Gaming Media and Social Effects

The scope of this book series is inter-disciplinary and it covers the technical aspect of gaming (software and hardware) and its social effects (sociological and psychological). This book series serves as a quick platform for publishing top-quality books on emerging or hot topics in gaming and its social effects. The series is also targeted at different levels of exposition, ranging from introductory tutorial to advanced research topics, depending on the objectives of the authors.

Yiyu Cai · Umesh Ramnarain · Jean Jieqiong Chen
Editors

Virtual and Augmented Reality Technology-Enhanced Learning

Research and Applications

Springer

Editors
Yiyu Cai
School of Mechanical and Aerospace Engineering
Nanyang Technological University
Singapore, Singapore

Umesh Ramnarain
University of Johannesburg
Pretoria, South Africa

Jean Jieqiong Chen
Shanghai Jiaotong University
Shanghai, China

ISSN 2197-9685 ISSN 2197-9693 (electronic)
Gaming Media and Social Effects
ISBN 978-981-96-2331-0 ISBN 978-981-96-2332-7 (eBook)
https://doi.org/10.1007/978-981-96-2332-7

This Springer imprint is published by the registered company Springer Nature Singapore Pte Ltd.
The registered company address is: 152 Beach Road, #21-01/04 Gateway East, Singapore 189721, Singapore

Foreword

When reading about the applications of technology in education, often the perspective is taken of a computer taking the role of a teacher or of a tool that students can use for educational tasks. Another often described use of technology in education is for (automated) assessment and of course for student administration and monitoring student progress.

While undoubtedly such applications of technology in education have their merits and usefulness, quite frankly they are also a bit boring. They make various roles in education more efficient, make it easier to process data, relieve tasks of teachers and administrators and can support students in many aspects. However, conceptually they do not add much news to how we see education.

In the current book, however, a different and more creative stance is taken. VARTeL applications offer the opportunity to engage in alternate worlds where you can be a dolphin trainer or take a deep dive into your intestines. Being able to immerse creates a whole new experience that can leave a lasting impression on learners.

Looking at the VR and AR applications that are presented in this book, we see the added value in the ability to immerse people in situations that they might encounter in real life, or into detailed views that would not be possible otherwise. The applications for special needs education are really helpful to prepare children for situations they encounter in real life. On top of that, the simulations can provide out-of-this-world experiences that can completely overturn perceptions. Who could imagine riding a protein lile a rollercoaster?!

The VARTeL conferences have been organized since 2012 and every time they were a great meeting place between designers, developers, educational researchers and end users. Over time, the books resulting from these conferences, of which this one is the latest, show a tendency of more advanced applications, more interactive engagement by students and especially more sophisticated educational designs, closer to learners' needs. As such they are a great example of international collaboration between researchers of various backgrounds, towards the goal of making education more effective and more engaging for students.

The conference of which this book is based was held just before the sudden entrance of generative AI systems such as ChatGPT. An interesting question is of course how these rapid developments will find its way to VARTeL. There are ample opprtunities. It should be not a big problem to have the AI create the virtual environments, and start dialogs with the learner. This book will not give the answer to this but the next conference and the resulting book certainly will.

Nevertheless, the current book will give you great insights in how to create, design or study the application of virtual and augmented reality in education. I wish you will draw much inspiration from the examples presented in the various chapters and feel challenged to apply VR and AR in your own work.

Utrecht, The Netherlands Wouter van Joolingen
July 2024

Contents

VARTeL Innovation

Yiyu Cai, Sui Lin Goei, Umesh Ramnarain, Jieqiong Chen, and Ingrid Winkler

Abstract Technology-enhanced Learning (TeL) is increasingly becoming popular worldwide. This chapter will start with a revisit to TeL in Singapore back in the early 2000s. Virtual and Augmented Reality Technology-enhanced Learning (VARTeL) will then be introduced. Emphasis will be placed on the discussion in relation to innovations on VARTel EduTech, VARTeL Space, VARTeL Pedagogy, VARTeL Content, and VARTeL Assessment. Efforts will be made in detailing VARTeL for K-12 Education in Singapore and Tertiary STEAM Education, particularly in Nanyang Technological University, as well as Public Education, and Special Needs Education. This will be followed by a sharing on International Collaboration and Globalization of VARTeL.

1 VARTeL: A Revisit

1.1 Background

In September 2003, an exhibition was launched at the Singapore Art Museum aiming to improve the public awareness of SARS pandemic [1]. As part of the one-year permanent exhibition on Convergence of Art, Science, and Technology, an immersive

---

Y. Cai (✉)
Nanyang Technological University, Singapore, Singapore
e-mail: myycai@ntu.edu.sg

S. L. Goei
Windesheim University of Applied Sciences, Zwolle, The Netherlands

U. Ramnarain
University of Johannesburg, Johannesburg, South Africa

J. Chen
Shanghai Jiao Tong University, Shanghai, China

I. Winkler
SENAI CIMATEC University Centre, Salvador, Brazil

Y. Cai et al. (eds.), *Virtual and Augmented Reality Technology-Enhanced Learning*, Gaming Media and Social Effects, https://doi.org/10.1007/978-981-96-2332-7_1

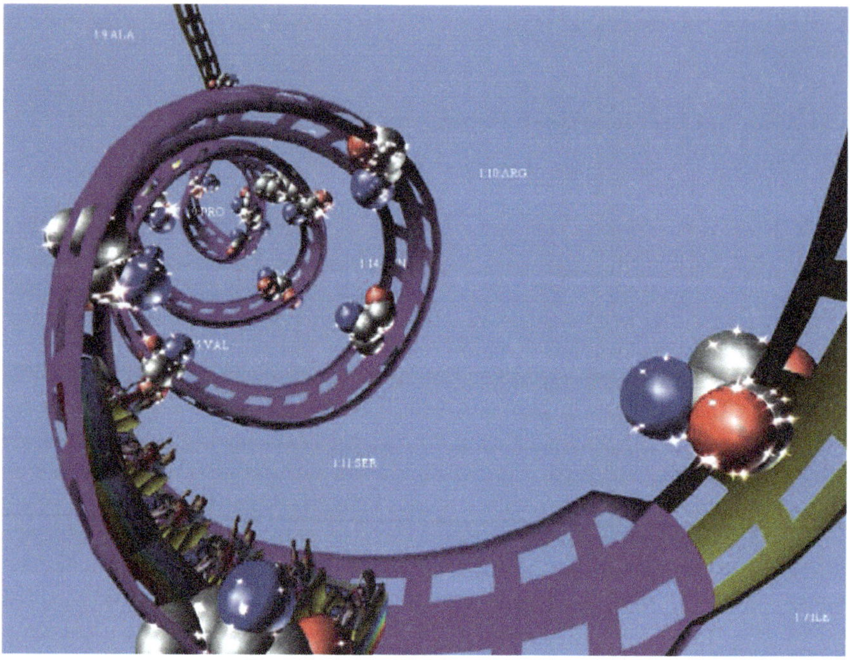

Fig. 1 The protein roller-coaster of a SARS virus [3]

and interactive VR protein roller-coaster (Fig. 1) was shown to visitors at Gallery #10, featuring the molecular structure for a main protease of the SARS virus [2]. Apart from the immersive and interactive VR games, protein music was also developed in the exhibit [3–5]. This marks the start of the journey for research and development at Nanyang Technological University (NTU) on Virtual and Augmented Reality Technology-enhanced Learning (VARTeL). In fact, VARTeL can be considered as a side product of the research project on Bio Visualization and Virtual Reality (2002–2006), funded by Singapore's Agency for Science, Technology and Research. Since then, VARTeL has been under continuous development in four main directions: (1) curriculum-based Science, Technology, Engineering, and Mathematics (STEM) Education; (2) Special Needs Education; (3) Public Education in science museums, libraries, etc.; and (4) VARTeL internationalization.

1.2 VARTeL Innovations

VARTeL is a specific type of Technology-enhanced Learning enabled by immersive and interactive VR and AR technology. Over the years, VARTeL has evolved through EduTech innovation, Learning Space innovation, curriculum and co-curriculum

Content innovation, Pedagogical innovation, and Assessment innovation. An interdisciplinary research team in NTU has worked hard to have the VARTeL innovation developed over years. Figure 2 illustrates VARTeL environments typically consisting of hardware, software, learning space, pedagogy, content, and assessment.

(1) **EduTech Innovation**

As an EduTech, VARTeL is initially built on VR technology and subsequently on AR technology due to historical reasons. Different tiers of VR, ranging from very high-end to very low-end, have been developed to support various VARTeL solutions. Normally, high-end VR is characterized by high quality and high cost, whereas low-end by lower quality and cost. Fortunately, rapid technological advancements have dramatically reduced the cost of VR hardware products and improved the quality of VR hardware production, leading to a substantial increase of VR installations for educational applications. In recent years, AR is being accepted at an accelerated rate for educational use due to its cost-effectiveness. For AR, VARTeL focuses on functional development based on different modalities such as location-based AR, model-based AR, book-based AR, image-based or note-based (handwriting) AR. While VR and AR have different focuses on EduTech development (Fig. 3), they do share some common enabling technologies such as modelling, visualization, interaction, and user interface. More detail can be found in the article published by Cai et al. [1]. Fundamentally, the development of VR and AR enabling technologies relies heavily on artificial intelligence, mathematical optimization, high-performance computing, data science, sensors, robotics, and mechatronics.

(2) **Learning Space Innovation**

The earliest version of VARTeL Space was based on a desktop design. It was supported by stereographic typically using CRT display, active shutter glasses, and emitters. The original design of this desktop VR was for the project of Bio Structure Visualization and Virtual Reality. Hwa Chong Institution (HCI), the first local secondary school in Singapore, adopted this Generation I design of VARTeL Space in 2004 for K-12 Education (Fig. 4a). The Opas Lab at The University of Toronto adopted this Generation I design as well for their cell and molecular research [6–11].

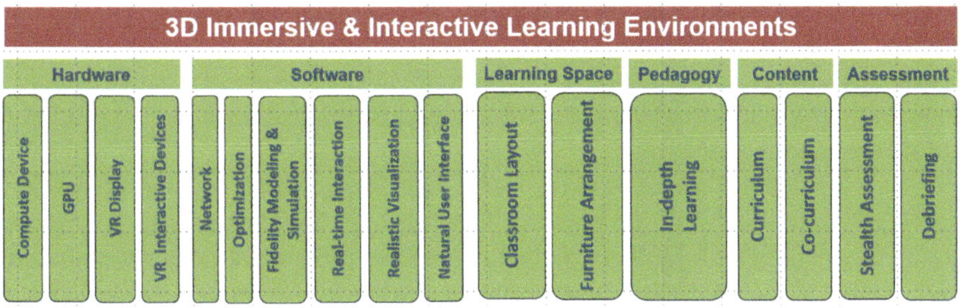

Fig. 2 3D immersive and interactive learning environments for VARTeL

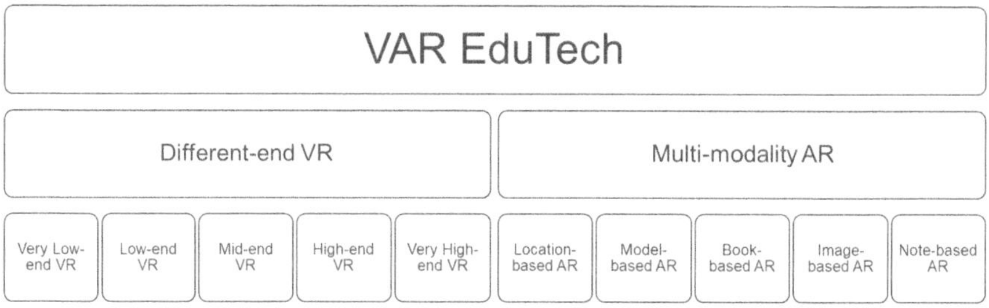

Fig. 3 VARTeL EduTech: different-end VR and multi-modality AR

Figure 4b showed a theatrette VR Space designed using a stereographic projection screen at NTU's Computer-aided Engineering Labs for engineering design and biology teaching. This Generation II design of VARTeL Space was adopted by several Singaporean schools, including HCI and River Valley High School (RVHS) in 2005 as well as Windesheim University of Applied Sciences (WUAS) in The Netherlands in 2014. Figure 4c depicted a Generation III design of VARTeL Space with multiple stereographic images synchronized and soft edge blended using clustering technology supported by 5 projectors and 5 computers. This design was a very high-end solution available with NTU's Institute for Media Innovation. Figure 4d was the Generation IV design of the VARTeL Space for team-based learning, first implemented in 2018 at The Hive, a new learning hub in NTU. This design was adopted by several schools, including Riverside Secondary School (RVSS) in Singapore, Renai School (RS) in China, and The University of Johannesburg (UoJ) in South Africa. All four generations of VARTeL Space were developed by the VARTeL team in NTU.

(3) **Pedagogical Innovation—In-Depth Learning**

In-depth Learning was first coined by the VARTeL team in the book—*3D Immersive and Interactive Learning* [1, 12]. It is a unique pedagogy of VARTeL featuring team-based learning, immersive visual learning, interactive hands-on learning, and fun learning. Simply speaking, VARTeL is designed to help students learn difficult topics involving abstract concepts, complex structures, or dynamic processes. In other words, VARTeL is not aimed to replace those well-established learning approaches that students and teachers traditionally feel comfortable with. Instead, In-depth Learning developed is a novel VARTeL approach to assisting students in learning challenging topics such as the abstract concept of skew lines and shortest distances, the complex structure of cells and organelles, and the dynamic process of electromagnetism (EM).

Figure 4d shows an example in the VARTel classroom where students are grouped into teams, each with 4–5 persons, for EM learning. Each team has a VR system assigned near their table for team-based learning. All team members can wear passive (polarizing) glasses to view the EM field in stereographics for immersive visual learning. Each member can take turns to become a master player manipulating the

(a) Desktop VARTeL Space (b) Theatrette VARTeL Space

(c) Immersive VARTeL Space (d) Team-based VARTeL Space

Fig. 4 VARTeL space innovation

haptic device to feel the EM force for interactive hands-on learning. This new and fun learning approach can be used for many difficult subjects in STEM education.

Gamification plays a vital role in In-depth Learning research, which can be seen in the protein roller-coaster and virtual pink dolphin projects. We have extended this research (Fig. 5) to include protein music or sonification [13] to the DNA dance [14] published in the journal Leonardo (MIT Press).

(4) **Content Innovation**

One of the challenges for VARTeL research and development involves a huge amount of effort to design content for curriculum or co-curriculum-based learning. The VARTeL team in NTU has worked hard over the years developing content [9, 15–17] for K-12 STEM Education as well as for Tertiary Education in Aerospace Engineering, Manufacturing Engineering, Petrochemical Engineering, Automobile Engineering, Civil and Construction Engineering, and Biomedical Science and Engineering (Fig. 6).

(5) **Assessment Innovation**

Recently, the VARTeL team in NTU is interested in assessment research. Specifically, they are investigating the novel stealth assessment technique as part of VARTeL. Stealth assessment can be considered as an in-process assessment technique capable of monitoring the progress of the learners without triggering either summative or formative assessments. Importantly, they are implementing stealth assessment in the VARTeL Space, and the results of in-process assessment may be used for debriefing.

(a) A protein music CD

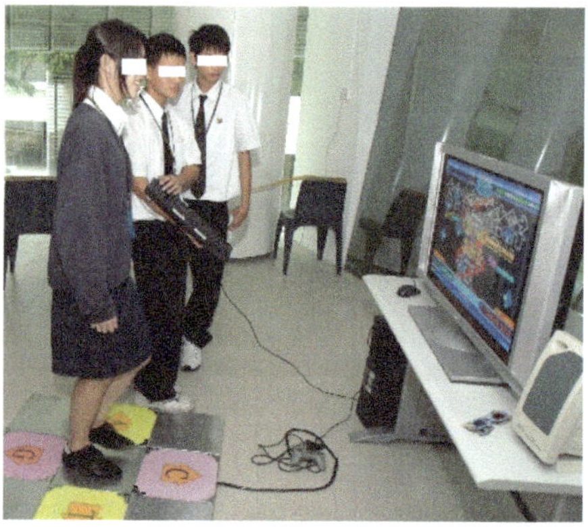

(b) DNA Dance Revolution

Fig. 5 Creative learning of biology

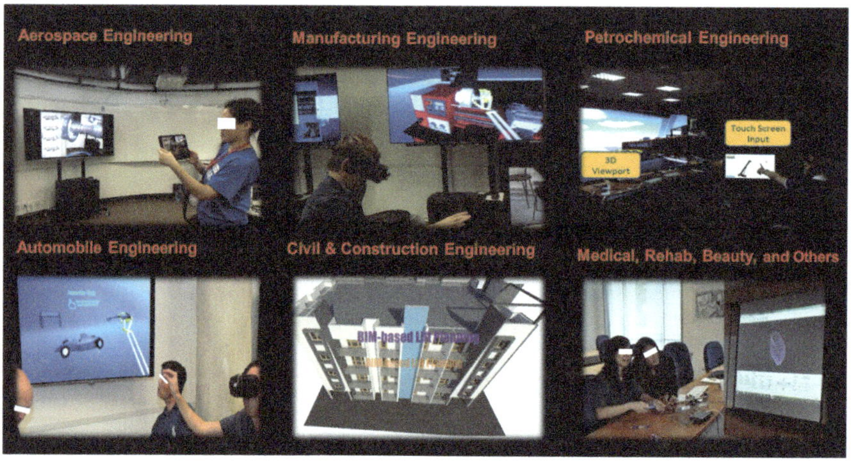

Fig. 6 Content innovation for curriculum and co-curricular-based learning

1.3 Organization of the Chapter

The rest of this chapter is organized as follows. Section 2 discusses VARTeL for K-12 Education in Singapore. Section 3 presents VARTeL for STEM Education in NTU. Section 4 focuses on VARTeL for Public Education in Singapore. Section 5 shares insights into VARTeL for Special Needs Education in Singapore. Section 6 describes the globalization of VARTeL for Education in Europe, Africa, China and Brazil. Section 7 introduces the rest of the book chapters which are mostly based on the papers presented at the 2022 VARTeL conference organized by University of Johannesburg in South Africa.

2 VARTeL for K-12 Education in Singapore

2.1 The Mobile Immersive Theatre

Supported by the Ministry of Information, Communication and the Arts under The Creative Community Singapore programme, the Mobile Immersive Theatre project aimed to promote VR technology for Science and Art Education in schools and community centres. From March 2006 to March 2008, the Mobile Immersive Theatre had organized 24 shows in 12 schools, 2 community centres, and 10 conferences/ events in Singapore (Table 1), attracting a very large audience.

The travelling shows were meaningful in the sense of (1) attracting a good number of schools interested in venturing into innovative education using VARTeL technology; and (2) engaging many secondary school students in pursuing VARTeL

Table 1 VR travelling shows using the mobile immersive theatre

#	Organization travelled	Date	Participants
	Nanyang Girls High School	22/3/2006	267
	West Spring Secondary School	18/5/06	178
	Hwa Chong Institution	11/5/2006	12
	Crescent Girls High School	4/8/2006	8
	Int'l Conf on Educational Technology 2006	7&8/9/2006	500+
	Choa Chu Kang Community Club	17/9/2006	300+
	National Junior College	14/10/2006	154
	Clementi Community Center	19/11/2006	300+
	TecXeprience 2006	22/11/2006	250+
	Greenfield 2006	27/11/2006	12
	Infocomm Clubs	28/11/2006	29
	Raffles Girls Secondary School	6 and 16/1/2007	20
	Maris Stella High School	30/1/2007	30
	Anglo-Chinese School (Independent)	02/3/2007	300+
	MOE Principals' forum 2007	07/5/2007	300+
	MOE West Zone School Cluster	01/6/2007	40
	MOE EXCELFEST	13/7/2007	80+
	National Junior College	24/8/2007	120+
	MOE South Zone Schools	10/10/2007	50+
	Raffles Institution	1/11/2007	200+
	Bedok Town Primary School	16/11/2007	40
	Singapore American School	17/11/2007	150+
	CREATIVE 24	1 and 2/12/2007	250+
	ACSI Life Science Symposium	3/3/2008	180+

related research. Nanyang Girls High School (NGHS) was the first school that hosted the Mobile Immersive Theatre. Several students from NGHS later joined NTU's Nanyang Research Programme (NRP) for VARTeL research. One of such research projects on protein sonification [13] was published in the journal *Leonardo* (MIT press).

2.2 FutureSchools @ Singapore

Funded by the National Research Foundation and supported by the Ministry of Education (MOE) and the Information Development Authority (IDA) of Singapore, the FutureSchools @ Singapore initiative involved approximately 10 primary and secondary schools. These schools each worked with an industrial consortium to promote a highly motivated digital learning lifestyle, which enabled anytime and anywhere learning to enhance knowledge construction and skill development; facilitated collaboration at local, national and government levels; and encouraged participation by all education stakeholders. HCI and Crescent Girls School (CGS) were two FutureSchools in Singapore selected by MOE and IDA to implement VR technology for K-12 education in collaboration with the NTU team.

(1) **FutureSchool @ Hwa Chong Institution**

HCI was the first secondary school in Singapore to have a desktop VR classroom (Generation I VARTeL Space) set up in 2005 (Fig. 3a). A VR Lab (Generation II VARTeL space) was developed as part of the FutureSchool @ HCI project, which was used for Biology and Mathematics Education, as well as for research project development. Several topics such as cell division (Fig. 7a) and solid geometry (Fig. 7b) were co-developed by HCI and NTU. Control and experimental comparisons were conducted to investigate the efficiency and performance improvement using VARTeL for Biology [18, 19] and Mathematics [20] Education. The VARTeL Space was also used for workshops and hands-on projects organized by the NTU team for HCI students under the HCI sabbatical programs [21]. The same VARTeL Space (Fig. 7c) was often used by the school to demonstrate student projects for local or international visitors. Some of the HCI students continued their VR interest by doing NRP work attached to the VARTeL team in NTU.

(2) **FutureSchool @ Crescent Girls School**

CGS was another FutureSchool in Singapore with an iiSPACE established for students to learn geography and physics (Fig. 8). The content developed included topics such as river formation, waterfall, volcano, plate tectonic for geography, and sound waves for physics. The iiSPACE is a Generation II VARTeL Space supporting immersive and interactive learning.

(a) Learning of cell division

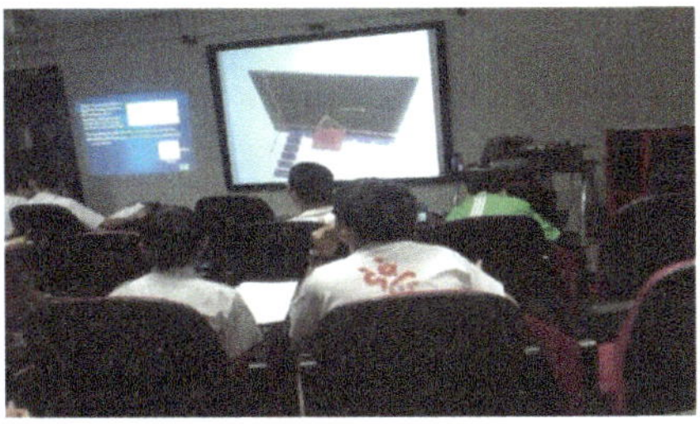

(b) Learning of solid geometry

(c) Sabbatical program on VR star warship design

Fig. 7 VARTeL @ Hwa Chong Institution

(a) iiSPACE

(b) Geography content

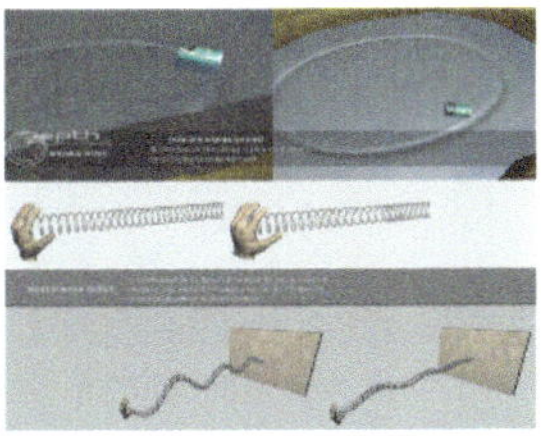

(c) Physics content

Fig. 8 VARTeL @ Crescent Girls School

2.3 VARTeL for MOE Schools

(1) **VARTeL @ River Valley High School**

In 2005, a Generation II VARTeL Space (Fig. 9a) was set up in the old campus of RVHS at Malan Road. The daVinci VR Lab (Fig. 9b) was established when the school relocated to its new premises at Boon Lay Way. Biology was a major subject conducted in this upgraded VARTeL Space. A good number of student projects (Fig. 9c–e) were developed in the lab. Some of these projects helped RVHS students clinch many awards including, the NRP Gold Award, the Gold Award at The Home Team Innovation Fest from the Ministry of Home Affairs, and the Merit Award from Singapore Information Technology Federation (SITF). RVHS represented Singapore to participate in the 2009 Asia–Pacific InfoComm Technology (APIT) Competition in Hong Kong [22]. In 2009, RVHS received the INERGY Award (Silver) from MOE for this educational innovation.

(2) **VARTeL @ Raffles Institution**

Raffles Institution (RI) is among the first batch of schools in Singapore embracing the VARTeL technology. Back in the mid-2000s, a group of students from RI attached to the VARTeL team in NTU and developed a simulation of the human blood system (Fig. 10a). Since 2019, RI and NTU have been collaborated, with funding support from MOE, to develop an AR app assisting RI students to learn the blood circulation system (Fig. 10b).

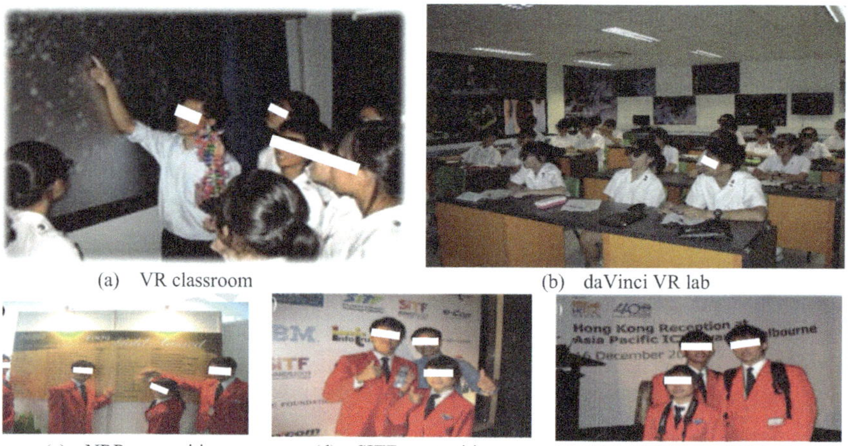

(a) VR classroom (b) daVinci VR lab

(c) NRP competition (d) SITF competition (e) APIT competition

Fig. 9 VARTeL @ River Valley High School [22]

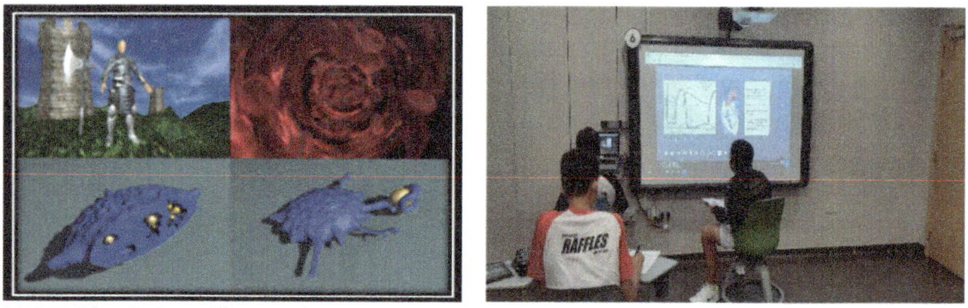

(a) VR Simulation of human immune system (b) Blood circulation system

Fig. 10 VARTeL @ Raffles Institution

(3) VARTeL @ National Junior College

National Junior College (NJC) has been an active follower of VARTeL innovation. They installed a Generation II VARTeL system in the school and shared the resource to a cluster of schools (S3). The VARTeL team also provided technical guidance to NJC students under their Science, Technology, and Research (STaR) programmes. Chan et al. [23] have described some of the STaR projects that utilise VR technology for the design and simulation, for example, UAVs, rockets, Formula One, and Singapore Flyer (Fig. 11a). NJC also used the VARTeL technology for their exchange program with Korea Science Academy (KSA). Figure 11b shows the VR photography project jointly developed by students participating in the NJC-KSA exchange.

(4) VARTeL @ Bukit Merah Secondary School and Kent Ridge School

Bukit Merah Secondary School (BMSS) and Kent Ridge Secondary School were two neighbourhood schools participating in the Media-in-Learning project, funded by Media Development Authority (MDA). Figure 12 shows the VARTeL Space established in BMSS for Biology Science learning.

(5) VARTeL @ Riverside Secondary School

RSS was the first secondary school in Singapore to adopt the Generation IV VARTeL Space. Three sets of VR systems were installed to students to engage in VARTeL activities (Fig. 13a). Pang et al. [24] described the use of VR technology for enzyme learning at RSS (Fig. 13b). Pang and Cai [25] also discussed transforming learning experiences through affordances of virtual and augmented reality.

(6) VARTeL @ Christ Church Secondary School

Over the past few years, the NTU team has made significant progress in AR technology development to enhance learning. Christ Church Secondary School (CCSS)

(a) STaR projects (b) NJC-KSA exchange

Fig. 11 VARTeL @ National Junior College

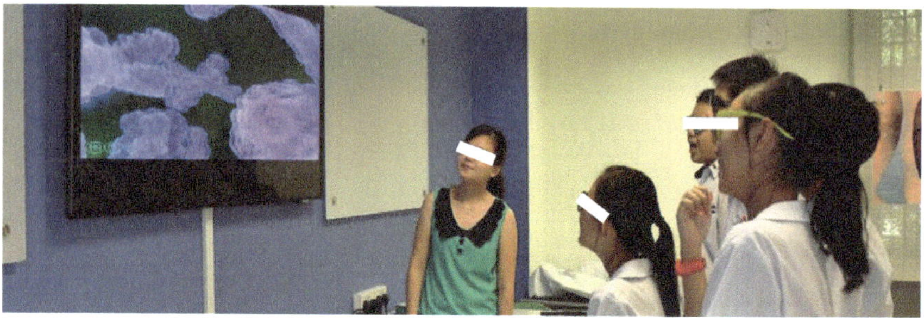

Fig. 12 VARTeL @ Bukit Merah Secondary School

(a) VARTeL space in RSS (b) Virtual enzyme

Fig. 13 VARTeL @ River Side Secondary

has worked closely with the NTU team in adopting VARTeL technology for Biology Education (Fig. 14). A YouTube video (https://youtu.be/q_Ca_lYp9n4) is available for sharing the details of this project.

(7) **VARTeL @ School of the Art**

During the COVID-19 pandemic, the School of The Art (SOTA) had tied up with NTU to organize VARTeL training courses for their students. Several workshops (Fig. 15a) were specifically designed to teach SOTA students on 3D modelling, which included the avatar modelling for AR applications (Fig. 15b). They applied

(a) VARTeL space in RSS (b) AR technology-enhanced biology education

Fig. 14 VARTeL @ Christ Church Secondary School

(a) VARTeL workshop

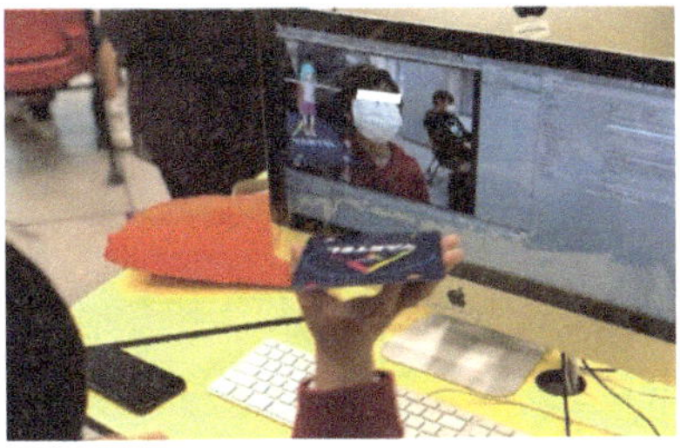

(b) AR avatar design

Fig. 15 VARTeL @ School of The Art

the skills and knowledge, in a major environmental protection project in collaboration with the World Wildlife Conservation Organization.

(8) **VARTeL @ Jurong West Secondary School**

Jurong West Secondary School (JWSS) was one of the schools in Singapore embracing very low cost VARTeL solution for learning back in the mid-2000s. JWSS students used cheap anaglyph (red and blue) filters to view 3D videos displayed on the conventional monitors that were generated by computers available at the school (Fig. 16). Wang et al. [26] have reported their findings from this research.

(9) **VARTeL @ Innova Primary School**

Students at the secondary school level could produce their own VARTeL contents after brief training via hands-on workshops or attachment programs. This was evidenced from the work done by students from HCI, RI, RVHS, NJC, etc. Would Primary school students be able to do the same? The answer was "yes" from Innova Primary School (IPS). Figure 17 shared some of the reflections from IPS students after a 5-day workshop on 3D modelling.

Fig. 16 VARTeL @ Jurong West Secondary School

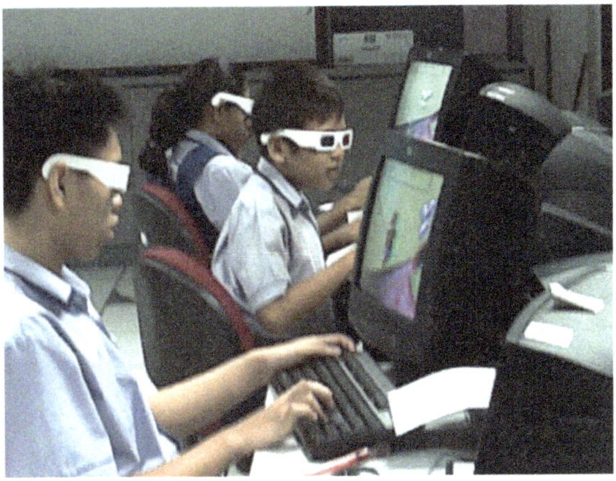

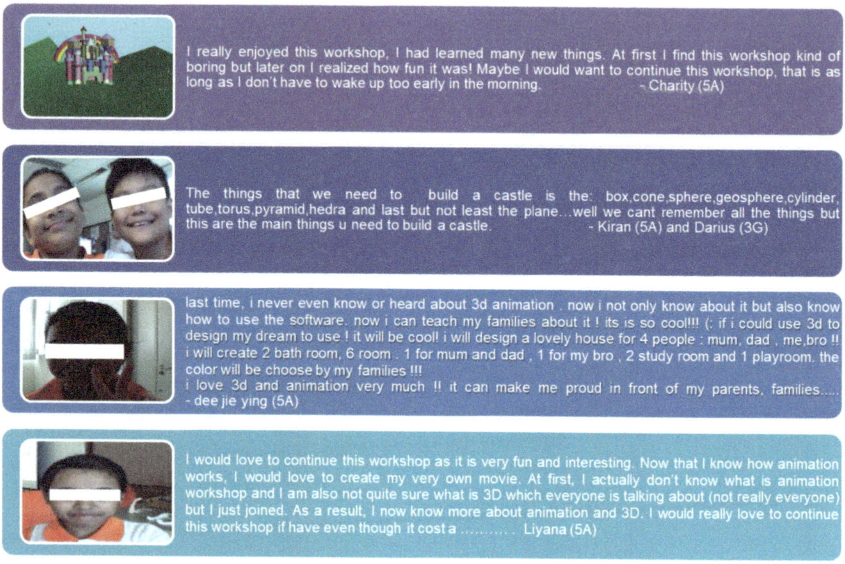

Fig. 17 VARTeL 3D modelling workshop @ Innova Primary School

(10) VARTeL @ Fuhua Primary School

Supported by the VARTeL team, Fuhua Primary School (FPS) focussed on the use of VARTeL for cultural exchange. Figure 18a showed FPS students demonstrating the AR dinosaurs at the School Open House. Figure 18b showed a student, from FPS's partner school in Chengdu, China, experimenting VARTeL technology to interact with the virtual Merlion in Singapore during the visit by the Singapore Minister of Education.

(11) VARTeL @ Springdale Primary School

Funded by the National Institute of Education (NIE), an AR learning project was developed for inquiry-based learning for primary school students. Designed by the

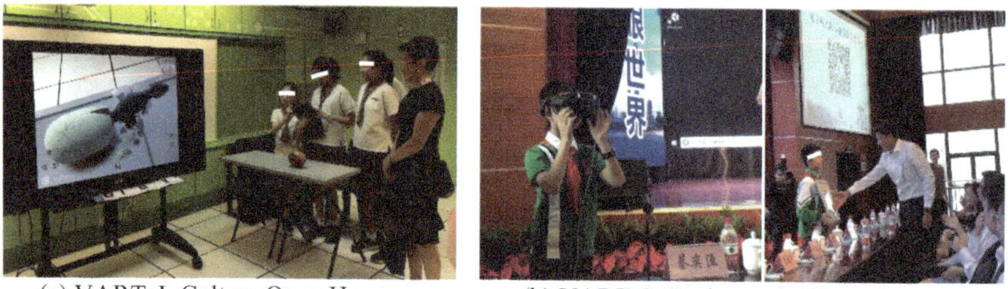

(a) VARTeL Culture Open House (b) VARTeL Exchange in Chengdu

Fig. 18 VARTeL @ Fuhua Primary School

(a) NJC

(b) RI (c) CGS (c) RVHS (d) BMSS

Fig. 19 Recognition and partnership

VARTeL in NTU, an AR learning trail and an AR application were used for Science Education in Springdale Primary School (SPS). Wen et al. [27] and Wen et al. [28] reported their research findings from this project in the *Interactive Learning Environments* (Sage) and *Education Technology Research Development* (Springer), respectively.

As a summary of this section, 12 selected schools in Singapore have their VARTeL associated works presented. Through this VARTeL journey, several partnerships had been developed (Fig. 19).

3 VARTeL for STEM Education in NTU

3.1 The Design of VARTeL Classroom for STEM Education

In 2017, a Generation IV VARTeL Space was set up in The Hive @ NTU for STEM education (Fig. 20a). This VARTel classroom, designed as an immersive and interactive learning environment, is the result of innovations in VAR EduTech, Learning Space, Pedagogy, Content, and Assessment [29]. Currently, the VARTeL Space is heavily used by many NTU students for STEM Education including Biology Science, Aerospace Engineering, Automobile Design Technology, and Mathematics (Fig. 20b–e). In addition, the VARTeL Space is used for students to learn Mechanical Engineering, Chemical Engineering, Technopreneurship, and Residential Education. Several publications with details relevant to VARTeL research and development are available [15, 30–35].

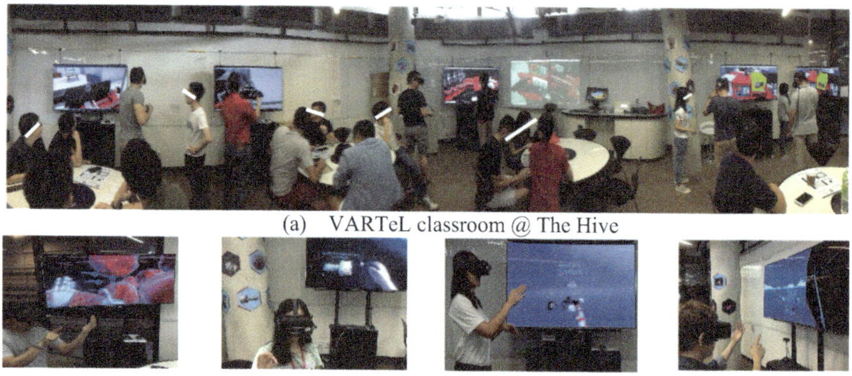

(a) VARTeL classroom @ The Hive

(b) Biology (c) Engineering (d) Technology (e) Mathematics

Fig. 20 VARTeL for STEM education @ Nanyang Technological University

3.2 *VARTeL Evaluation Research*

Supported by the MOE Academic Fund and the Tertiary Research Fund, several projects were developed to evaluate the situational interest, self-efficacy, and affordances of the VARTeL innovation [36]. Wong et al. [37] investigated the VARTeL environments and their benefits to students from the perspective of a technology-rich and outcome-focussed learning environment inventory. More works on VARTeL evaluation are available as chapters in this book.

3.3 *VARTeL @ National Institute of Education*

Apart from the Springdale Primary School project [27, 28], two additional projects funded by NIE under the I3G scheme have been developed to explore the use of VARTeL technology for teacher education in the areas of organic chemistry and cell culturing (Fig. 21). A new project funded by ERFP scheme was recently awarded to investigate VARTeL solution for Teaching Fundamental Movement Skills in Primary School: Technology-Enhanced Pedagogy for Improving Throwing in Children.

3.4 *VARTeL @ School of Chemistry, Chemical Engineering, and Biotechnology*

Singapore is a global hub for chemical engineering. NTU's Interdisciplinary School of Chemistry, Chemical Engineering, and Biotechnology is responsible for training quality future engineers for the industry. Distillation, one of the most fundamental

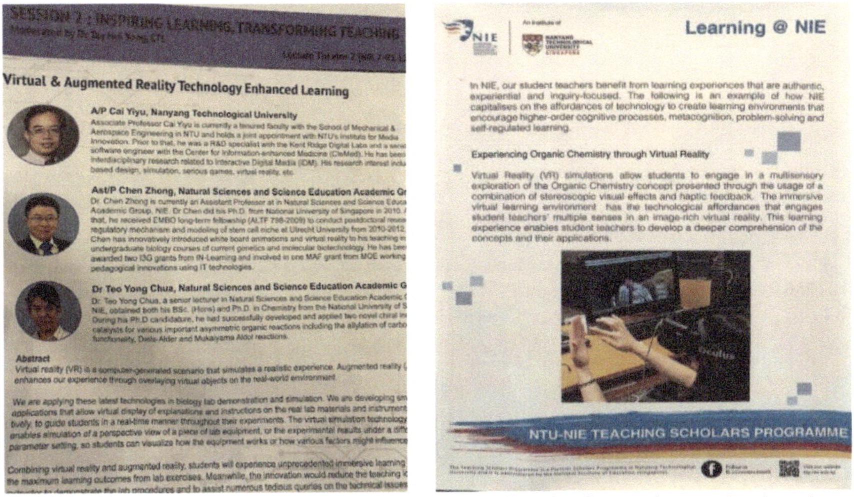

(a) NTU-NIE VARTeL collaboration (b) Cell culturing and organic chemistry

Fig. 21 VARTeL @ National Institute of Education

processes, is not easy to teach and learn as a module. With a scale-down pilot plant available in the school, traditional way of learning distillation process is not only time-consuming and costly but also sometimes risky for a large cohort of students to conduct physical experiments. An alternative VARTeL solution has been developed in NTU [38] with two parts: a VR version for classroom learning (Fig. 22a) and an AR version for individual learning to use their mobile device (Fig. 22b). Details of this research project and development were published in the journal *Computer in Chemical Engineering* [39].

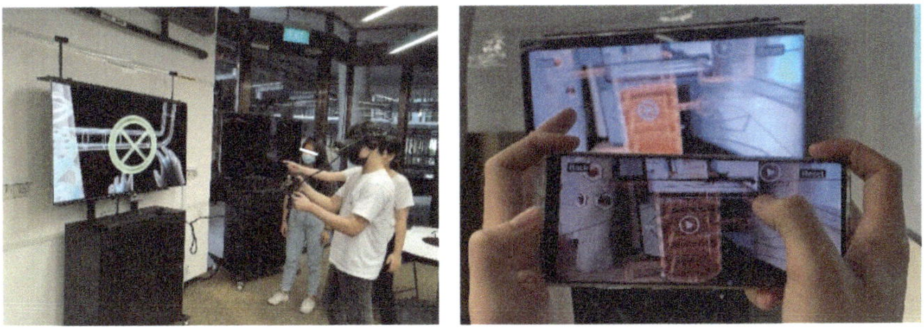

(a) VR learning of distillation (b) AR learning of distillation

Fig. 22 VARTeL for distillation @ School of Chemistry, Chemical Engineering, and Biotechnology

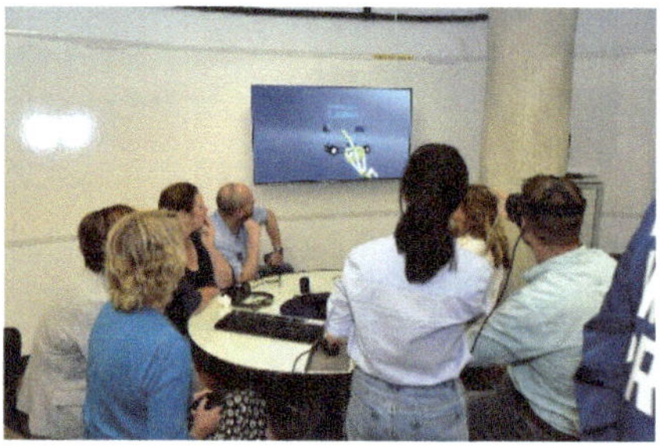

(a) VARTeL for Technopreneurship Education

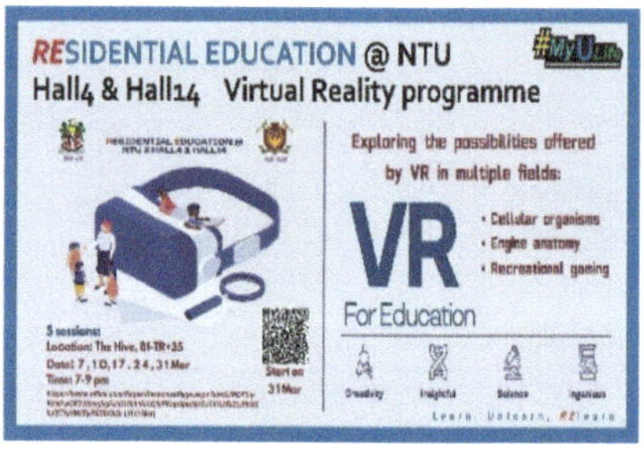

(b) VARTeL for Residential Education

Fig. 23 VARTeL for technopreneurship and residential education @ Nanyang Technopreneurship Centre

3.5 VARTeL for Technopreneurship and Residential Education

Nanyang Technopreneurship Centre offers a Master of Science Degree in Technopreneurship and Innovation. Every year, two classes of students come to the VARTeL classroom @ The Hive to examine the latest trends and key issues in VAR EduTech. The participants discuss the potential commercialization of EduTech inventions and innovations (Fig. 23a).

NTU is actively promoting Residential Education for students staying in campus. The VARTeL classroom in The Hive is also offering VR programs to residents from different hostels every semester (Fig. 23b).

4 VARTeL for Public Education in Singapore

4.1 VARTeL @ Singapore Art Museum

To improve the public awareness of SARS pandemic, a VR game was developed by the VARTeL team in NTU for Public Education in Singapore Art Museum. Specifically, a protein roller-coast (Fig. 24a) was designed using novel techniques based on protein structure modelling, visualization, interaction and sonification [3, 4]. For any protein in the Protein Data Bank, a unique approach developed could be used to automatically generate its primary structure (amino acid sequence) and secondary structure (ribbon or helices). The secondary structure, with helices and amino acids represented by roller-coaster tracks and balls, respectively, was then visualized through graphical rendering. The roller-coaster was interactively controlled by gamers, with VR devices, to allow close examination of protein structures and functions. Protein sonification was further developed by mapping the amino acid sequence into music nodes (Fig. 24b).

(a) Protein roller-coaster modelling, visualization, and interaction.

A – alanine C – cysteine	I – isoleucine K – lysine	Q – glutamine R – arginine
D – aspartate	L – leucine	S – serine
E – glutamate	M – methionine	T – threonine
F – phenylalanine	N – asparagine	V – valine
G – glycine	P – proline	W – tryptophan
H – histidine		Y – tyrosine

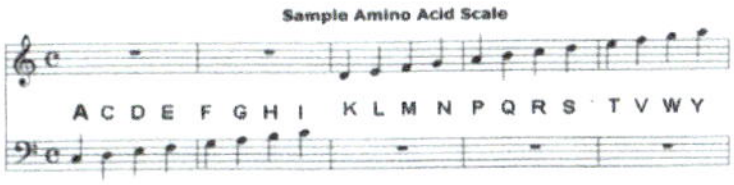

(b) Protein sonification

Fig. 24 Protein modelling, visualization, interaction, and sonification [2, 3]

4.2 VARTeL @ Science Centre Singapore

Science Centre Singapore (SCS) often organizes permanent exhibitions (typically lasting 5 years) to promote STEM education on topics of current importance. VARTeL technology was accepted by SCS around 2004 after the successful exhibition of protein VR roller-coaster at the Singapore Art Museum. Figure 25a showed a VR walkthrough game for a dengue virus protein exhibited in the Life Science Corridor of the DNA Learning Lab [2, 3]. Virtual cells, interactive DNA, and protein rendition were three VARTeL works developed for SCS's permanent exhibition on Genomics (Fig. 25b–d). T4-phage and Influenza were two VARTeL works designed for the Virus Exhibition (Fig. 24e–f). Seeing Your Hearing (Fig. 25g) and Foetus Development (Fig. 25h) were two VARTeL works created for the Sound Exhibition and the Human Body Exhibition, respectively.

4.3 DinoWorld VR Travelling Show

In July and August 2009, twelve high-resolution 3D anaglyph photos were designed and rendered for a travelling show in two shopping malls in Singapore: Causeway Point in July and CentrePoint in August. This DinoWorld travelling show was designed primarily for young kids to have fun experiences using low-cost VARTeL technology (Fig. 26).

4.4 National Science and Art Competition

Supported by the Ministry of Information, Communication and the Arts, a National Science and Art Competition was held in 2008 to promote STEM Education in this

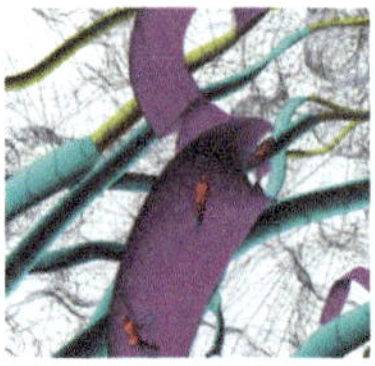
(a) VR walkthrough of a dengue virus

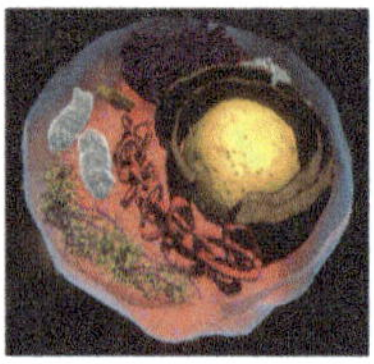
(b) Virtual cell

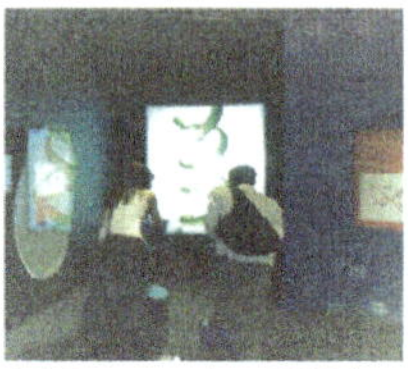
(c) Interactive DNA

(d) Protein rendition

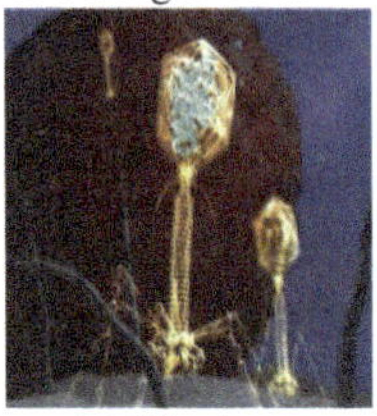
(e) T4-phage

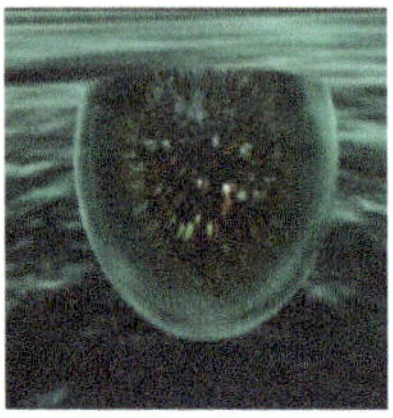
(f) Flu virus

(g) Seeing your hearing

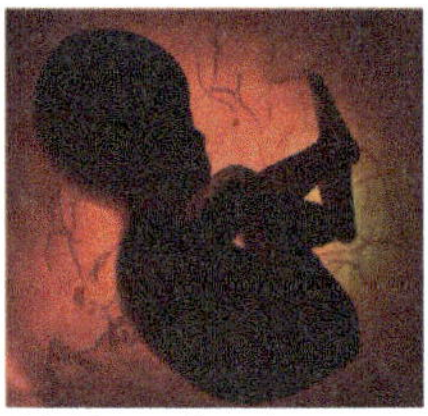
(h) Fetus development

Fig. 25 Permanent exhibitions of VARTeL @ Science Centre, Singapore

Fig. 26 DinoWorld VARTeL travelling shows at shopping malls

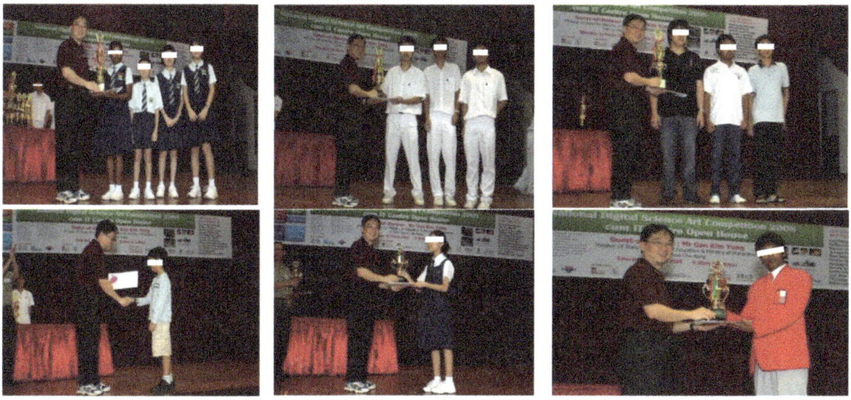

Fig. 27 Prize giving ceremony of The National Science and Art Competition

island country. The competition featured four categories: primary schools, secondary schools, tertiary learning institutions, and open category, each with Gold, Silver, and Bronze Awards (Fig. 27). In addition to cash prizes for winning individuals, Fuji Xerox provided colour printers, as prizes, for the Best Teams in each category. The Minister of Education was the Guest of Honour at the event.

4.5 VARTeL @ Choa Chu Kang Community Club

The prize-winning works, created by the contestants, were exhibited at the Choa Chu Kang Community Club attracting old and young participants (Fig. 28a, b). Several training courses including Kids' 3D Animation Course (Fig. 28c) were later offered in the Club.

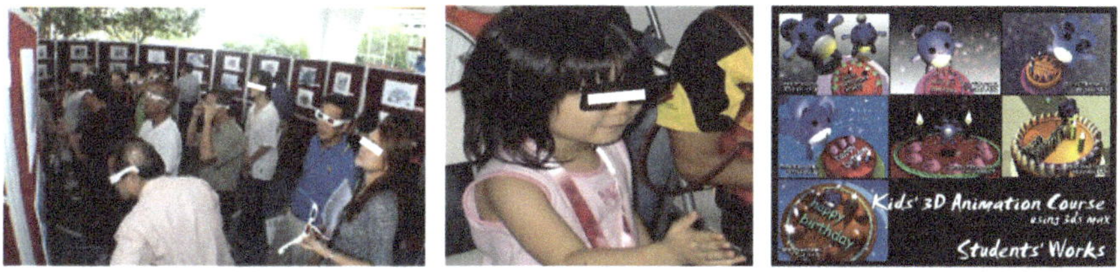

(a) Exhibition (b) Young Participants (c) Kids' Animation Course

Fig. 28 VARTeL@ CCK Community Club

4.6 VARTeL @ National Library Board

Supported by the National Library Board (NLB), VARTeL technology was developed in the newly renovated Choa Chu Kang Community Library (CCKCL) to promote innovative learning among the new generation of library users. An AR booklet and an AR learning trail, focussing on sustainability, were implemented by the NTU team in collaboration with NLB. The AR booklet (Fig. 29a) features about 10 books authored by local writers on topics like Butterflies of Singapore, Native Orchids of Singapore, Birds of Singapore, and so on. The AR learning trail consists of 8 stops (Fig. 29b) including the last wild tiger in the Choa Chu Kang area and the sambar deer (Fig. 29c).

5 VARTeL for Special Needs Education in Singapore

Special Needs Education (SNE) is one of the important VARTeL application fields. Supported by the Temasek-Trust funded Millennium Foundation, the Virtual Pink Dolphin (Fig. 30a) was a flagship project aimed at developing VARTeL solutions to improve the learning of children with autism spectrum disorder (ASD) [1, 40]. In collaboration with The Underwater World Singapore and AWWA School, the NTU team started this project, with a seed grant from The Institute for Media Innovation, through designing the Virtual Pink Dolphins game implemented in a high-end immersive room (Fig. 30b). A dolphin lagoon and three pink dolphins [41, 42] were modelled to allow children with ASD to act as dolphin trainers using their gestures and body language (Fig. 30b). This project later became an international effort [43, 44]. Numerous SNE associated VARTeL projects (Fig. 30c) were developed by many students and researchers from the NTU team [45].

(a) An AR booklet (b) An AR learning trail (c) New learning experience @ CCKCL

Fig. 29 VARTeL @ CCK Community Library

(a) Virtual Pink Dolphins (b) Virtual Pink Dolphin game (c) VARTeL projects for SNE

Fig. 30 VARTeL for Special Needs Education

5.1 VARTeL for Special Needs Education @ AWWA School

To develop a low-cost VARTeL solution for SNE, the NTU team worked very closely with AWWA School [46, 47] for the customization of the pink dolphin game in assisting the academic learning of children with ASD. Figure 31a–d showed some of the VR games designed for children with ASD to learn living and job skills. For the project "Learning Through 3D Technologies", AWWA School was the recipient of the MOE—NCSS Innovation Award 2014 (Fig. 31e). The students and teachers were visited by President of The Republic, Dr. Tony Tan, during the Lunar New Year (Fig. 31f).

5.2 VARTeL Executive Function Learning @ Institute of Mental Health

Supported by the National Health Group under the Rehabilitation Research Grant scheme, this project was a collaboration between NTU and the Institute of Mental Health (IMH). A cognitive rehabilitation programme (CRP) was developed to improve executive functions among children with ASD focussing on daily living skill learning in a real-life environment. Three serious AR games (Fig. 32a) were designed for interventions at home, school, and supermarket settings. Figure 32b showed a game designed for supermarket route planning [48] and Fig. 32c showed another game for school bag packing [49].

6 Globalization of VARTeL

VARTeL was originated in NTU, Singapore. Over the years, through international collaboration, VARTeL has gradually been accepted by learning institutions in Asia, Europe, and Africa.

(a) VR shower learning (Teoh et al, 2021) (b) VR road crossing learning (Zhang et al, 2021) (c) VR living skill learning (Chung et al, 2021; Hoe et al, 2021)

(d) VR cleaning job skill training (Chong et al, 2021) (e) VR supermarket shopping training (Goh, Cao, Cai, 2021) (f) VR emotion learning (Chua et al, 2017; Goh, Cao, Chen, Cai, 2021)

(g) VR gesture training (Chan et al, 2021) (h) Innovation award (f) President Dr Tony Tan visited AWWA

Fig. 31 VARTeL @ AWWA School

(a) Three CRP games (b) An AR game for supermarket route planning (c) An AR game for bag packing

Fig. 32 VARTeL for executive function learning @ Institute of Mental Health

6.1 VARTeL @ The Netherlands

Around 2010, Prof Sui Lin Goei, a visiting professor from Windesheim University of Applied Sciences and VU University (Amsterdam), was attached to the Special Needs Education Group at NIE. She became interested in VARTeL after several discussions with the NTU team. She was invited to join the Virtual Pink Dolphin project and other VARTeL related research projects.

(1) **VARTeL @ Windesheim University of Applied Sciences (WUAS)**

About three years later, a book titled "Simulations, Serious Games and their Applications" [50] was published by Springer. After another three years, a second book on "Simulations and Serious Games for Education" [51] was published by Springer. Trooster [52] from WUAS investigated the use of virtual worlds and serious gaming in education. This effort was soon joined by other WUAS faculty on the interdisciplinary and international collaboration between NTU and WUAS towards an inquiry-based pedagogy [53]. A VARTeL Space was set up in WUAS for the uses of STEM teaching and SNE research (Fig. 33).

(2) **VARTeL @ Utrecht University**

Utrecht University (UU) is the second Dutch university collaborating with NTU on several VARTeL projects. Profs. van Joolingen and Matimba, worked closely with the NTU team in developing and integrating an AR app for teaching and learning about enzymes [54]. Part of the collaboration outcomes were published in two Springer books [55, 56].

Prof. van Joolingen visited the VARTeL team almost every year before the Covid-19 (Fig. 34a). His Ph.D. student Susanne Jansen came to NTU for a six-month physical attachment (Fig. 34b) for a joint project development. The idea was to

Fig. 33 A VARTeL space @ Windesheim University of Applied Sciences

design an avatar serving as a biological teacher for Biology Education (Fig. 34c, d). A VARTeL app was developed as part of her Ph.D. thesis [57].

(3) **VARTeL @ Gymnasium Novum, Hague**

The partnership and collaboration between Singapore and Holland continued expanding from VARTeL research to K-12 STEM education. Gymnasium Novum (GN), in Hague, was the first Dutch school involved in the VARTeL collaboration. Students and teachers from this school used the specially designed avatar, for learning and teaching of the human digestive system (Fig. 35).

(4) **VARTeL @ Hollandse School, Singapore**

Following the Dutch education system, Hollandse School is an international school for Dutch pupils in Singapore. The original campus of this school, established 100 years ago is now part of the Shangri-la Hotel in Singapore (Fig. 36a). A virtual campus of the old school, designed by the NTU team, was used not only for the school's 100-anniversary ceremony but also for school history education (Fig. 36b). A VARTeL Space was set up, as part of the Maker Space, in this school (Fig. 36c).

(a) Prof. van Joolingen in NTU

(b) Jansen in NTU's VARTeL Classroom

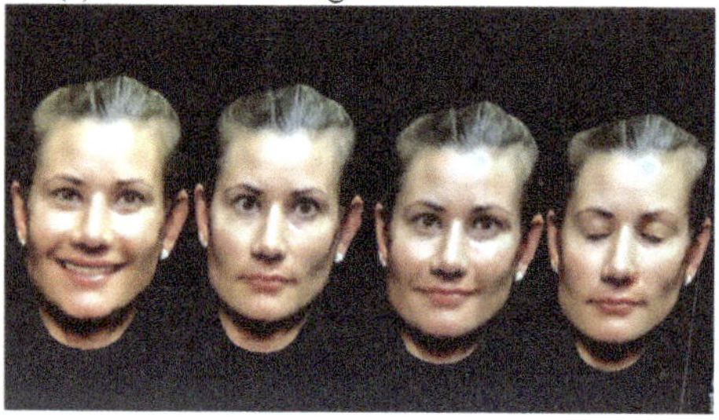

(c) Facial modeling of the virtual teacher

(d) An avatar

Fig. 34 Collaboration and academic exchange of VARTeL projects between UU and NTU

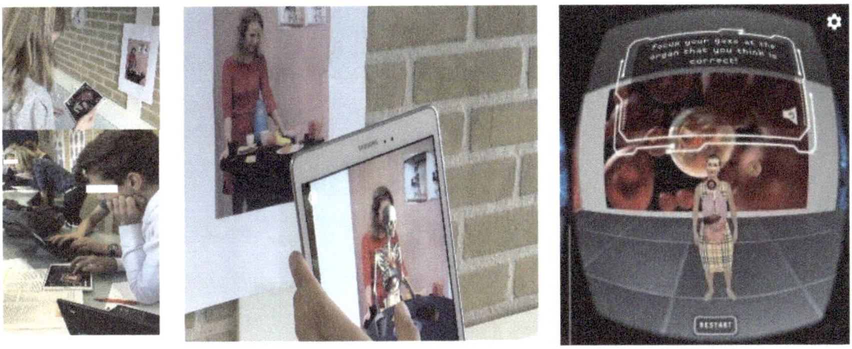

(a) GN students (b) AR for digestive system learning (c) An avatar bio teacher

Fig. 35 VARTeL @ GN

(a) The old campus (b) The virtual campus modelled for the 100-anniversary ceremony (c) The VARTel Maker Space

Fig. 36 VARTeL @ Hollandse School

6.2 *VARTeL @ PR China*

Back in 2005, the NTU team was invited to present their first VR exhibition in China. Over the years, several collaborations have been developed to promote VARTeL innovation in Public Education, and SNE Education.

(1) **VARTeL @ Shanghai International Art Expo**

An immersive and interactive Virus Theme Park was exhibited in the Oriental Pearl Tower (Zero Metre Hall) as part of the 2005 Shanghai International Art Expo in May 2005. About 20 different viruses related to HIV, Dengue, Mad Cow and so on were modelled by the NTU team for immersive visualization and interaction. The viral exhibition designed based on extreme game concept with The Virus Theme Park (Fig. 37a) such as roller-coaster [3], attracted around 70,000 visitors of different age groups for this week-long show (Fig. 37b, c).

(2) **VARTeL @ China National Science and Technology Museum**

Soon after the Shanghai exhibition, the VARTeL team in NTU was invited to do an immersive and interactive show at Beijing's China Science & Technology Museum

(a) Virus Theme Park (b) The Immersive & Interactive Show (c) The Visitors

Fig. 37 VARTeL @ Oriental Pearl Tower

(Hall C) in Oct 2025. X-games featuring various virus and protein structures were developed (Fig. 38). Specifically, the secondary structures of the virus proteins were now designed based on the patterns and shapes of the Great Wall [2, 4].

(3) **Singapore-Shanghai Exchange**

The National Science and Art Competition was held in Singapore in March 2008. The prize-winning works were invited for an exhibition in Shanghai in May 2008 (Fig. 39). This Singapore-Shanghai Exchange Programme was sponsored by the Lee Foundation.

(4) **VARTeL @ Zhejiang Science & Technology Museum**

The Mobile Immersive Theatre was adopted by Zhejiang Science and Technology Museum for travelling shows across the province to promote creative learning through VARTeL innovations.

(5) **VARTeL @ Suzhou Renai School**

Renai School is a special needs school located in the Suzhou Industrial Park. A VARTeL Classroom was set up in the school with the help from NTU, allowing students to learn communication skills through the Pink Dolphin Play (Fig. 40a).

Fig. 38 VARTeL @ China Sciences and Technology Museum

Fig. 39 VARTeL @ Singapore-Shanghai Exchange

Research was carried out using the VARTeL platform at Renai School. Ge and Fan [58] explored Social Development for Children with Autism using Kinect Gesture Games, while Wu and Zheng [59] investigated Motion Sensing Games for Children with ASD. Students from Renai School in China and AWWA schools in Singapore met online via Skype meeting (Fig. 40b) before their physical meeting in Suzhou in Oct 2018 (Fig. 40c, d).

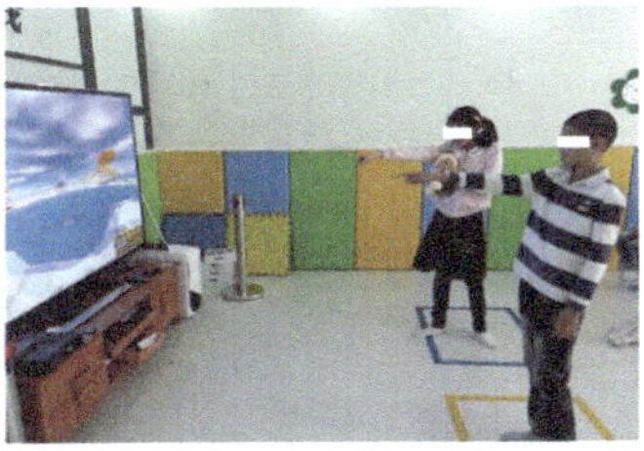

(a) A VARteL Space in Renai

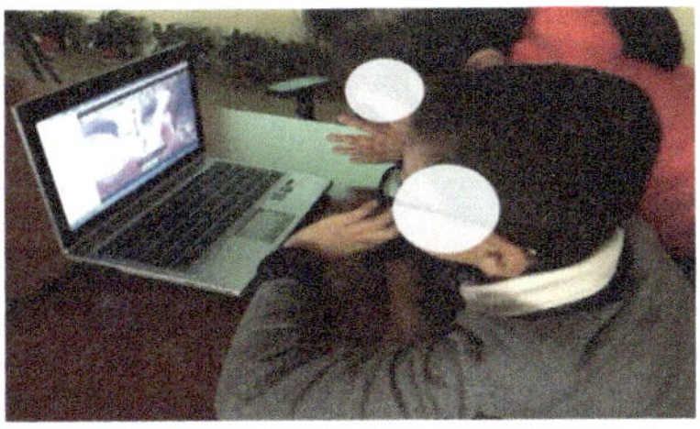

(b) Skype Meeting for students from Renai and AWWA

(c) Face-to-face exchange in Suzhou

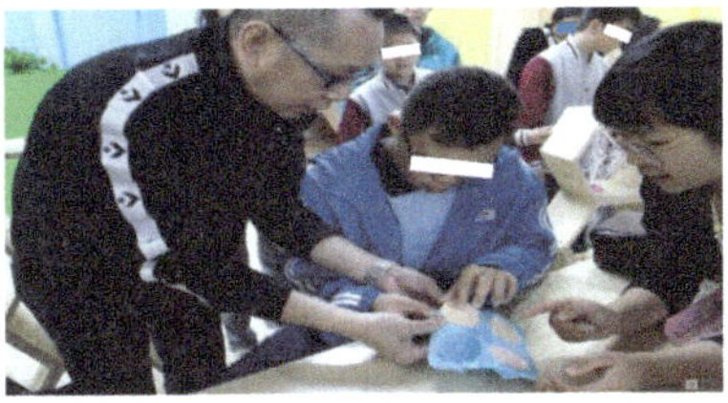

(d) AWWA students in Renai

Fig. 40 VARTeL @ Suzhou Renai School

6.3 VARTeL @ University of Johannesburg, South Africa

In collaboration with NTU, a VARTeL Space was established in 2021 at the Faculty of Education at the University of Johannesburg (UoJ) (Fig. 41), South Africa. This space was used to support teacher education preparing them for the Future of Classroom. Prof Ramnarain and his team investigated South African Pre-service Teachers' Use of Simulations in Virtual Physical Sciences Learning [60], and discussed issues and perspectives on Virtual and Augmented Reality in STEM education [61].

6.4 VARTeL Network, Conferences and International Partnership

In 2012, The first Asia-Europe Symposium on Simulation and Serious Games was held in NTU. The symposium travelled to The Netherlands in 2014 [51], China in 2016 [56], Finland in 2018 [55], Online in 2020 [62], South Africa in 2022. Each conference has one book published by Springer. Figure 42a was the 2014 meeting in Zwolle, Holland. The next conference will be back to NTU, Singapore in Dec 2024.

In Dec 2016, an international network was formed to promote partnership and collaboration in relation to VARTeL (Fig. 42b). Through the VARTeL network, Renai School in China and AWWA School in Singapore established a sister-school partnership (Fig. 40). In 2023, Renai School paid a return visit to AWWA in Singapore. The Education School of WUAS and Renai School signed an agreement between the two organizations for international collaboration (Fig. 42c). A Singapore-Holland Forum (Fig. 42d) was held in NTU every year except during the Covid-19 outbreak for principals and researchers from the two countries to share the latest development in education and EduTech innovation. Principals, school leaders, and researchers in Special Needs Education came together from different countries to learn from each other and explore international collaboration (Fig. 32e). Selected research outcomes from the collaboration were jointly presented at leading conference like Third International Conference on Immersive Learning [63]. Recently, Prof Ingrid Winkler from Brazil (Fig. 32f) joined this network with initial collaboration kicked start in metaverse and digital twin [17, 37].

7 About the Book

This book consists of two parts. Part 1 is edited by Yiyu Cai and Jean Chen, and Part 2 by Umesh Ramnarain.

Focusing on VARTeL Research, Part 1 has five chapters contributed from Nanyang Technological University, Singapore. Chan et al. describe, in Chap. 2, on the Effectiveness of VARTeL Environments in Enhancing Students' Self-Efficacy. Chung et al.

Fig. 41 VARTeL @ University of Johannesburg

discuss, in Chap. 3, on Optimizing Learning Through Game Design: An Essential Role in Educational Games. Phang et al. share their work, in Chap. 4, on Stealth Assessment for VR-based Skill Training Utilizing Hand Motion Detection. Song et al. present their research, in Chap. 5, on VARTeL for STEAM Education.

Part 2 on VARTeL Applications has twelve chapters. Patel et al. from India and Singapore present, in Chap. 6, their collaborative efforts on Virtual Reality Enabled Robotic Surgery Training through TeleSurgery. Tan et al. from Singapore and UK

(a) VARTeL conference

(b) VARTeL network

(c) WUAS-Renai partnership

(d) Singapore-Holland education forum

(e) Singapore-Holland-China SNE roundtable

(f) Brazil-Singapore Partnership

Fig. 42 VARTeL network and partnership

discuss, in Chap. 7, on Tree Pruning Simulation and Tree Growth Modelling. Ang et al. also from Singapore and UK describe, in Chap. 8, their Educational Metaverse for Learning of CPU Scheduling Algorithms in Operating Systems Subject. Asino et al. from USA share, in Chap. 9, on Designing Immersive Experiences Grounded in Culturally Relevant Teaching and Learning Strategies. Asino et al. from USA share, in Chap. 10, on Evaluating Learning and Engagement in Immersive Space. Potgieter et al. from South Africa describe, in Chap. 11, Using A Hierarchy of Data Sources to Represent Cultural Heritages Sites. Luo et al. from China present, in Chap. 12, An Interactive VR/Desktop-3D System for Nasal Sinus Anatomy Education. Luo et al. from China discuss, in Chap. 13, The Effectiveness of the Force Feedback Technology Application in the Immersive Physics Learning. Myburgh et al. from South Africa describe their efforts, in Chap. 14, towards universally accessible interactive augmented reality in education: A case study of developing a photoelectric effect experiment in interactive augmented reality. Alexdottir from UK presents, in Chap. 15, Phantom Touch in VR and its Potential uses in the Education Field. Zeng et al. from Republic of China present, in Chap. 16, Immersive virtual reality for science education. Finally, Mitchell et al. detail, in Chap. 17, A Cyber Physical Architecture for Symbiotic Multi-Robot Fleet Management.

Acknowledgements The first author would like to thank the following agencies or organizations for their support to this research: (1) The Science Engineering Research Council of The Agency of Science, Technology and Research (A*STAR); (2) Ministry of Info Communication and The Art under The Creative Community Singapore programme; (3) National Research Foundation, Ministry of Education and Information Development Authority under The FutureSchools @ Singapore Initiative; (4) Media Development Authority under The Media-in-Learning programme; (5) MOE Academic Fund; (6) MOE Tertiary Research Fund; (7) Temasek Trust-funded Singapore Millennium Foundation; (8) National Institute of Education under I3G and ERFP schemes; (9) The

Institute for Media Innovation; (10) National Library Board; (11) Fuji Xerox Pte Ltd; (12) Underwater World Singapore; (13) AWWA School; (14) Hwa Chong Institution; (15) River Valley High School; (16) National Junior College; (17) Raffles Institution; (18) Suzhou Industrial Park Renai School; (19) The Institute for Mental Health; (20) National Health Group Under Rehabilitation Grant; (21) Utrecht University; (22) Windesheim University of Applied Sciences; (23) University of Johannesburg; (24) University of Toronto; and (24) Lee Foundation.

Many people offered valuable help during the VARTeL journey. Special thanks should go to Dr. B Lu, Dr. Z Fan, Dr. H Wan, Prof M Opas, Mr. BK Ngoi, Ms. Serena Li, Madam KL So, Mr. M Wu, Mr. Russelle Loh, Mr. KM Kwek, Ms. Ruby Seah, Prof. N Thalmann, Prof. D Thalmann, Prof. J Zheng, Mr. Ryan Ba, Dr. YK Aung, Ms. Jasemine Tay, Mr. Nay, Dr. CW Hon, Dr. Sandra Tan, Dr. HN Gwee, Mr. Joseph Tan, Mr. Nick Chan, Prof. W van Joolingen, Dr. Susanne Jansen, Prof. Wen Yun, Prof. CK Looi, Prof. V Kong, Prof. L Wu, Dr Z Chen, Mr. Zhi Xiong Tay, Dr. Daniel Chua.

References

1. Cai, Y., Tay, C.T., Ngo, B.K.: Introduction to 3D immersive and interactive learning. In: Cai, Y. (ed) 3D Immersive and Interactive Learning. Springer, Singapore (2013). https://doi.org/10.1007/978-981-4021-90-6_1
2. Cai, Y., Lu, B., Fan, Z., Chan, C.W., Lim, K.T., Qi, L., Li, L.: Proteins, immersive games and music. Leonardo **39**(2), 135–137 (2006). https://doi.org/10.1162/leon.2006.39.2.135
3. Cai, Y., Lu, B., Fan, Z., Chandrasekaran, I., Lim, K.T., Chan, C.W., Jiang, Y., Li, L.: Bio-edutainment: learning life science through X gaming. Comput. Graph. **30**(1), 3–9 (2006). https://doi.org/10.1016/j.cag.2005.10.003
4. Cai, Y., Lu, B., Zheng, J., Li, L.: Immersive protein gaming for bio edutainment. Simulation Gaming **37**(4), 466–475 (Sage Publisher) (2006). https://doi.org/10.1177/1046878106293677
5. Lu, B., Lim, K.T., Zheng, J., Cai, Y.: Learning molecular biology by VR playing. In: Proceedings of the 2004 ACM SIGGRAPH International Conference on Virtual Reality Continuum and Its Applications in Industry, June 2004, pp. 168–172 (2004). https://doi.org/10.1145/1044588.1044622
6. Guan, Y., Cai, Y., Lee, Y.T., Opas, M.: An automatic method for identifying appropriate gradient magnitude for 3D boundary detection of confocal image stacks. J. Microsc. **223**(1), 66–72 (2006). https://doi.org/10.1111/j.1365-2818.2006.01600.x
7. Guan, Y., Cai, Y., Opas, M., Xiong, Z., Lee, Y.T.: A VR enhanced collaborative system for 3D confocal microscopic image processing and visualization. Int. J. Image Graphics **6**(02), 231–250 (2006). https://doi.org/10.1142/S0219467806002215
8. Guan, Y., Cai, Y., Zhang, X., Lee, Y.T., Opas, M.: Adaptive correction technique for 3D reconstruction of fluorescence microscopy images. Microsc. Res. Tech. **71**(2), 146–157 (2008). https://doi.org/10.1002/jemt.20536
9. Indhumathi, C., Cai, Y., Cao, C., Lu, B., Zheng, J.: Virtual reality prototyping of biomolecules. Virtual Phys. Prototyping **2**(1), 37–49 (2007). https://doi.org/10.1080/17452750601170316
10. Indhumathi, C., Cai, Y., Guan, Y., Opas, M.: An automatic segmentation algorithm for 3D cell cluster splitting using volumetric confocal images. J. Microsc. **243**(1), 60–76 (2011). https://doi.org/10.1111/j.1365-2818.2010.03482.x
11. Indhumathi, C., Cai, Y., Guan, Y., Opas, M.: 3D boundary extraction of confocal cellular images using higher order statistics. J. Microsc. **235**(2), 209–220 (2009). https://doi.org/10.1111/j.1365-2818.2009.03203.x
12. Cai, Y. (editor).: 3D Immersive and Interactive Learning, Springer (2013). https://doi.org/10.1007/978-981-4021-90-6
13. Shi, X., Cai, Y., Chan, C.W.: Electronic music for biomolecules using short music phrases. Leonardo **40**(2), 137–141 (2007). https://doi.org/10.1162/leon.2007.40.2.137

14. Cai, Y., Chen, J., Chan, W.H., Tan, S.Y.: DNA dance revolution. Leonardo (2024). https://doi.org/10.1162/leon_a_02525
15. Cai, P., Chandrasekaran, I., Cai, Y., Chen, Y., Wu, X.: Simulation-enabled vocational training for heavy crane operations. In: Cai, Y., Goei, S., Trooster, W. (eds) Simulation and Serious Games for Education. Gaming Media and Social Effects. Springer, Singapore (2017). https://doi.org/10.1007/978-981-10-0861-0_4
16. Cai, Y., Teo, Q.C.E., Cao, Q., Ba, R.A., Taib, S.F.B.M., Chen, J.: Virtual and augmented reality technology help learning of biology in secondary schools. In: 13th Annual International Conference on Education and New Learning Technologies (EDULEARN21), 5–6 July 2021, pp. 9752–9756 (2021). ISBN 9788409312672. https://doi.org/10.21125/edulearn.2021.1968
17. Sim, J.K., Xu, K.W., Jin, Y., Lee, Z.Y., Teo, Y.J., Mohan, P., Huang, L., Xie, Y., Li, S., Liang, N., et al.: Designing an educational metaverse: a case study of NTUniverse. Appl. Sci. **2024**(14), 2559 (2024). https://doi.org/10.3390/app14062559
18. Tan, S.: Gender differences in learning molecular biology using virtual learning environments. In: Cai, Y., Goei, S. (eds) Simulations, Serious Games and Their Applications. Gaming Media and Social Effects. Springer, Singapore (2014). https://doi.org/10.1007/978-981-4560-32-0_14
19. Tan, S., Waugh, R.: Use of virtual-reality in teaching and learning molecular biology. In: Cai, Y. (eds) 3D Immersive and Interactive Learning. Springer, Singapore (2013). https://doi.org/10.1007/978-981-4021-90-6_2
20. Gwee, H.N.: Effects of virtual-reality elements on spatial visualization skills of secondary three students in Singapore. In: Cai, Y. (eds) 3D Immersive and Interactive Learning. Springer, Singapore (2013). https://doi.org/10.1007/978-981-4021-90-6_3
21. Tan, J., Ngo, B.K., Chandrasekaran, I., Cai, Y.: Cultivating STEM education through 3D sabbaticals. In: Cai, Y. (eds) 3D Immersive and Interactive Learning. Springer, Singapore (2013). https://doi.org/10.1007/978-981-4021-90-6_5
22. Chow, B.H.: The virtual-reality campuses project. In: Cai, Y. (eds) 3D Immersive and Interactive Learning. Springer, Singapore (2013). https://doi.org/10.1007/978-981-4021-90-6_4
23. Chan, N.C.L., Ang, S.S., Goh, A.Y.: A 3D synergized discovery and learning journey: a case study of national junior college. In: Cai, Y. (eds) 3D Immersive and Interactive Learning. Springer, Singapore (2013). https://doi.org/10.1007/978-981-4021-90-6_6
24. Pang, C.G., Devi, S., Wong, D., Cai, Y., Ba, R.: The use of immersive virtual reality technology to deepen learning in singapore schools. In: Cai, Y., van Joolingen, W., Veermans, K. (eds) Virtual and Augmented Reality, Simulation and Serious Games for Education. Gaming Media and Social Effects. Springer, Singapore (2021). https://doi.org/10.1007/978-981-16-1361-6_5
25. Pang, C.G., Cai, Y.: Transforming learning experiences through affordances of virtual and augmented reality. In: Cai, Y., Mangina, E., Goei, S.L. (eds) Mixed Reality for Education. Gaming Media and Social Effects. Springer, Singapore (2023). https://doi.org/10.1007/978-981-99-4958-8_6
26. Wang, C.J., Low, A.S.L., Woo, C.S.: 3D system for learning gains in the lower secondary normal (technical) stream. In: Cai, Y. (eds) 3D Immersive and Interactive Learning. Springer, Singapore (2013). https://doi.org/10.1007/978-981-4021-90-6_7
27. Wen, Y., Lai, C., He, S., Cai, Y., Looi, C.K., Wu, L.: Investigating primary school students' epistemic beliefs in augmented reality-based inquiry learning. Interactive Learn. Environ. 1–18 (2023). https://doi.org/10.1080/10494820.2023.2214182
28. Wen, Y., Wu, L., He, S., et al.: Integrating augmented reality into inquiry-based learning approach in primary science classrooms. Educ. Tech. Res. Dev. **71**, 1631–1651 (2023). https://doi.org/10.1007/s11423-023-10235-y
29. Cai, Y.: Virtual reality technology enhanced learning. In: 9th International Conference on Education and New Learning Technologies, Barcelona, Spain, 3–5 July 2017 (2017). https://doi.org/10.21125/edulearn.2017.1058
30. Ba, R., Cai, Y., Guan, Y.: Augmented reality simulation of cardiac circulation using APPLearn (Heart). In: 2018 IEEE International Conference on Artificial Intelligence and Virtual Reality (AIVR), Taichung, Taiwan, 2018, pp. 241–243 (2018). https://doi.org/10.1109/AIVR.2018.00055

31. Ba, R., Huang, L., Taib, S.F.B.M., Chen, J., Aung, Y.K., Cai, Y.: AR simulation of cardiac circulation. In: Cai, Y., van Joolingen, W., Veermans, K. (eds) Virtual and Augmented Reality, Simulation and Serious Games for Education. Gaming Media and Social Effects. Springer, Singapore (2021). https://doi.org/10.1007/978-981-16-1361-6_12
32. Ba, R., Xie, Y., Zhang, Y., Taib, S.F.B.M., Cai, Y., Walker, Z., Chen, Z., Tan, S., Chow, B.H., Lim, S.M., Pang, D., Goei, S.L., Matimba, H.E.K., van Joolingen, W.: Virtual reality enzymes: an interdisciplinary and international project towards an inquiry-based pedagogy. In: Cai, Y., van Joolingen, W., Walker, Z. (eds) VR, Simulations and Serious Games for Education. Gaming Media and Social Effects. Springer, Singapore (2019). https://doi.org/10.1007/978-981-13-2844-2_5
33. Ng, E., Bakr, A.A., Cai, Y.: Integration of Wearable Devices in VR-Enhanced Aircraft Maintenance Training. In: Cai, Y., Mangina, E., Goei, S.L. (eds) Mixed Reality for Education. Gaming Media and Social Effects. Springer, Singapore (2023). https://doi.org/10.1007/978-981-99-4958-8_12
34. Xie, Y., Zhang, Y., Cai, Y.: Virtual reality engine disassembly simulation with natural hand-based interaction. In: Cai, Y., van Joolingen, W., Walker, Z. (eds) VR, Simulations and Serious Games for Education. Gaming Media and Social Effects. Springer, Singapore (2019). https://doi.org/10.1007/978-981-13-2844-2_11
35. Yu, Z., Huang, L., Ooi, C.H., Gong, Y., Cai, Y.: Immersive and interactive VR for lifting training of tower cranes. In: Cai, Y., Mangina, E., Goei, S.L. (eds) Mixed Reality for Education. Gaming Media and Social Effects. Springer, Singapore (2023). https://doi.org/10.1007/978-981-99-4958-8_9
36. Lee, T., Wen, Y., Chan, M.Y., Azam, A.B., Looi, C.K., Taib, S., Cai, Y., et al.: Investigation of virtual and augmented reality classroom learning environments in university STEM education. Interact. Learn. Environ. 1–16 (2022). https://doi.org/10.1080/10494820.2022.2155838
37. Wong, J.Y., Azam, A.B., Cao, Q., Huang, L., Xie, Y., Winkler, I., Cai, Y.: Evaluations of virtual and augmented reality technology-enhanced learning for higher education. Electronics **13**, 1549 (2024). https://doi.org/10.3390/electronics13081549
38. Gunawan, P., Kwan, J., Cai, Y., Yang, R.: Augmented reality application for chemical engineering unit operations. In: Cai, Y., van Joolingen, W., Veermans, K. (eds) Virtual and Augmented Reality, Simulation and Serious Games for Education. Gaming Media and Social Effects. Springer, Singapore (2021). https://doi.org/10.1007/978-981-16-1361-6_4
39. Gao, S., Lu, Y., Ooi, C.H., Cai, Y., Gunawan, P.: Designing interactive augmented reality application for students directed learning of continuous distillation process. Comput. Chem. Eng. 169 (2023). https://doi.org/10.1016/j.compchemeng.2022.108086
40. Cai, Y., Chia, N.K.H., Thalmann, D., Kee, N.K.N., Zheng, J., Thalmann, N.M.: Design and development of a virtual dolphinarium for children with autism. In: IEEE Transactions on Neural Systems and Rehabilitation Engineering, vol. 21, no. 2, pp. 208–217, March 2013 (2013). https://doi.org/10.1109/TNSRE.2013.2240700
41. Liu, W.R., Cao, Q., Cai, Y.: Serious game design for virtual dolphin-assisted learning. In: Cai, Y., Cao, Q. (eds) When VR Serious Games Meet Special Needs Education. Gaming Media and Social Effects. Springer, Singapore (2021). https://doi.org/10.1007/978-981-33-6942-9_6
42. Lu, D., Cao, Q., Cai, Y.: Virtual pink dolphins and lagoon. In: Cai, Y., Cao, Q. (eds) When VR Serious Games Meet Special Needs Education. Gaming Media and Social Effects. Springer, Singapore (2021). https://doi.org/10.1007/978-981-33-6942-9_5
43. Cai, Y., Cao, Q.: VR serious games for special needs education on living skills training. In: 13th Annual International Conference on Education and New Learning Technologies (EDULEARN21), 5–6 July 2021, pp. 7629–7634 (2021). ISBN 9788409312672. https://doi.org/10.21125/edulearn.2021.1551
44. Cai, Y., Chiew, R., Fan, L., Kwek, M.K., Goei, S.L.: The virtual pink dolphins project: an international effort for children with ASD in special needs education. In: Cai, Y., Goei, S., Trooster, W. (eds) Simulation and Serious Games for Education. Gaming Media and Social Effects. Springer, Singapore (2017). https://doi.org/10.1007/978-981-10-0861-0_1

45. Cai, Y., Cao, Q. (eds.).: When VR Serious Games Meet Special Needs Education, Springer, Singapore (2021). https://doi.org/10.1007/978-981-33-6942-9
46. Cai, Y., Chiew, R., Nay, Z.T., Indhumathi, C., Huang, L.: Design and development of VR learning environments for children with ASD. Interact. Learn. Environ. **25**(8), 1098–1109 (2017). https://doi.org/10.1080/10494820.2017.1282877
47. Lu, A., Chan, S., Cai, Y., Huang, L., Nay, Z.T., Goei, S.L.: Learning through VR gaming with virtual pink dolphins for children with ASD. Interact. Learn. Environ. **26**(6), 718–729 (2018). https://doi.org/10.1080/10494820.2017.1399149
48. Mohd Taib, S.F.B., Zhang, Y., Cai, Y., Goh, T.J.: Supermarket route-planning game: a serious game for the rehabilitation of planning executive function of children with ASD. In: Cai, Y., van Joolingen, W., Walker, Z. (eds) VR, Simulations and Serious Games for Education. Gaming Media and Social Effects. Springer, Singapore (2019). https://doi.org/10.1007/978-981-13-2844-2_10
49. Aziz, M.A.B.A., Cao, Q., Cai, Y.: Design of a home bag-packing serious game for children with ASD. In: Cai, Y., Cao, Q. (eds) When VR Serious Games Meet Special Needs Education. Gaming Media and Social Effects. Springer, Singapore (2021). https://doi.org/10.1007/978-981-33-6942-9_9
50. Cai, Y., Goei, S.L. (eds.): Simulations, Serious Games and Their Applications. Springe, Singapore (2013). https://doi.org/10.1007/978-981-4560-32-0
51. Cai, Y., Goei, S.L., Trooster, W. (eds.): Simulation and Serious Games for Education. Springer, Singapore (2016). https://doi.org/10.1007/978-981-10-0861-0
52. Trooster, W.: The use of virtual worlds and serious gaming in education. In: Cai, Y., Goei, S. (eds) Simulations, Serious Games and Their Applications. Gaming Media and Social Effects. Springer, Singapore (2014). https://doi.org/10.1007/978-981-4560-32-0_8
53. Gomes, T.D.P., Goei, S.L., van Joolingen, W., Cai, Y.: VR biology, an interdisciplinary and international student project towards an inquiry-based pedagogy. In: Proceedings of the 3rd Asia-Europe Symposium on Simulation and Serious Gaming, December 2016, pp. 169–171 (2016). https://doi.org/10.1145/3014033.3014044
54. van Joolingen, W., Goei, S.L., Matimba, H., Ba, R.K.T.A., Pedro Gomez Dias, T.: Developing and integrating an augmented reality app for teaching and learning about enzymes. In: Cai, Y., van Joolingen, W., Veermans, K. (eds) Virtual and Augmented Reality, Simulation and Serious Games for Education. Gaming Media and Social Effects. Springer, Singapore (2021). https://doi.org/10.1007/978-981-16-1361-6_11
55. Cai, Y., van Joolingen, W., Veermans, K. (eds.): Virtual and Augmented Reality, Simulation and Serious Games for Education. Springer (2021). https://doi.org/10.1007/978-981-16-1361-6
56. Cai, Y., van Joolingen, W., Walker Z. (eds.): VR, Simulations and Serious Games for Education. Springer, Singapore (2019). https://doi.org/10.1007/978-981-13-2844-2
57. Jansen, S., Taib, S.F.B.M., Cai, Y., van Joolingen, W.R.: Applying prior meta-modeling knowledge to a VR model of a biological process. In: Cai, Y., Mangina, E., Goei, S.L. (eds) Mixed Reality for Education. Gaming Media and Social Effects. Springer, Singapore (2023). https://doi.org/10.1007/978-981-99-4958-8_4
58. Ge, Z., Fan, L.: Social development for children with autism using kinect gesture games: a case study in Suzhou Industrial Park Renai School. In: Cai, Y., Goei, S., Trooster, W. (eds) Simulation and Serious Games for Education. Gaming Media and Social Effects. Springer, Singapore (2017). https://doi.org/10.1007/978-981-10-0861-0_8
59. Wu, C., Zheng, Q.: Motion sensing games for children with autism spectrum disorder. In: Cai, Y., van Joolingen, W., Walker, Z. (eds) VR, Simulations and Serious Games for Education. Gaming Media and Social Effects. Springer, Singapore (2019). https://doi.org/10.1007/978-981-13-2844-2_6
60. Ramnarain, U., Penn, M.: An investigation of South African Pre-service teachers' use of simulations in virtual physical sciences learning: process, attitudes and reflections. In: Cai, Y., van Joolingen, W., Veermans, K. (eds) Virtual and Augmented Reality, Simulation and Serious Games for Education. Gaming Media and Social Effects. Springer, Singapore (2021). https://doi.org/10.1007/978-981-16-1361-6_7

61. Ramnarain, U., Mavuru, L., Ramaila, S., Dhurumraj, T., Penn, M., Cai, Y., Rodil, K.: Issues and perspectives on virtual and augmented reality in STEM education: (panel paper). In: Conference of the Southern African Association for Research in Mathematics, Science and Technology Education 2022, Cambridge, United Kingdom (2022)
62. Cai, Y., Mangina, E., Goei, S.L. (eds.): Mixed Reality for Education. Springer (2023). https://doi.org/10.1007/978-981-99-4958-8
63. Cai, Y., Walker, Z., van Joolingen, W., Goei, S.L., Tan, S.H.: Immersive learning of biomolecules. In: Proceedings of the iLRN: 3rd International Conference on Immersive Learning, pp. 156–161. iLRN, Coimbra, Portugal (2017). https://doi.org/10.3217/978-3-85125-530-0-22
64. Chan, S.M.Y., Cao, Q., Chen, J., Cai, Y.: Evaluation of serious games for special needs education. In: Cai, Y., Cao, Q. (eds) When VR Serious Games Meet Special Needs Education. Gaming Media and Social Effects. Springer, Singapore (2021). https://doi.org/10.1007/978-981-33-6942-9_7
65. Chong, S., Cao, Q., Cai, Y.: Game-assisted vocational training. In: Cai, Y., Cao, Q. (eds) When VR Serious Games Meet Special Needs Education. Gaming Media and Social Effects. Springer, Singapore (2021). https://doi.org/10.1007/978-981-33-6942-9_8
66. Chua, L., Goh, J., Nay, Z.T., Huang, L., Cai, Y., Seah, R.: ICT-enabled emotional learning for special needs education. In: Cai, Y., Goei, S., Trooster, W. (eds) Simulation and Serious Games for Education. Gaming Media and Social Effects. Springer, Singapore (2017). https://doi.org/10.1007/978-981-10-0861-0_3
67. Chung, Y.P.J., Cao, Q., Cai, Y.: Design of a virtual home for special needs children to learn life skills. In: Cai, Y., Cao, Q. (eds) When VR Serious Games Meet Special Needs Education. Gaming Media and Social Effects. Springer, Singapore (2021). https://doi.org/10.1007/978-981-33-6942-9_3
68. Goh, J.H., Cao, Q., Cai, Y.: Design of a VR supermarket serious game. In: Cai, Y., Cao, Q. (eds) When VR Serious Games Meet Special Needs Education. Gaming Media and Social Effects. Springer, Singapore (2021). https://doi.org/10.1007/978-981-33-6942-9_11
69. Goh, Z.H.J., Cao, Q., Chen, J., Cai, Y.: iPad serious game to aid children with special needs in emotion learning. In: Cai, Y., Cao, Q. (eds) When VR Serious Games Meet Special Needs Education. Gaming Media and Social Effects. Springer, Singapore (2021). https://doi.org/10.1007/978-981-33-6942-9_10
70. Hoe, G., Cao, Q., Chen, J., Cai, Y.: Learning life skills through gaming for children with autism spectrum disorder. In: Cai, Y., van Joolingen, W., Veermans, K. (eds) Virtual and Augmented Reality, Simulation and Serious Games for Education. Gaming Media and Social Effects. Springer, Singapore (2021). https://doi.org/10.1007/978-981-16-1361-6_6
71. Teoh, S.E.M., Cao, Q., Cai, Y.: Learning to take a shower through VR serious gaming. In: Cai, Y., Cao, Q. (eds) When VR Serious Games Meet Special Needs Education. Gaming Media and Social Effects. Springer, Singapore (2021). https://doi.org/10.1007/978-981-33-6942-9_2
72. Zhang, Q., Cao, Q., Cai, Y.: Learning to cross roads through VR playing. In: Cai, Y., Cao, Q. (eds) When VR Serious Games Meet Special Needs Education. Gaming Media and Social Effects. Springer, Singapore (2021). https://doi.org/10.1007/978-981-33-6942-9_4

Effectiveness of Virtual and Augmented Reality Learning Environments in Enhancing Students' Self-efficacy

Ming Yi Chan, Abu Bakr Azam, Lihui Huang, Yuan Xie, Siyi Li, and Yiyu Cai

Abstract This Chapter presents the results of an experimental study of the effects of Virtual and Augmented Reality (VAR) learning environments on students' academic performance and self-efficacy (SE). In the study, SE was measured across three dimensions, course-specific SE, content-specific SE, and problem-specific SE. VAR simulations were used to teach engineering students about abstract concepts and hands-on experience. Quantitative data on students' academic performance and SE were collected using Likert questionnaires administered before and after the VAR teaching activities. Focus group discussions were conducted as a qualitative manner to gain an in-depth understanding of student's overall learning experience in the VAR environment. The results have shown that VAR learning activities had a positive effect on students' academic performance and SE. Across three SE dimensions, problem-specific SE had the greatest improvement. The implications of the results are then discussed for further research in the field of VAR technology-enhanced learning.

1 Introduction

1.1 Background

Since Albert Bandura's introduction of the concept of self-efficacy (SE), it has become an important topic in educational research. Bandura defined SE as the perceived capability to perform a target behavior [1–3]. He suggested that the level of SE can be used to predict whether a task will be initiated, the amount of effort that will be exerted, and the level of determination to complete the task when faced with challenges [1]. Extensive studies were done to further the explanation of SE to explain behavioral differences in the areas of academic achievements [4–6], physical and cognitive performance [7], health [8], and mental health [9]. In particular, SE

M. Y. Chan · A. B. Azam · L. Huang · Y. Xie · S. Li · Y. Cai (✉)
Nanyang Technological University, Singapore, Singapore
e-mail: myycai@ntu.edu.sg

Y. Cai et al. (eds.), *Virtual and Augmented Reality Technology-Enhanced Learning*, Gaming Media and Social Effects, https://doi.org/10.1007/978-981-96-2332-7_2

has been extensively studied as it plays a significant role in influencing important academic outcomes (e.g., students' motivation and achievement).

In the context of education, SE refers to students' beliefs about their academic capabilities to achieve academic success [10]. Several studies revealed the importance of SE and its effect on academic achievement. Chemers and Garcia found that students with a higher level of SE have greater academic expectations and display a greater commitment to academic pursuits [11]. Other researchers have demonstrated that SE is a strong predictor of learning-related emotions and academic success [4, 5]. In contrast, students with low SE attribute their failure to their low abilities. They are also more likely to avoid and abandon difficult tasks.

Understanding the relevance of SE, researchers have explored instruments that measure this construct. At the most generic level of measurement, the more trait-like generality dimension of SE is assessed, which is termed general SE (GSE) [12]. GSE captures individual differences in the confidence of their abilities in meeting task demands in a broad array of contexts [13]. The most commonly used GSE measure is the General Self-Efficacy Scale [13]. This scale is a self-report measure of GSE consisting of 17 items that are designed to measure "a general set of expectations that the individual carries into new situations" [14]. Scholars have also developed instruments that assess students' GSE for more generalized academic behaviors without specifying a particular subject matter. Despite the accumulation of research in GSE, academia has continued to question the validity of this approach. This approach is incongruent with Bandura's theoretical view of the SE construct. Bandura suggested that GSE measures "bear little or no relation either to efficacy beliefs related to particular activity domains or to behavior" [3]. To put it simply, SE measurement should be specific to the context instead of regarded as a trait-like construct [15]. In the academic setting, students' SE is mostly construed as specific: course-specific, content-specific, and problem-specific.

1.1.1 Course-Specific Self-efficacy

Course-specific SE investigates students' confidence in their abilities with regard to a specific subject matter. For instance, an early study examined the relationship between SE and classroom academic performance for seventh graders [16]. A self-report measure of student SE was administered, and performance data were obtained from classroom assignments. Sample items in the survey read "I expect to do very well in this class" and "I think I will receive a good grade in this class". The survey included items that assess students' perception of their capability to master course material and attain a high grade in a specific domain. Later studies also used a similar methodology to measure academic SE at the course-specific level [7, 17–19].

1.1.2 Content-Specific Self-efficacy

Content-specific SE concerns ones' confidence in their understanding of a specified subject domain. Survey items are designed to assess students' confidence in their capability to respond correctly to items testing specific content knowledge. For example, Bong developed a SE instrument in connection with a specific content domain to predict college students' performance and future enrollment intentions [20]. Five items were posed about students' confidence in mastering key contents taught related to instructional technology (IT). An example reads "I'm confident that I can successfully solve problems on the definitions of IT" [20]. Another example of this measure is the mathematics SE scales developed by Lopez and his colleagues, items were designed to measure students' confidence in geometry and advanced algebra [21].

1.1.3 Problem-Specific Self-efficacy

On the other hand, problem-specific SE concerns ones' confidence in their abilities to apply the content knowledge to solve specific problems. Written descriptions of problems or tasks are presented in place of the actual problems students must solve. For instance, "How confident are you that you can successfully solve equations containing square roots?" or "How confident are you that you can successfully read a given passage and come up with appropriate title words?" [22]. Students then rate the degree of their confidence in solving each problem correctly without solving the problem. In practice, there are studies that utilized all three SE constructs for comparison [20, 22].

Given the relevance of SE in academic achievement, it is critical to develop SE beliefs in the classroom. In this regard, VAR can play a crucial role in developing students' SE. VAR are immersive technologies wherein sensory experiences, (e.g., visual, auditory and touch stimuli) are artificially created, enabling users to manipulate the objects within the virtual environment [23]. VAR solutions are deployed to create an immersive and interactive learning environment, enabling students to learn experientially. This hands-on approach to learning has positive cognitive and pedagogical benefits, for example, draws student interest and increases motivation, which ultimately enhances overall educational outcomes [24]. In particular, the ability to provide a visual presentation of 3D models and abstract concepts makes VAR a valuable tool for teaching engineering applications that require visualization and treatment of objects [25]. By offering more intuitive, interactive, and effective learning experiences, students gain valuable experience in performing tasks. In which, mastery experiences contribute to developing students' confidence and increasing SE [26].

Despite the growing enthusiasm among educators around using VAR as an educational tool, few scholars have attempted to investigate the effects of VAR on students' SE specifically. A pilot study evaluated VAR-based learning against a more traditional technology, PowerPoint (PPT) [27]. Participants completed a short five-minute anatomy demonstration using VAR or PPT. SE scores were significantly higher for using VAR compared to using PPT. While the study shows that VAR has a pronounced immediate effect on SE, the longer-term effects are unknown. It is possible that a longer intervention could produce different results. Another research compared the effects of VAR versus traditional videos on SE [17]. Middle-school students learned about humans' impact on the ocean through VAR or traditional videos [17]. They found that students in the VAR condition reported higher SE than the ones who were shown traditional videos. In the study, SE was measured in self-reporting questionnaires with items assessing students' SE on generalized academic behaviors without specifying a particular subject matter. This approach only considers SE at a general level, which is incongruent with Bandura's theoretical view of the SE construct [1]. The measurement of SE should be specific to the context instead of regarded as a trait-like construct. In the academic setting, it is important to consider course-specific, content-specific SE, and problem-specific SE.

1.2 Objective

To ensure the validity of the SE measurement, this Chapter aims to examine the effectiveness of VAR in enhancing students' academic performance and SE along three SE dimensions. More specifically, this study aims to answer the following research questions:

1. What are the effects of VAR learning environments on students' academic performance?
2. What are the effects of VAR learning environments on students' SE along three SE dimensions (e.g., course-specific SE, content-specific SE, and problem-specific SE)?

Insights from this study will help to provide a better understanding of the process of learning in VAR learning environments and direct the proper use of VAR in education to enhance students' academic performance and SE.

1.3 Scope

Our study sample comprised 93 students from School of Mechanical and Aerospace Engineering (MAE) at Nanyang Technological University (NTU), Over the course of 5 weeks, VAR applications were used to teach students about engineering applications, helping them grasp abstract engineering concepts and gain hands-on experience. Students' SE was measured across three dimensions (course-specific SE, content-specific SE, and problem-specific SE). SE measurements were obtained from Likert scale self-reporting questionnaires, supplemented by qualitative data retrieved from focused group discussions.

2 Literature Review

2.1 The Relationship Between Self-efficacy and Academic Performance

In the educational context, students' SE refers to individuals' beliefs in their ability to perform academic tasks successfully to reach their personal goals [10]. Students with a high level of SE have high confidence in their capabilities. They are more likely to invest a high degree of effort in their studies and be intrinsically motivated. Students with low SE, on the other hand, believe that they lack the ability to succeed. Consequently, they are less likely to make a concerted effort and tend to give up quickly when difficulties arise. As a result, students with a low level of SE have low aspirations which can impede academic performance and become part of self-fulfilling prophecies of failure [28].

SE was found to be a good predictor of academic scores, which is consistent and supportive of the theoretical relationship between SE beliefs and academic performance. A growing body of literature explored the relationship between students' SE and their academic performance. Many studies have shown that there is a positive relation between SE and academic performance, independent of students' gender, domains, and disciplines. Several studies examined the significance of SE beliefs in predicting academic success in the sciences and engineering fields [29–31]. Other studies have further established how SE beliefs could account for variance in educational outcomes in other educational levels and disciplines [11, 32–34].

2.2 Development of Self-efficacy

Incorporating strategies to raise students' SE in class is crucial as SE is learned, and not inherited. Previous studies examined the theoretical formulations of SE and were able to explain the nature, sources, and psychological processes involved in the development of SE beliefs [3, 6]. It was found that levels of SE are not constant, and its development is closely intertwined with a person's interpretations of former and current experiences. Learners obtain information to appraise their SE from four main sources. These include mastery experiences, vicarious experiences, verbal persuasion, and emotional state. Lent suggested the most effective way of developing students' SE is through mastery experiences [35]. Performing a task successfully boosts our sense of SE while failing to complete the task can erode SE. When a student experiences setbacks and overcomes these challenges, he or she becomes resilient and subsequently acquires the essential experience needed to overcome difficult tasks. Another source is vicarious experience, observing a peer similar to oneself succeed at a task can strengthen observers' beliefs that they too possess abilities to succeed. Verbal persuasion, on the other hand, is the third source of SE that relies on teachers verbally persuading students that they possess the capabilities to master given activities and motivate students to make a greater effort. The last source suggests that emotional states can affect one's judgment in their capabilities. Positive emotions can translate to an increase in confidence and contribute to strong performances [23].

Notably, there is a growing interest in integrating VAR into classroom activities to improve students' SE. With VAR, students gain experience in performing the task and observe others performing the task in simulations. This highlights the potential of using an integrated curriculum of VAR for effective SE development through mastery experience and vicarious experience.

2.3 Virtual and Augmented Reality Learning Environment

The use of VAR in education is no longer a novelty and is continuously transforming classroom experiences. VAR applications are compatible with head-mounted displays (HMD). With the rapid development of HMD, VAR is becoming easier to use and more affordable. As such, an increasing number of institutions have started to incorporate VAR technologies in their curriculum. A survey finds that over two-thirds of higher education institutions have engaged in some level of VAR deployment [37]. For example, undergraduate students at Arizona State University learn about the impacts of climate change through interactive virtual field trips to the polar region [38]. Astronomy students at Purdue University study 3D models of astronomical objects in a virtual, collaborative environment using VAR technologies [39].

To date, the theoretical foundations of using VAR in education are often discussed with reference to the constructivism theory [40, 41]. The constructivist philosophy states that "learning" is the outcome of a process of construction and not a mechanism of representation, it comes about through active, authentic, cooperative, and reflective learning activities [42]. As proposed by Jonassen on constructive learning environments, they should offer reproductions of reality that can provide cases beyond predetermined instructional sequences; offer multiple representations of reality and support the construction of knowledge in accordance with the context [42]. In this philosophy of learning, VAR could reproduce the complexity of reality in an appealing and interesting way through the use of simulations, 3D imagery, and advanced audio-visual effects. This may help learners to value the importance and relevancy of the content taught, which could result in increased motivation and participation [43]. In the VAR learning environments, students can navigate and interact with virtual objects. VAR is therefore exceptionally well-suited for encouraging experimentation-based learning. This is in accordance with constructivists' beliefs that students can learn better when they are actively engaged in constructing knowledge in a "learning-by-doing" setting. As seen, the characteristics of VAR learning environments support this learning theory.

In addition to the theoretical framework discussed, extensive research uses experimental methods to examine the effects of VAR on students' learning outcomes. VAR learning activities were introduced in the classroom, and various aspects of students' performances were subsequently measured. This approach allows researchers to investigate how VAR can provide specific benefits to students and verifies the theoretical findings. For example, it was found that the use of interactive physics education books with VAR successfully enhanced students' enthusiasm and motivation [44]. Another research discovered that gamification learning supported by VAR induced deeper concentration [45]. Further studies reported that VAR applications improved collaborative learning by providing the opportunity for effective communication [46, 47]. While the benefits of VAR in education have been studied, there is insufficient attention paid to the effects of VAR on students' SE specifically.

2.4 *Role of VAR in Enhancing Student's Self-efficacy*

VAR learning environments have significant potential for SE development, especially in teaching engineering applications. The popularity of VAR in engineering can be attributed to the ability to simulate real-world experiences where students can interact with objects to acquire the necessary skills [48]. In the virtual environment, students gain hands-on learning experiences that sometimes would not otherwise be possible due to physical and resource limitations. For instance, students can practice car detailing or dissembling an aircraft engine in the virtual environment which are costly operations in the real world. The more of these experiences, the more students will develop SE and the confidence to solve engineering problems in real life. In addition, when students observe their peers successfully completing a learning activity in the virtual environment, they believe they too possess the ability. Students gain mastery experiences and vicarious experiences in the VAR learning environment which are positive strategies for fostering students' SE.

However, few studies have examined the impact of VAR on students' SE. Parmar conducted an experiment to evaluate the effects of VAR programming practice on middle-school students' learning achievements [49]. Results from the experiments show that immersive VAR is greatly effective in developing students' confidence in programming. Another research on SE compared the effects of VAR versus traditional videos [4]. Middle-school students learned about humans' impact on the ocean through VAR or traditional videos. They found that students in the VAR condition reported higher SE than the ones who were shown traditional videos. In the study, SE was measured in a self-reporting questionnaire with items assessing generalized academic behaviors. This approach only considers general SE. To ensure the validity of SE measurement, the SE construct should be discipline-specific. However, since content taught in middle-school is relatively basic and generic, it could be challenging to assess content-specific SE and problem-specific SE.

The majority of studies on the impact of VAR on students' SE were conducted at the middle-school level, few studies focused on tertiary education. A pilot study in 2019 evaluated VAR-based learning against a more traditional technology,

PowerPoint (PPT) in delivering human gross anatomy laboratory sessions to first-year osteopathic medical students [11]. Participants completed a short five-minute, anatomy identification demonstration using VAR or PPT. The results showed that SE was significantly higher for using VAR compared to using PPT. While the study demonstrated that VAR has a pronounced immediate effect on SE, the longer-term effects are unknown. It is possible that a longer intervention could produce different results.

While prior studies have confirmed the positive impact of VAR on SE for middle-school students, there lacked research within tertiary education. Furthermore, SE is often considered as a single general construct in these studies. Students' general SE was measured without specifying a particular subject matter. Motivated by such observation in the current field, the present study examines the effectiveness of VAR in enhancing university students' SE along three SE dimensions. By focusing on course-specific SE, content-specific SE and problem-specific SE, the present study aims to provide a more comprehensive understanding of the effects of VAR on SE. As VAR technologies continue to evolve and become more accessible, insights from this study can contribute to a better understanding of the potential benefits of VAR for students' learning outcomes in higher education.

3 Methodology

The research aimed to examine the effectiveness of VAR in enhancing students' academic performance and SE. The study adopted a within-group experimental design. Quantitative data on students' academic performance and SE were collected using Likert questionnaires administered before and after the VAR teaching activities. Focus group discussions were conducted as a qualitative approach to gain an in-depth understanding of student's overall learning experience in the VAR environment. The observed results were compared, and conclusions were drawn regarding the impact of VAR on academic performance and SE.

3.1 Participants

The research sample consisted of a total of 93 students from two tutorial groups for the module "MA4831 Computer Aided Engineering". In each group, there were 43 and 48 undergraduate students, respectively. Students were from the School of Mechanical and Aerospace Engineering majoring in mechanical or aerospace engineering. The students were selected from the second semester of the undergraduate program in the 2022/2023 academic year. Over the course of 5 weeks, students completed VAR learning activities in groups with the guidance of the professor and teaching assistants. As the module is a major prescribed elective, it is not mandatory for students to enroll in the module. It is thus likely that students enrolled in the module would have some prior interest in VAR.

3.2 VAR Learning Environment Set-Up

The study was conducted in a tutorial room at Hive which was adapted to support VAR learning activities. Specifically, it was equipped with a projector, six individual high-resolution displays, and six sets of HMDs. The classroom layout can be seen in Fig. 1. The VAR applications were developed using Unity 3D and were compatible with different head-mounted displays (HMD). The HMD used was Meta Quest 2. It is currently (as of early 2023) the best-selling HMD in the market [50]. As seen in Fig. 2 [51], Meta Quest 2 has a compact and ergonomic design that increases balance and comfort. Wearing comfort is one of the key considerations in selecting HMDs for educational purposes as HMDs are known to cause discomfort and potentially nausea, after prolonged use. The hand control used was Meta Quest 2 Touch Controllers (Fig. 3) [51]. The controllers allow users to use their hands and fingers to manipulate and interact with virtual objects. They are equipped with motion sensors that track the position and movement of the user's hands and fingers, enabling accurate and responsive hand-tracking in the virtual environment.

Fig. 1 VAR classroom layout @ The Hive

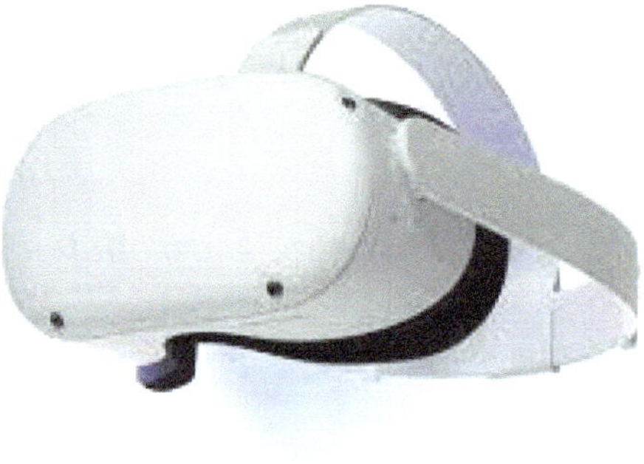

Fig. 2 Meta Quest 2 used in the experiment [51]

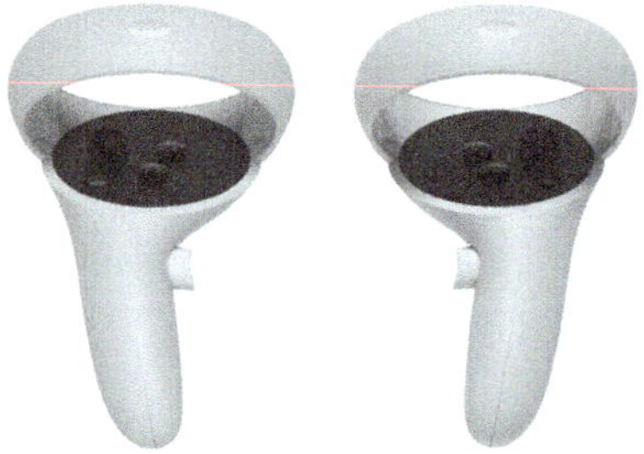

Fig. 3 Meta Quest 2 touch controllers [51]

3.3 Lesson Structure

Each session was one hour in length, and sessions were conducted over the course of five weeks, occurring once per week. In each session, a brief introduction of the lesson content was given by the professor followed by a demonstration of the VAR learning activity conducted by teaching assistants. The class was then divided into groups of five to six, the VAR activity was played alternatively between group members, and the VAR activity was conducted with one student at a time due to the limited number of headsets available. The student wearing the HMD could move freely in physical reality while receiving multisensory information from the VAR environment. The length of each VAR activity was 3–5 min. To ensure every student experienced VAR learning during the lesson period, the duration of each VAR activity was no more than 7 min. To optimize learning outcomes and better cater to student's needs, the lesson structure was designed to be flexible and adaptable, allowing for modifications and adjustments as necessary.

3.4 Lesson Design and Planning

For this study, the lesson plan for the module "MA4831 Computer Aided Engineering" was redesigned. This project adopted the user-centered design (UCD) framework, in which the lesson plan focused on the students and their needs in each phase of the design process [52]. Based on student responses to the survey and classroom observations from previous years, the lesson plan was revamped to enhance the quality of teaching.

3.4.1 Course Curriculum

The curriculum was explicitly geared toward addressing the critical competency gaps identified among engineering students. These gaps include technical knowledge, problem-solving skills, communication skills, and teamwork. The curriculum incorporated coursework and virtual experiences that focus on developing these competencies. Six activities were designed, each teaching different engineering applications. During the span of 5 weeks, students learn about the applications of VAR, the operation of lathe and milling machines, aircraft hydraulic systems, smart crane lifting operations, and aircraft engine disassembly process through the VAR activities. By emphasizing the development of these core competencies, the curriculum aimed to nurture well-rounded students who are not only technically proficient but also able to communicate effectively and work collaboratively to solve complex problems. The detailed lesson plan is shown in Table 1.

3.4.2 Team-Based Learning

The lesson design incorporated teamwork-based objectives which required participation from members of the group who were not using the HMD to increase student collaboration and participation. When a student (master player) was wearing the HMD, the other students in the same group observed the process in real-time from the display using 3D glasses (Fig. 4). As students stayed engaged, the course curriculum was effectively covered in a structured and organized manner, allowing students to gain a comprehensive understanding of the subject matter while learning in groups. Students can take turn to serve as a master player.

Furthermore, students who were not wearing the HMD could refer to the group displays, tutorial guides, or demonstration videos (Fig. 5) offering assistance to the master player wearing the HMD in completing the tasks. This approach promoted collaboration among the group members by encouraging students to engage in discussions on the best approaches to tackle the challenges presented. Through guided group discussions, students compared their individual experiences in the virtual environment, and learn from each other's experiences to improve their understanding of the tasks presented.

Table 1 Lesson plan

Week	Topic	Lesson outline
1	Introduction	In this lesson, students participated in two basic introductory VAR activities, namely, 3D virtual cell and F1 car assembly **Objectives**: To familiarize students with the VAR capabilities before performing more complex tasks in the virtual environment in subsequent sessions **Remarks**: Basic hand gestures that are required to navigate the VAR environment (e.g., pinch and open palm) were introduced
2	Lathe and Mill	As an extension to the module "MA2071 Laboratory Experiments", students operated virtual machines to practice lathe and milling **Objectives**: To gain hands-on experience in operating lathe and milling machines and learn the standard operating procedures (SOP) of lathe and mill **Remarks**: Meta Quest 2 touch controllers were used to create more realistic sensations in operating machines
3	Hydraulics system	Students were tasked to dissemble components of aircraft hydraulic systems in the virtual environment **Objectives**: To learn the functionalities of the different components in the aircraft hydraulics system through 3D visualization and gain hands-on experience in disassembly **Remarks**: More complex hand gestures (e.g., rotate and grab) were required to manipulate tools in the virtual environment
4	Smart crane lifting and aircraft engine disassembly	The fourth lesson comprised of two VAR learning activities. Students were introduced to smart crane lifting where they completed crane lifting planning operations in the crane simulation. Afterward, students participated in the aircraft turbojet engine disassembly simulation **Objectives**: To identify different types of crane operations and gain hands-on experience in removing parts from an aircraft engine **Remarks**: As the smart crane lifting activity did not require the use of HMDs, students worked in groups by viewing the simulations on the displays while referring to the tutorial guide on their personal devices
5	Concluding session	For the last session, students reflected on their learning experiences for the past four weeks and were encouraged to provide feedback on the lesson. They could choose to redo any of the previous activities **Objectives**: Allow students to revisit VAR activates to strengthen understanding on the selected topics **Remarks**: The majority of the students replayed the hydraulic system activity as it was identified as the most challenging activity

Overall, this approach was effective in enhancing student collaboration and teamwork, which were identified as critical competency gaps among engineering students.

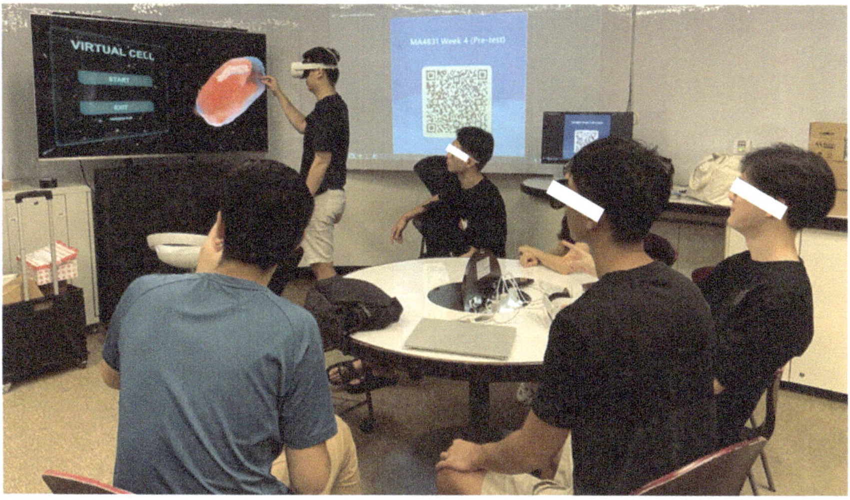

Fig. 4 Teamwork-based group learning in the VAR environment

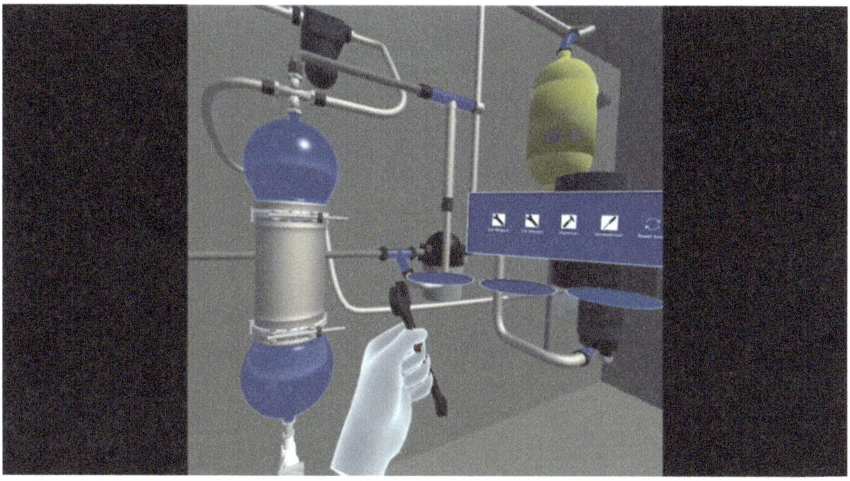

Fig. 5 Demonstration video used in the experiement

3.4.3 Level of Difficulty

By gradually increasing the level of difficulty of the VAR learning activities, students were tasked to apply their knowledge and skills in increasingly complex scenarios. As the weeks progressed, students were required to perform more advanced hand gestures to complete the VAR learning activities.

In the first session, students were introduced to basic hand gestures that are required to navigate in the virtual environment. Some examples include pointing

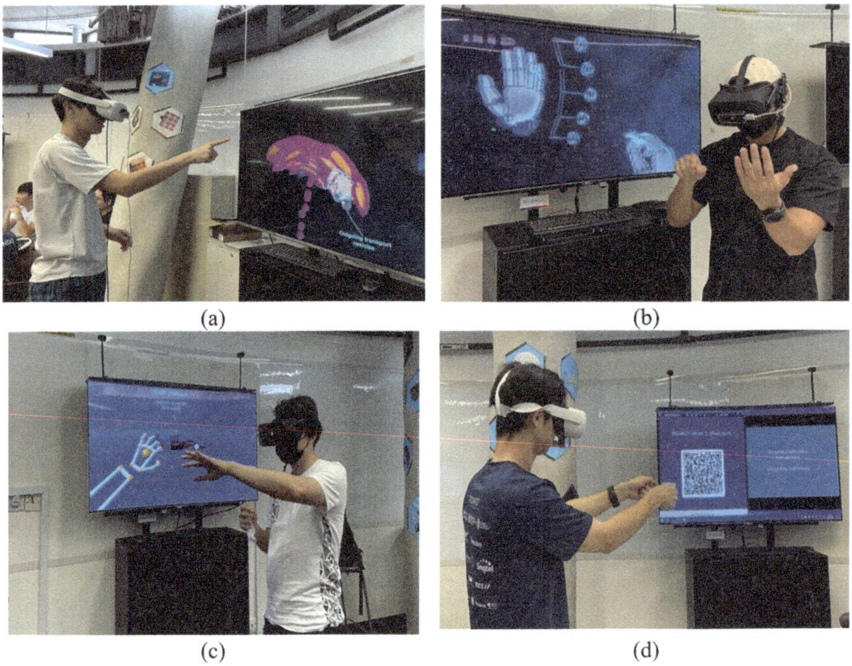

Fig. 6 Immersive and interactive learning using hand gestures

their index finger at the desired target to select an object (Fig. 6a) and opening their palm to select the menu option (Fig. 6b). Mastering these hand gestures was essential for students to perform more complex tasks in subsequent weeks, such as disassembling objects or manipulating machinery, with greater accuracy and efficiency. In the following weeks, more complex hand gestures were required to complete the tasks. Some examples include swiping sideways to rotate the orientation (Fig. 6c) and rotating the wrench to loosen the nut and bolt (Fig. 6d). This gradual increase in complexity helped build students' confidence and competence in learning in the virtual environment.

4 Data Collection

4.1 Data Collection Points

Informed consent was obtained prior to the start of the program to permit data collection. Participation in the study included attending the VAR learning activities and completing two different sets of surveys, the first one assessing students' knowledge, and the second one assessing students' SE. All surveys were conducted online using

Table 2 Data collection schedule

Instruments	Week				
	1	2	3	4	5
Knowledge assessment		X	X	X	
Self-efficacy survey	X			X	
Focus group discussions				X	X

Microsoft Forms. In the last two sessions, students participated in focus group discussions to share their experiences in the VAR learning environment. The schedule for data collection is shown in Table 2.

4.2 Measures and Instrumentation

4.2.1 Knowledge Assessment

The main objective is to evaluate the effectiveness of VAR in enhancing students' understanding of the topics discussed. Before the commencement of the VAR learning activity on Weeks 2, 3 and 4, a pre-test was conducted to assess students' prior knowledge of that week's featured topic. Directly after the VAR activity, a post-test with identical questions as the pre-test was administered. The results were then compared to evaluate students' learning gains in the session. The pre and post-test consisted of 5 multiple-choice questions (Appendix A), assessing the key concepts introduced in each week's VAR activity. Some examples of items include: "Laser scanning is commonly used in construction and building…" and "Which of the following is/are incorrect in robotization of cranes?". The validity of the test items was verified by the professor and teaching assistants and the level of difficulty was calibrated to ensure consistency across weeks.

4.2.2 Self-efficacy Survey

Students' SE was measured primarily using self-reported surveys. Before the commencement of the first session, students completed the SE survey. Similarly, students filled in an identical SE survey after the completion of all the VAR activities. The results were then compared to evaluate the effectiveness of the VAR learning environment in enhancing students' SE. The survey comprised three sections, namely, course-specific, content-specific, and problem-specific SE Scale. All three surveys employed a 5-point Likert rating scale which ranged from 1 (Strong Disagree) to 5 (Strongly Agree). This survey format has been used extensively in the field of educational research to measure students' SE [20].

The first session consisted of 5 items (Table 3) to measure course-specific SE. The survey items were adapted from the course-specific SE scale used by Bong [20] and were specific to engineering courses. Survey item SE_1 measured students' perception of their ability to excel in the specific module "Computer Aided Engineering". While survey items SE_2, SE_3, SE_4.and SE_5 measured the extent to which they believed they would excel in all engineering courses.

The second session consisted of 5 items (Table 4) from the content-specific SE Scale, which enabled data to be obtained about SE specific to the content taught in each week's activities. Students indicated the extent to which they believed they could understand the specific concepts about VAR, lathe and milling, hydraulic systems, aircraft engines and crane lifting operations. The survey items addressed each learning objective independently.

The third session consisted of five items (Table 5) from the problem-specific SE Scale. However, the way it was administered was different from the first two SE surveys. Five real-world engineering problems were presented to students in the form of a short video for 50 s through the projector. The problems presented are shown in Fig. 7. Written descriptions of the problems along with pictorial illustrations that mimicked problems faced in the real world were shown to the students. Each problem was shown for a duration of 10 s, such that students would have sufficient time to recognize the types of problems presented, yet not have time to devise a solution. Students would have encountered similar problems during the VAR activities, and thus, gained hands-on experience in solving such issues.

While interacting with the VAR learning content, students would have rated their confidence in solving the problems on their personal devices. To ensure the time

Table 3 Course-specific SE scale

SE_1	I can master the content in this module
SE_2	I can master the content in even the most challenging engineering course
SE_3	I can do a good job on almost all my engineering coursework
SE_4	I can learn the content taught in my engineering-related courses
SE_5	I can earn a good grade in my engineering-related courses

Table 4 Content-specific SE scale

Can do an excellent job on tasks and problems on …	
SE_6	Fundamental understanding of the VAR technology
SE_7	Understand the functionalities of lathe and milling machines
SE_8	Describe aircraft hydraulics systems
SE_9	List the various stages of the aircraft engine disassembly process
SE_10	The working mechanism of crane lifting operations

Table 5 Problem-specific SE scale

Can do an excellent job on tasks and problems on …	
SE_11	Navigate the VAR environment using Meta Quest 2
SE_12	Using lathe and milling machines to manufacture the product below
SE_13	Remove the accumulator form the aircraft hydraulic system
SE_14	Identify the key components of a turbojet engine
SE_15	Develop a crane-lifting plan with appropriate parameters

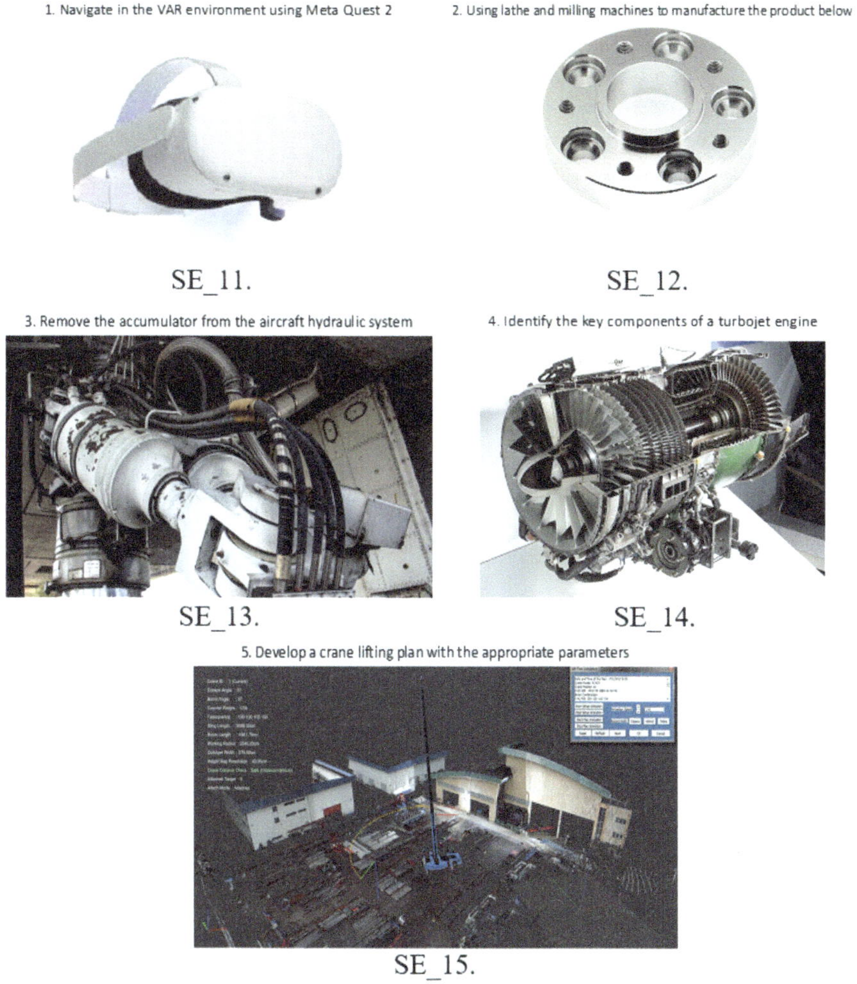

Fig. 7 Problem-specific SE video

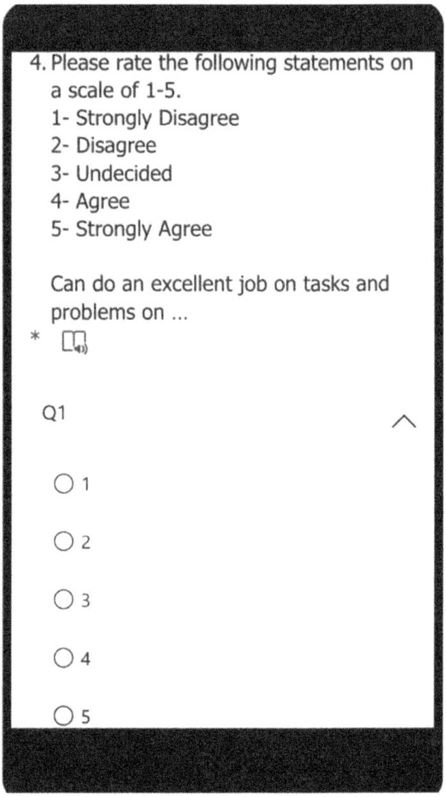

Fig. 8 Problem-specific SE Scale displayed on personal devices

restriction for each problem was strictly adhered to, students were not allowed to re-watch the video, and the problem statements were not shown to them in the survey as shown in Fig. 8.

4.2.3 Focus Group Discussions

While the self-reported SE instruments have high validity and reliability for measuring students' SE [20], the use of qualitative data could enhance and enrich the understanding of students' learning experiences and how VAR could affect students' SE. For this reason, the study included focus group discussions after the completion of all the VAR activities. A total of 30 students were randomly selected to participate in the discussion. Responses to questions listed in Table 6 were collected. Follow-up questions were asked to prompt students to elaborate on their individual experiences and provide insights into how to improve the program.

Table 6 Focus group discussion questions

FGD1	What are your favorite and least favorite aspects of the experience?
FGD 2	Do you prefer learning engineering concepts through VAR learning activities or traditional video lectures?
FGD 3	Do performing tasks (e.g., crane lifting operations, lathe and milling) in the simulations help you acquire the necessary skills to perform a similar task in real word?
FGD 4	Do you feel confident in performing similar tasks in real-world?
FGD 5	Would you recommend other modules to utilize VAR technologies?

5 Results and Discussion

The statistical analysis software platform IBM-SPSS for Windows v.28 and Excel were used to conduct the statistical analysis.

5.1 The Effects of VAR on Student's Academic Performance

The first research question aimed to examine the effects of VAR learning environments on students' academic performance. The results of the pre-and post-knowledge tests administered on Week 2, Week 3 and Week 4 were analyzed to identify the learning gains. To ensure data robustness, only students who completed both the pre- and post-knowledge tests for all three weeks were included in the study sample. After the initial data screening process, 58 participants were included in the analysis. The data was then analyzed using repeated measures ANOVA. The results of the pre-and post-knowledge tests are presented in Fig. 9. The means and standard deviations of the respective learning gains are reported in Table 7.

The results revealed that VAR had a positive effect on students' academic performance. As seen in Table 7, the mean learning gains were positive for every week. This implies that students' knowledge improved after the VAR learning activities, which reflects the effectiveness of the learning process.

However, there was a notable difference in the mean values of the learning gains between Week 2 ($M = 0.345$, SD = 1.42) and Week 3 ($M = 0.862$, SD = 1.32). In general, students showed greater improvement in the later weeks. The difference in learning gains could be due to students' increased familiarity with the VAR learning environment. That is, before they can focus on learning engineering concepts through the VAR, students must first learn how to navigate the virtual environment so as to achieve a more effective learning experience. Similarly, it was discovered that students who had prior exposure to VAR technology showed significantly better learning outcomes than those who were new to the technology [53]. These findings

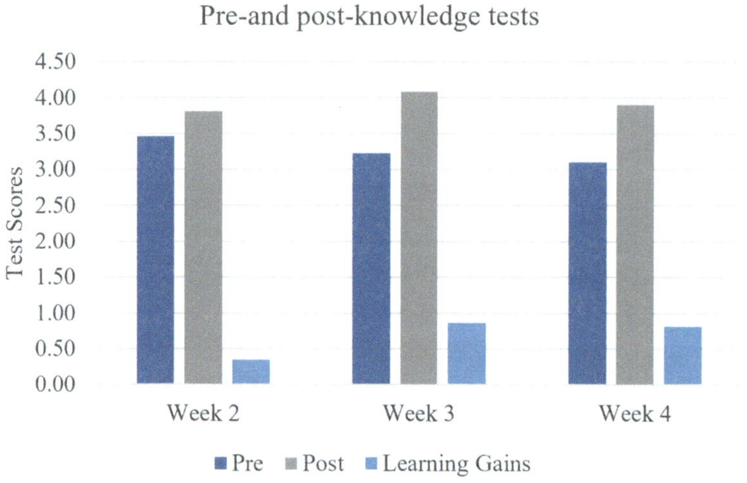

Fig. 9 Results of pre-and post-knowledge tests

Table 7 Means and standard deviations of learning gains

Descriptive statistics			
	Mean	Std. deviation	*N*
Week 2	0.3448	1.4210	58
Week 3	0.8621	1.3172	58
Week 4	0.8103	1.1766	58

suggest that providing students with ample practice and training to familiarize themselves with the VAR learning environment is critical for enhancing the effectiveness of VAR as a learning tool.

5.2 *The Effects of VAR on Student's Self-efficacy*

The second research question aimed to examine the effects of the VAR learning environment on students' SE. The results of the pre-and post-SE tests administered on Week 1 and Week 4 were compared to identify the SE gains. To ensure data robustness, only students who completed both the pre-and post-SE tests were included in the study sample. After the initial data screening process, 55 participants were included in the analysis. The results of the pre-and SE tests are presented in Fig. 10. Firstly, descriptive analyses were conducted to identify the means of course-specific SE, content-specific SE, and problem-specific SE from the pre-SE test and post-SE test (Table 8). Following which, a series of paired samples t-tests were conducted to determine the differences in mean SE between the pre-and post-tests. The Cronbach's alpha coefficient for the SE tests was 0.921, indicating good internal consistency and reliability.

As seen in Fig. 8, the pre-SE test shows that students had lower problem-specific SE compared to course-specific SE and content-specific SE before the lessons. The above results suggest that while students are confident in their ability to master theoretical concepts and principles, they are less confident in applying them to real-world engineering problems. This could be due to engineering modules often placing heavy emphasis on theories rather than application.

Paired samples t-tests were then used to test whether the mean differences between the pairs of pre- and post-SE measurements were significant. The results for course-specific SE are summarized in Table 9. The mean course-specific SE increased considerably after the VAR sessions. Mean course-specific SE was found to have increased from the pre-test ($M = 3.53$, SD = 0.796) to the post-test ($M = 4.03$, SD

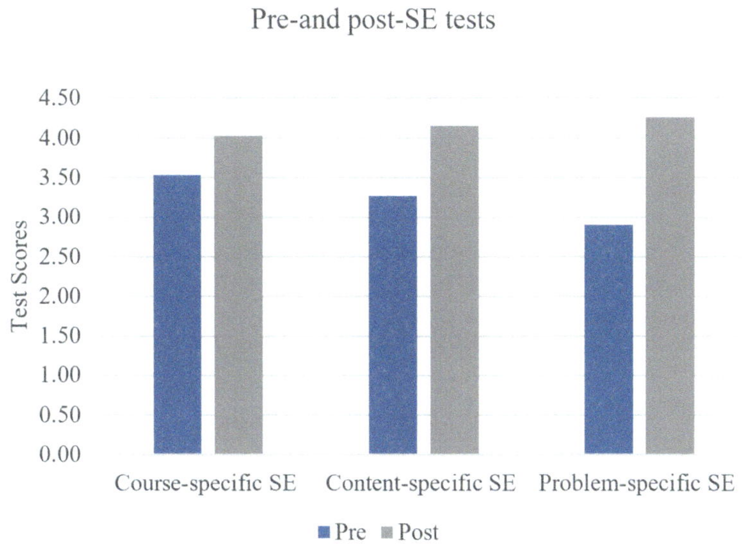

Fig. 10 Results of pre-and post-SE tests

Table 8 Means and standard deviations of SE

Descriptive statistics			
	Mean	Std. deviation	N
Course-specific SE (pre)	3.531	0.7956	55
Course-specific SE (post)	4.033	0.8967	55
Content-specific SE (pre)	3.269	0.9792	55
Content-specific SE (post)	4.153	0.9092	55
Problem-specific SE (pre)	2.898	1.1944	55
Problem-specific SE (post)	4.262	0.8027	55

= 0.897). This increase was statistically significant ($t = 3.52, p < 0.001$). In particular, the scores for survey items SE_2, SE_3, SE_4.and SE_5 increased in the post-SE test. As these items measured the extent to which students believed they would excel in all engineering modules, the results demonstrate that students' learning experience in one module could have a positive spill-over effect on their SE in other engineering courses.

The above results for course-specific SE are consistent with the findings in a similar study conducted in 2019 at NTU [54]. The study evaluated the effects of VAR-based learning on enhancing engineering students' course-specific SE. The results showed that students' course-specific SE increased significantly after participating in the VAR learning activities. In the study, mean course-specific SE was found to have increased from the pre-test ($M = 3.52$, SD = 0.722) to the post-test ($M = 4.11$, SD = 0.727) [54]. The increase was found to be statistically significant ($t = 6.84$, $p < 0.001$). These findings are similar to the present results on course-specific SE gains. Therefore, the current study confirms that VAR learning environment has a positive effect on students' course-specific SE, as is expected from prior research.

The results for content-specific SE are summarized in Table 10. The mean content-specific SE increased significantly after the VAR sessions. The mean content-specific SE increased considerably after the VAR sessions. Mean content-specific SE was found to have increased from the pre-test ($M = 3.27$, SD = 0.979) to the post-test ($M = 4.15$, SD = 0.909). This increase was statistically significant ($t = 5.841$, $p < 0.001$). Specifically, survey item SE_8 "Describe aircraft hydraulics systems" had the highest gain. This item assessed students' confidence in their knowledge of the aircraft hydraulic system, which was taught in Week 3's VAR learning activities. Correspondingly, the mean learning gain was highest in Week 3. This shows that there may be a connection between students' SE and learning gains.

Table 9 Paired sample T-test for course-specific SE

Paired samples statistics							
Mean	Std. deviation	Std. error mean	95% confidence interval of the difference		*t*	*df*	2-sided p
			Lower	Upper			
0.5018	1.0561	0.1424	0.2163	0.7873	3.524	54	< 0.001

Table 10 Paired sample T-test for content-specific SE

Paired samples statistics							
Mean	Std. deviation	Std. error mean	95% confidence interval of the difference		*t*	*df*	2-sided p
			Lower	Upper			
0.8836	1.1220	0.1513	0.5803	1.1869	5.841	54	< 0.001

Table 11 Paired sample T-test for problem-specific SE

Paired samples statistics							
Mean	Std. deviation	Std. error mean	95% confidence interval of the difference		*t*	*df*	2-sided p
			Lower	Upper			
1.3636	1.2585	0.1697	1.0234	1.7039	8.036	54	< 0.001

The results for problem-specific SE are summarized in Table 11. Mean problem-specific SE was found to have increased from the pre-test (M = 2.90, SD = 1.19) to the post-test (M = 4.26, SD = 0.803). This increase was statistically significant (t = 8.036, $p < 0.001$).

The gain in SE is more apparent for problem-specific SE than content-specific SE and course-specific SE. This suggests that the VAR learning environment is most effective in developing students' confidence in their problem-solving skills. One reason for this difference in gain could be due to the VAR games being centered around real-world engineering problems, which could help students develop critical thinking skills and problem-solving skills.

Given that content-specific SE did not increase as much as problem-specific SE, some possible enhancements could be considered in the game design process. These enhancements include providing short video lectures during the game to explain the engineering concepts used; and adding pauses in the game to give students time to consolidate their knowledge.

5.3 *Focus Group Discussions*

Data collected through interviews were recorded and transcribed. Five open-ended questions examining students' learning experiences were discussed.

Item FGD1 inquired about students' overall learning experience in the VAR environment. Three categories, namely, visualization, hands-on learning, and immersive experience, were derived while analyzing students' favorite aspects of the experience. Another three categories, namely, difficult to navigate, inaccurate leap motion, and discomfort, were derived while analyzing students' least favorite aspects of the experience. The results are presented in Table 12.

Item FDG2 asked students to compare the effectiveness of VAR learning activities and traditional lectures. While most students agreed that VAR learning activities provide an immersive hands-on learning opportunity, they expressed that traditional lecture is more effective in teaching engineering principles. Hence, through the integration of these two approaches, such as supplementing a traditional lecture with a VAR simulation, educators can create a comprehensive hybrid learning experience that combines the best of both worlds.

Table 12 FGD1 discussion

Category	Description
Visualization	VAR helped to visualize complex designs and concepts
Hands-on learning	Through the simulations, students practiced skills in a safe and controlled environment
Immersive experience	Students were fully engaged due to the immersive nature
Difficult to navigate	Students faced difficulties navigating in the virtual environment
Inaccurate leap motion	Hand-tracking was inaccurate and unresponsive in certain situations
Discomfort	Some students experienced motion sickness and headaches

Items FDG3 and FDG4 inquired about the effectiveness of VAR in acquiring new skills and the effects on students' SE. Since items FDG3 and FDG4 were similar, the same four categories were identified: accumulate experience, develop practical skills, immediate feedback, and sense of achievement. At the most basic level, VAR learning activities offered students a unique opportunity to accumulate experience in performing engineering tasks. By doing so, they can develop the practical skills required to perform similar tasks in the real world. Moreover, VAR provided immediate feedback to students, allowing them to identify areas for improvement and refine their approaches. This feedback loop helped students build confidence in their problem-solving abilities and many expressed they felt a sense of accomplishment after completing each task.

Item FDG5 inquired about the applicability of VAR in other engineering modules. Many students suggested that VAR technology would be more effective for modules that require visualization, such as engineering graphics and aircraft design.

6 Conclusion

To reiterate, this study evaluated the effectiveness of VAR in enhancing students' academic performance and SE along the three SE dimensions (e.g., course-specific SE, content-specific SE, and problem-specific SE). Students participated in VAR classroom activities over 5 weeks, whereby their academic performance and SE were assessed using surveys and focus group discussions. During the sessions, students actively participated in the learning process. As students gained hands-on experience in performing engineering tasks in virtual settings, their confidence in performing a similar task or solving a similar problem in the real world increased. Upon the completion of all the VAR activities, students' knowledge and SE scores improved significantly. The results have shown that VAR classroom activities have a positive effect on students' academic performance and SE. This is consistent with Chen's and Queiroz's findings that VAR has a pronounced effect on academic performance and SE [17, 27].

Across the three SE dimensions, problem-specific SE had the greatest improvement, followed by content-specific SE and course-specific SE. However, the increase in problem-specific SE and content-specific SE did not translate to an equal magnitude increase in course-specific SE. This suggests that VAR is more effective in teaching specific hands-on engineering applications as opposed to engineering theories and principles. This could be because students' course-specific SE is dependent on many other factors such as students' prior achievements, numeracy skills, and intrinsic motivation. Nonetheless, these results suggest that VAR possesses much potential in its application in education to enhance academic performance and SE.

6.1 Contributions

6.1.1 Self-efficacy Instrument

While researchers have investigated the effect of VAR on students' SE, few scholars attempted to measure SE in specific domains. Although the existing body of research demonstrated that VAR is effective in enhancing students' general SE, the mechanism through which VAR enhances students' SE is not known.

On the other hand, course-specific SE, content-specific SE, and problem-specific SE are well-established SE constructs and are often studied in educational contexts to better understand how students perceive their abilities and how this perception affects their learning outcomes. Measuring students' SE along the three SE dimensions is a widely used approach in educational research, particularly in the field of educational psychology.

The current study expanded on the scope of previous research in 2019 [54] by introducing the three dimensions SE model as the SE instrument. By measuring students' SE in different dimensions, this project enables the comparison of the effectiveness of VAR in enhancing various aspects of SE and helps to identify the crucial elements of an effective educational VAR game. Therefore, this study integrated the established methodology used in educational research into the field of VAR technology research to enhance understanding of the effects of VAR on students' SE.

6.1.2 Lesson and Experiment Design

In this study, the lesson plan for the "MA4831 Computer Aided Engineering" module was redesigned using the UCD framework. The revised lesson plan incorporated coursework and virtual experiences that focused on addressing the critical competency gaps identified among engineering students. As VAR is in the early stages of being adopted as a teaching tool in NTU, this lesson design provides a roadmap for educators to follow in integrating VAR learning activities into the traditional curriculum. Furthermore, this experimental approach can be adapted to evaluate the effects of the VAR learning environment on other aspects of student's performance, such as students' motivation and attention span.

6.2 Limitations

6.2.1 Longer Intervention

First, the time constraint due to limited class time and a large pool of participants was a significant challenge to incorporate VAR learning activities into lessons. Since each student was only allocated around 5 min per class to use the headset, the VAR learning activities were carefully designed to ensure every student could complete the tasks within the time limit. Thus, students' exposure to the virtual learning environment was limited, which could reduce the effectiveness of the intervention. For instance, some students may require more time to get familiarized with the virtual environment and hand gestures. Future research can focus on investigating if a longer intervention would increase the effectiveness of VAR in enhancing students' SE and academic performance.

6.2.2 Long-Term Effects

Second, since the knowledge tests and SE tests were administrated right before and after the VAR learning activities, this study only measured the immediate effect. The improvement in academic performance and the increase in SE due to the intervention may fade out over time. It is not uncommon for students' SE to depreciate over time, especially when faced with repeated failures or setbacks. Future research is needed to examine the long-term effects of using VAR in education, beyond the short-term gains.

6.2.3 Gender Control

Lastly, this Chapter did not include controls for gender as the sample was majorly composed of male students. Thus, the results discussed are more relevant and applicable to male students than female students. Since female students may respond differently than males to immersive technology, and gender biases can impact how people perceive and interact with virtual environments. Further research can investigate the impact of VAR on female students specifically. Other factors such as prior experience with VAR and age may impact the effectiveness of VAR in enhancing academic performance and SE. By controlling for demographic variables, the advantages and drawbacks of employing VAR for various educational levels can be studied.

6.3 Implications

This research provides evidence that VAR simulations led to improvement in academic performance and increments in self-reported SE. The findings demonstrated that VAR simulations could be an effective learning experience and could hold great potential to strengthen the link between theory and practice in engineering education. Concurrently, it was found that VAR simulations are most effective in enhancing problem-specific SE as compared to content-specific SE and course-specific SE. These findings present some important implications for the applications of VAR in education.

First, VAR can be applied across multiple engineering modules for facilitating exposure to unique engineering scenarios and are particularly effective in developing students' problem-solving skills. The VAR simulations should mimic real-world engineering problems to provide students with a realistic learning experience. Moreover, it is crucial to incorporate hands-on elements into the VAR games for students to interact with virtual objects and be familiar with the motion. The game should be designed in a way that offers multiple pathways to reach the solution, allowing students to have the flexibility to experiment with different methodologies to solve the problem. By varying the level of difficulty of the VAR activities, it provides students with a greater sense of challenge and engagement as they progress in their knowledge and skills.

Second, VAR learning activities could be introduced as early as freshman year, such that students can develop the skills needed to effectively learn in the virtual setting. For many students, the experience of learning in an immersive environment may be overwhelming at first, and they may need time to adjust to the new technology. With trial and error, students can develop the skills to navigate and interact with objects in the virtual space. Furthermore, introducing VAR to freshmen can be advantageous because they are often more open to new experiences and technologies. By introducing the technology at an early stage, it can help to create a strong foundation of knowledge regarding VAR and pave the way for future learning opportunities with VAR.

Third, while VAR technologies have been shown to be effective pedagogical tools, they should not replace traditional teaching methods entirely in engineering education. Through the focus group discussions, the majority of students expressed a preference for learning about theories and concepts in lectures rather than through immersive technologies. Therefore, while the VAR learning environment offers a range of benefits, such as interactivity and engagement, it may not be the preferred learning method for complex engineering principals. Some students feedback that the VAR learning environment was overwhelming or distracting, preferring to read and digest the concepts at their own pace through more traditional methods. Besides, traditional teaching methods such as lectures, seminars, and tutorials offer the opportunity for direct interaction with the professor, allowing students to ask questions and receive immediate personalized feedback. For these reasons, VAR could be used as a supplementary pedagogical tool to complement traditional teaching methods by providing additional hands-on practice sessions that can help to develop practical skills.

6.4 Recommendations for Future Research

As interest in this area is expected to grow, these findings open up avenues for further research in the field of the application of VAR in education. Broadly, there is a need to continue examining the effectiveness of VAR in education using different methodologies and extend its application. The present study could be extended in the following directions.

With NTU's vision of utilizing digital technology solutions to enhance learning experiences, VAR can be integrated into a wider range of modules and disciplines. Recognizing the benefits of interdisciplinary education and the role of technology in education, NTU can offer students an immersive and interactive learning experience in unrestricted electives and common core modules. Through VAR simulations, students from all backgrounds can gain a deeper understanding of different fields and learn about the applications of the technology. Students can also benefit from collaboration with peers from different disciplines, thereby promoting cross-disciplinary learning and enhancing their problem-solving skills. To further enhance the effectiveness of VAR teaching, future research can be conducted to examine the effects of the VAR learning environment on enhancing non-engineering students' academic performance and SE. Ultimately, the effectiveness of VAR in education varies depending on the specific needs and goals of each student and discipline.

To summarize, the present study has shown the benefits of the VAR learning environment on students' academic performance and SE. Further research addressing the areas mentioned above could provide a more comprehensive understanding of the potential benefits and limitations of the use of VAR in education. As VAR is in the early stages of being adopted as a pedagogical tool in higher education, the benefits of using VAR in education have yet to be fully realized.

References

1. Bandura, A.: Self-efficacy: toward a unifying theory of behavioral change. Psychol. Rev. **84**(2), 191–215 (1977)
2. Bandura, A.: The explanatory and predictive scope of self-efficacy theory. J. Soc. Clin. Psychol. **4**(3), 359–373 (1986)
3. Bandura, A.: Self-Efficacy: The Exercise of Control. W.H. Freeman and Co, New York (1997)
4. Pintrich, P.R., Schunk, D.H.: Motivation in Education: Theory, Research, and Applications. Merrill Prentice Hall, Upper Saddle River (2002)
5. Usher, E.L., Pajares, F.: Sources of self-efficacy in school: critical review of the literature and future directions. Rev. Educ. Res. **78**(4), 751–796 (2008)
6. Wilde, N., Hsu, A.: The influence of general self-efficacy on the interpretation of vicarious experience information within online learning. Int. J. Educ. Technol. Higher Educ. **16**(1) (2019)

7. Horcajo, J., Santos, D., Higuero, G.: The effects of self-efficacy on physical and cognitive performance: an analysis of meta-certainty. Psychol. Sport Exerc. **58**, 102063 (2022)
8. O'Leary, A.: Self-efficacy and health. Behav. Res. Ther. **23**(4), 437–451 (1985)
9. Zhou, C., Yue, X.D., Zhang, X., Shangguan, F., Zhang, X.Y.: Self-efficacy and mental health problems during COVID-19 pandemic: a multiple mediation model based on the health belief model. Personality Individ. Differ. **179**, 110893 (2021)
10. Schunk, D.H.: Self-efficacy and academic motivation. Educ. Psychol. **26**, 207–231 (1991)
11. Chemers, M.M., Hu, L.T., Garcia, B.F.: Academic self-efficacy and first year college student performance and adjustment. J. Educ. Psychol. **93**(1), 55–64 (2001)
12. Luszczynska, A., Scholz, U., Schwarzer, R.: The general self-efficacy scale: multicultural validation studies. J. Psychol. **139**(5), 439–457 (2005)
13. Chen, G., Gully, S.M., Eden, D.: Validation of a new general self-efficacy scale. Organ. Res. Methods **4**(1), 62–83 (2001)
14. Dunn, S., Jacobs, B., Rogers, R.W.: The self-efficacy scale: construction and validation. Psychol. Rep. **51**, 663–671 (1998)
15. Pajares, F.: Self-efficacy beliefs in academic settings. Rev. Educ. Res. **66**(4), 543–578 (1996)
16. Pintrich, P.R., De Groot, E.V.: Motivational and self-regulated learning components of classroom academic performance. J. Educ. Psychol. **82**(1), 33–40 (1990)
17. Queiroz, A.C., Fauville, G., Herrera, F., Leme, M.I., Bailenson, J.N.: Do students learn better with immersive virtual reality videos than conventional videos? A comparison of media effects with middle school girls. Technol. Mind Behav. **3**(3) (2022)
18. Leach, C.W., Queirolo, S.S., DeVoe, S., Chemers, M.: Choosing letter grade evaluations: the interaction of Students' achievement goals and self-efficacy. Contemp. Educ. Psychol. **28**(4), 495–509 (2003)
19. Juang, L., Vondracek, F.W.: Developmental patterns of adolescent capability beliefs: a person approach. J. Vocat. Behav. **59**(1), 34–52 (2001)
20. Bong, M.: Role of self-efficacy and task-value in predicting college students' course performance and future enrollment intentions. Contemp. Educ. Psychol. **26**(4), 553–570 (2001)
21. Lopez, F.G., Lent, R.W., Brown, S.D., Gore, P.A.: Role of social–cognitive expectations in high school students' mathematics-related interest and performance. J. Couns. Psychol. **44**(1), 44–52 (1997)
22. Bong, M.: Predictive utility of subject-, task-, and problem-specific self-efficacy judgments for immediate and delayed academic performances. J. Exp. Educ. **70**(2), 133–162 (2002)
23. Dick, E.: The Promise of Immersive Learning: Augmented and Virtual Reality's potential in education (2023). https://itif.org/publications/2021/08/30/promise-immersive-learning-augmented-and-virtual-reality-potential/
24. İbili, E.: Effect of augmented reality environments on cognitive load: pedagogical effect, instructional design, motivation and interaction interfaces. Int. J. Progress. Educ. **15**(5), 42–57 (2019)
25. Radianti, J., Majchrzak, T.A., Fromm, J., Wohlgenannt, I.: A systematic review of immersive virtual reality applications for higher education: design elements, lessons learned, and research agenda. Comput. Educ. **147**, 103778 (2020)
26. López Ríos, O., Lechuga López L.J., Lechuga López, G.: A comprehensive statistical assessment framework to measure the impact of immersive environments on skills of higher education students: a case study. Int. J. Interactive Des. Manuf. **14**(4), 1395–1410 (2020)
27. Chen, C., Zhang, L., Luczak, T., Smith, E., Burch, R.F.: Using Microsoft hololens to improve memory recall in anatomy and physiology: a pilot study to examine the efficacy of using augmented reality in education. J. Educ. Technol. Dev. Exchange **12**(1) (2019)
28. Margolis, H., Mccabe, P.P.: Improving self-efficacy and motivation. Interv. Sch. Clin. **41**(4), 218–227 (2006)
29. Lent, R.W., Brown, S.D., Larkin, K.C.: Self-efficacy in the prediction of academic performance and perceived career options. J. Couns. Psychol. **33**(3), 265–269 (1986)
30. Jamil, N.L., Mahmud, S.N.: Self-efficacy relationship on science achievement amongst National Secondary School Students. Creat. Educ. **10**(11), 2509–2527 (2019)

31. Zakariya, Y.F.: Improving students' mathematics self-efficacy: a systematic review of intervention studies. Front. Psychol. **13** (2022)
32. Liem, A.D., Lau, S., Nie, Y.: The role of self-efficacy, task value, and achievement goals in predicting learning strategies, task disengagement, peer relationship, and achievement outcome. Contemp. Educ. Psychol. **33**(4), 486–512 (2008)
33. Bartimote-Aufflick, K., Bridgeman, A., Walker, R., Sharma, M., Smith, L.: The study, evaluation, and improvement of university student self-efficacy. Stud. High. Educ. **41**(11), 1918–1942 (2016)
34. Obrentz, S.B.: Predictors of science success: the impact of motivation and learning strategies on college chemistry performance (2022). https://scholarworks.gsu.edu/cgi/viewcontent.cgi?article=1078&context=epse_diss
35. Brown, S.D., Lent, R.W.: A Social Cognitive View of Career Development and Counselling. in Career Development and Counseling: Putting Theory and Research to Work. Wiley, Hoboken (2021)
36. Usher, E.L., Pajares, F.: Sources of self-efficacy in mathematics: a validation study. Contemp. Educ. Psychol. **34**(1), 89–101 (2009)
37. Fineman, B.: VR/AR in Research and Education Survey (2018). https://meetings.internet2.edu/media/medialibrary/2018/10/25/20181016-fineman-metaverse.pdf
38. Valentine, K.: Team awarded NSF grant to teach virtual explorers about permafrost, arctic climate change (2023). https://news.asu.edu/20210719-team-awarded-2m-nsf-grant-teach-virtual-explorers-about-permafrost-and-arctic-climate
39. Steff, B.: Flying among the stars: Purdue University professor using virtual reality to teach astronomy (2023). https://www.purdue.edu/newsroom/releases/2021/Q2/flying-among-the-stars-purdue-university-professor-using-virtual-reality-to-teach-astronomy.html
40. Greening, T.: Building the constructivist toolbox: AN exploration of cognitive technologies. Educ. Technol. **38**(2), 28–35 (1998)
41. Spector, J.M., Anderson, T.M.: Integrated and holistic perspectives on learning, instruction and technology: understanding complexity. Springer, Dordrecht (2002)
42. Jonassen, D.H., Rohrer-Murphy, L.: Activity theory as a framework for designing constructivist learning environments. Educ. Tech. Res. Dev. **47**(1), 61–79 (1999)
43. Chen, C.J.: Theoretical bases for using virtual reality in education. Themes Sci. Technol. Educ. **2**(1–2), 71–90 (2009)
44. Dünser, A., Walker, L., Horner, H., Bentall, D.: Creating interactive physics education books with augmented reality. In: Proceedings of the 24th Australian Computer-Human Interaction Conference (2012)
45. Iwata, T., Yamabe, T., Nakajima, T.: Augmented reality go: extending traditional game play with interactive self-learning support. In: 2011 IEEE 17th International Conference on Embedded and Real-Time Computing Systems and Applications (2011)
46. Wang, H.Y., Lin, T.J., Tsai, C.C., Duh, H.B.L., Liang, J.C.: An investigation of students' sequential learning behavioral patterns in mobile CSCL learning systems. In: 2012 IEEE 12th International Conference on Advanced Learning Technologies (2012)
47. Kamarainen, A.M., Metcalf, S., Grotzer, T., Browne, A., Mazzuca, D., Tutwiler, M.S., Dede, C.: EcoMOBILE: integrating augmented reality and probeware with environmental education field trips. Comput. Educ. **68**, 545–556 (2013)
48. Tai, K.H., Hong, J.C., Tsai, C.R., Lin, C.Z., Hung, Y.H.: Virtual reality for car-detailing skill development: learning outcomes of procedural accuracy and performance quality predicted by VR self-efficacy, VR using anxiety, VR learning interest and flow experience. Comput. Educ. **182**, 104458 (2022)
49. Parmar, D.: Evaluating the Effects of Immersive Embodied Interaction on Cognition in Virtual Reality (2023). https://tigerprints.clemson.edu/all_dissertations/1987
50. Deng, I.: Facebook owner meta in talks to bring Quest 2 VR headset to China (2023). https://www.scmp.com/tech/big-tech/article/3211014/meta-talks-sell-quest-2-virtual-reality-headset-china-after-failing-bring-facebook-country-report

51. Meta.: Meta quest 2: immersive all-in-one VR headset | meta store (2023). https://www.meta.com/quest/products/quest-2/
52. Young, G., Stehle, S., Walsh, B., Tiri, E.: Exploring virtual reality in the higher education classroom: using VR to build knowledge and understanding. JUCS J. Univ. Comput. Sci. **26**(8), 904–928 (2020)
53. Makransky, G., Andreasen, N.K., Baceviciute, S., Mayer, R.E.: Immersive virtual reality increases liking but not learning with a science simulation and generative learning strategies promote learning in immersive virtual reality. J. Educ. Psychol. **113**(4), 719–735
54. Lee, T., Wen, Y., Chan, M.Y., Azam, A.B., Looi, C.K., Taib, S., Ooi, C.H., Huang, L.H., Xie, Y., Cai, Y.: Investigation of virtual and augmented reality classroom learning environments in university STEM education. Inter. Learn. Environ. 1–16 (2022)

Optimizing Learning Through Game Design: An Essential Role in Educational Games

Jin Wen Chung, Abu Bakr Azam, Yuan Xie, Siyi Li, Lihui Huang, and Yiyu Cai

Abstract With the use of educational games in traditional classrooms becoming increasingly common as a tool to aid educators in teaching, this study aims to explore the importance of game design on enhancing learning experiences, as well as the benefits of integrating virtual reality educational games into a flipped classroom model. Specifically, the study focuses on the redevelopment of a hydraulic system game with key considerations such as the gameplay sequences, color and sound usage, understanding cognitive loads, and stage design. The game is then integrated into a traditional classroom setting as a form of assessment, instead of using the traditional pen and paper. Walkthrough video guides area also used during the lesson, as referenced from the flipped classroom model, to help with independent learning. Students' engagement levels and learning outcomes are then assessed through a survey using the 5-point Likert scale, feedback on the game design is received through this survey. The study's findings suggest that effective game design can help to improve learning outcomes, promote student engagement and collaboration. The study then concludes that this pedagogy approach of integrating virtual reality educational games into a flipped classroom model could be a potential path forward for education in the future.

1 Introduction

1.1 Background

In the current digital age, video games have captured the attention of today's society and became an integral part of our daily lives. With technology driving innovation at an unprecedented pace, the world of gaming has evolved to encompass various parts of society, such as the educational and medical sectors [1]. From fascinating mobile games to immersive virtual reality (VR) experiences, the gaming industry

J. W. Chung · A. B. Azam · Y. Xie · S. Li · L. Huang · Y. Cai (✉)
Nanyang Technological University, Singapore, Singapore
e-mail: myycai@ntu.edu.sg

Y. Cai et al. (eds.), *Virtual and Augmented Reality Technology-Enhanced Learning*, Gaming Media and Social Effects, https://doi.org/10.1007/978-981-96-2332-7_3

continues to extend the limitations of what is possible, and in doing so, draws in people from all walks of life. Educational games are one genre of games that have become increasingly popular in recent years and have been recognized as a beneficial and useful form of teaching strategy [2–4].

Games that go beyond entertainment with the intentions to teach or train are also known as serious games [5]. These video games are intricate, complex creations that consist of a variety of important components, including narrative elements such as storylines and characters, and interactive mechanics such as gameplay and user interfaces. The most important component, however, would be game design. Its role is to integrate all these different components together and ensure that they work seamlessly to create an enjoyable gaming experience [6]. When applied effectively, a well-designed educational game can help educators create an immersive and interactive learning environment, which in turn allows for a better learning experience while also maintaining engagement and motivation [7, 8].

This process of adding game-like elements into non-game contexts, also known as gamification, is a popular application of game design in education [9]. As the use of educational games become more popular, and more relevant as a teaching and learning strategy [10], it is therefore imperative to further research on the fundamentals of game design so that game designers can create educational games that are both meaningful and beneficial to the player.

1.2 Objective and Scope

The objective of this study is to evaluate the importance of game design, how it can affect a students' learning experiences, as well as the integration of VR educational games into a traditional classroom setting. Through examination of existing literature as well as experimentations, this study seeks to shed light on the importance of game design in educational games.

The scope of this project covers the redevelopment of a hydraulic aircraft game. Since the original game was built using Unity Engine, it will be the main game engine that will be used. Microsoft's Mixed Reality Toolkit (MRTK) will also be used in conjunction with Unity as it contains pre-build assets that can help accelerate the development process. MRTK also supports game development in VR and AR, which is the area of interest in this project. Blender will be the main modeling software that will be used to design and create the hydraulic parts of the system, as it is user-friendly and free. The testing of the game will also be done by residential students and class participants, on both the Oculus Rift along with the LEAP motion detector, as well as the Oculus Quest headset.

2 Literature Review

2.1 Education 4.0 and Serious Games

With the fourth industrial revolution came a rapid advancement in technology. Artificial intelligence (AI), machine learning (ML), and the Internet of Things (IoT) are topics that have been widely talked about as to how it will impact the future of jobs and education [11]. Education 4.0, therefore, refers to the digital integration of such technologies into the teaching and learning process. To address the needs of the industrial revolution, higher learning institutions will have to continuously integrate new and innovative methods with modern technologies to enhance the teaching and learning process [12].

Education 4.0 also incorporates the principle of experiential learning, where students learn by doing hands-on experiences [13], one example of this would also be educational games also known as serious games. These games are designed with its focus on education instead of entertainment [14]. One benefit of serious games is that it offers a unique and innovative approach to education, which can be crucial in keeping the students engaged during the learning process. Unlike traditional classroom lectures, serious games are able to take control of a player's active neural pathway within the brain, assisting in more effective learning as well as realizations [15]. Serious games have also been proposed as a productive tool for nurturing the development of necessary twenty-first century skills, skills that may be required for individuals to thrive and excel today [16].

There are various examples of how serious games are being applied to the different working sectors. Games such as Economia is a game used by the European Central Bank to teach employees about financial concept [17], games that help preschoolers with autism to learn emotion identifying skills [18], so on and so forth. Although the purpose of the game may be different for each case, the overarching game design for serious games may not differ that much from one another.

Exceptional serious games designs can be classified into 3 criteria, which is the serious aspects, the game aspects, and the balance between them. For the serious aspect, high-quality serious games have a clear characterizing goal, which is a goal that refers to the desired outcome a game designer and educator wishes the players to achieve, and therefore strive to motivate the players to achieve this outcome [19]. The game device used to play the game can also directly influence the characterizing goal of a serious game, if it was a game that required body movements, then an Xbox Kinect or VR games would be the ideal option [5]. In the game aspect of a serious game, game designers must ensure that the game itself is fun and enjoyable to play [19]. According to Raph Koster's Theory of Fun for Game Design [20], he suggests that having fun is mainly about practicing and learning, rather than simply achieving mastery. When the player is able to recognize and understand patterns for learning, the brain registers this feedback as a level of enjoyment. Additionally, serious games should also ensure proper flow for the most optimal experience [21]. The optimal flow state is said to be achieved when the difficulty and rewards are proportional, leading to a state of enjoyment, feeling of timelessness, and serious focus for the players [22].

Finally, a successful serious game will have to strike a good balance between the serious and game aspects. The characterizing goal should be seamlessly embedded into the game play and impossible to avoid, all while maintaining player immersion [23].

2.2 Gamification Concept

Gamification, as previously mentioned, is the process of adding game-like elements into non-game contexts. Serious games, on the other hand, are games designed with a main purpose of achieving serious goals, games such as surgical training (health), games that feature social justice issues (change), and educational games. Unlike serious games, gamification is not limited to learning and can be used in any context [24].

Gamification can be seen as a type of design process [25], this process can typically comprise of three stages: learning foundation, preparation, prototyping, and playtesting [26]. The preparation phase focuses intensively on the topic or domain of interest, ideas in this stage are evaluated by its technical feasibility, market opportunities, budget, and artistic considerations. In the next stage, a prototype of the game is used as a representation of the game idea to test the game's feasibility and to make improvements on it, focusing on how the game elements should fit together [27]. It is also in this stage where the designers will develop gamification concepts [28]. Finally, the playtesting stage is a stage that is constantly repeated throughout the game design process, as a game designer continuously improves the game after each iteration [27].

However, in context for a learning-game design process, it is necessary to establish the learning foundations prior to the preparation stage. These foundations are the learning objectives and goals, and they form the basis for conceptualization of the game [29]. From a study conducted by Palomo-Duarte [30], who gamified a German language course, students have benefited from language improvement as well as increase vocabulary acquisition. When the gamification process is implemented and designed properly, it can result in several benefits, therefore demonstrating the usefulness of gamification in education.

2.3 Game-Based Learning

A game can be described as a form of leisure or recreation, usually with rules, goals, and challenges, with the aim of providing entertainment [31]. In recent years, there has been a noticeable rise in using games for educational purposes [32], whether it be through gamification or by using serious games. It is then important to understand how game-based learning works, and how educators and game designers can utilize it to their advantage.

Game-based learning refers to the utilization of games to create a learning environment, in which the game content and gameplay work together to improve a player's knowledge and skills. The game is designed to include challenges that, upon solving, will provide players with a sense of accomplishment [33]. Research has suggested that game-based learning can have positive behavioral outcomes in education, such as higher engagement and participation rates [34, 35].

While game-based learning has its benefits, it also possesses flaws that need to be considered. From a study done by Russell Francis in 2006 [16], it was noted that incorporating complex games into a traditional classroom duration of 1 h can prove to be very challenging. The study also talks about the importance of maintaining educational focus on the activity, instead of getting carried away with playing the game.

Despite these flaws, Russell Francis acknowledges the importance of game-based learning environments and emphasizes the need for educators to explore ways to integrate games as part of an activity, when compared with simply focusing on the game itself. By using teaching methods that are suited for game-based learning, educational game developers can continually refine and improve their games.

2.3.1 Virtual Reality Integration in Classroom-Based Education

While many studies have focused on the uses and benefits of using VR as a form of educational games, few have discussed as to how these games are integrated into the traditional classroom settings [36]. It is worth noting that when VR educational games are implemented poorly in classrooms, they may result in a mere duplication of the traditional instructional learning experience [37]. Instead, a better way to incorporate VR into learning is to use it to create unique experiences that can help enhance a students' understanding of the learning context [38], rather than simply using it as a novel educational tool.

Some learning models that have incorporated VR, or other technologies, into traditional classrooms would also be known as hybrid learning. Hybrid learning involves the incorporation of educational games inside and outside of the classroom to motivate and keep the students engaged [39]. Hybrid learning makes use of the flipped classroom model, a teaching approach that uses videos as an instructional tool. These videos typically contain relevant materials that will be covered during class and are easily accessible for the students to review at any time through the internet. Moreover, the students can review the videos multiple times at their own pace and convenience, allowing for students to better comprehend the topic at hand and some flexibility in learning [40]. This flipped learning approach can help to promote effective teamwork [41], as well as student engagement [42]. Using this model, students are first tasked with completing a pre-class learning assignment, and afterward, participate in application activities that are related to the pre-class learning assignment during class time [41]. This project will attempt to make use of this teaching approach, integrating instructional videos and VR games as applicational activities into the traditional classroom scene.

3 Hydraulic System Game

Before the start of this project, it should be mentioned that there was already a hydraulic system game developed by previous students. The game was developed using Unity engine (2018) and MRTK (version 2.6). However, since the development of the game, both Unity and Microsoft have released newer versions with enhanced features and utilities that can help game designers in creating higher quality games. The game itself also has design areas for improvement, which will be further discussed in the report. Given the team's belief in the potential of this game to enhance student learning, it has been decided that rebuilding the hydraulic system game using the latest, and more advanced, versions of Unity (2021) and MRTK (version 2.8) would be the path forward.

The most suitable game device for this project would be the VR headsets. These headsets can facilitate learning through gesture, or motion-based, interactions allowing for students to not only be engaged, but also better understand the lessons more effectively.

3.1 Serious Aspect and Game Aspect

It is evident that educational games must establish clear characterizing goals to effectively achieve their desired learning outcomes. As such, the goals for this project are to enhance the students' knowledge of the hydraulic system in aircrafts, to understand how fluid flows within the system, and to teach them the process of removing a component from the system. To motivate and help the students in achieving these goals, this project aims to develop a hydraulic system game centered around the Learjet 24 aircraft, featuring game stages that will introduce various components of the hydraulic system to the students, interactive features that can enable them to better understand fluid dynamics within the system, and a tutorial and test stage on how to remove an accumulator from the hydraulic system. The game is carefully designed to have a balance between achieving the learning outcomes and providing a fun and engaging experience. The aim is to ensure that the game is both entertaining to play, and also able to effectively convey the intended learning outcomes.

Table 1 Components used from MRTK

MRTK component	Uses
NearInteractionGrabbable	This component allows the user to grab any game object equipped with a collider
ObjectManipulator	This component allows the developer to toggle scripts during object manipulation, which can prove to be useful in certain cases
ToolTip	This component helps to display in game text, as well as important information for the user
Dock	This component is used as a set-up for other game objects to be docked
Dockable	This component allows a game object to be placed at the dock

3.2 Virtual Reality and Mixed Reality Toolkit

VR has been a popular medium across various fields, such as the health and educational fields [43]. One of the key advantages of VR is the ability to promote experiential and situated learning, where students can learn by using their bodies in a natural way and engaging with the environment [44]. Additionally, using VR for simulation training is a promising approach [45], and often contributes to a more effective learning experience. VR also offers an immersive learning environment that can be controlled by educators and game designers, reducing distractions while allowing for higher focus on the topic at hand [46].

The MRTK is an open-source cross-platform toolkit developed by Microsoft for creating mixed reality applications in Unity. With pre-built assets and prefabs, MRTK can offer game designers a variety of time-saving features [47]. By utilizing those features, developers can hasten the development process with little effort. This project made use of several prefabs and UX building blocks from MRTK for the development of this project, such as the button prefab, experimental docking system, tooltips, panels, and the hand menu, as shown in appendix A.

In addition to prefabs and UX features, MRTK also offers components that are ready to be used and incorporated with objects in Unity. This project made particular use of specific components to accelerate the development process, and to create an interactive and unique experience for the students or players. Here is a table of the components that were used in the project (Table 1).

3.3 Game Design, Past Versus Present

This section of the report will discuss the improvements in game design made to the older version of the hydraulic system game. It also addresses the reason why these modifications were necessary, if needed, and shows how it positively affects a students' learning experience.

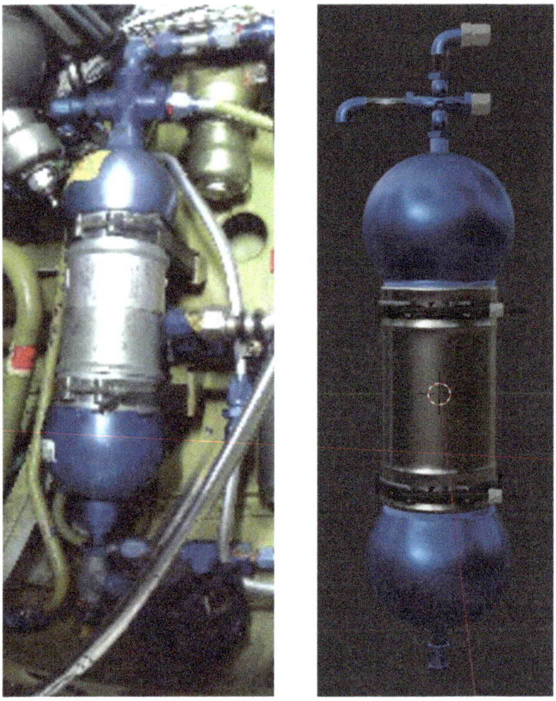

Fig. 1 Comparison between the real accumulator and the modeled accumulator

3.3.1 Hydraulic Components

The hydraulic components in the game were modeled using Blender, doing so allow game designers to segment different Unity components and materials into specific parts of the game objects conveniently. For some components, there was no need for redesigning or remodeling, hence they were used directly from the previous version of the game. These components were designed to be as accurate as possible to the components in the hydraulic system, so that players are able to recognize them in real life too. For the hydraulic system there were five main components to be designed, the reservoir, accumulator, pressure relief valve, auxiliary pump, and the pressure filter (Figs. 1, 2, 3, 4 and 5).

3.3.2 Gameplay Sequence

Gameplay sequence is a very important factor in creating a successful game design [48]. It can help the player to understand which game segment they are currently in, therefore allowing them to better understand the goals of the task at hand. Proper gameplay sequencing allows a player to progress gradually and linearly toward the end goal, resulting in a smooth and engaging game experience, also often referred to as being in the state of flow [22].

Fig. 2 Comparison between the real pressure relief valve and the model relief valve

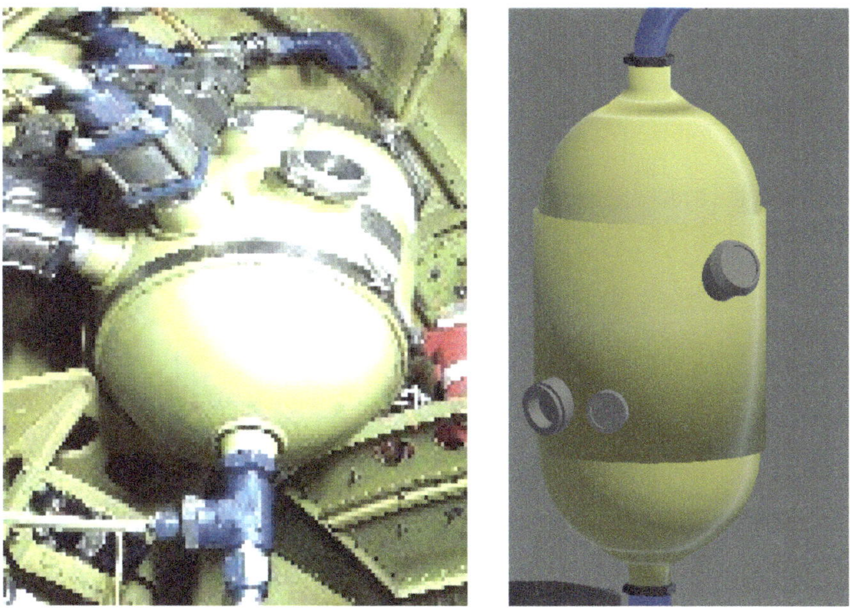

Fig. 3 Comparison between the real reservoir and the model reservoir

Fig. 4 Comparison between the real auxiliary pump and the model pump

The gameplay sequence of the older version of the hydraulic game was as such: Introduction, followed by tutorial tasks, and finally a test segment. This sequence is ideal for the hydraulic system game, and therefore will be retained in the current version of the hydraulic game.

3.3.3 Game Segments

This portion of the report will focus on the individual segments of the gameplay and examine areas of improvements made from the previous version of the game to the current one. Although the gameplay sequence for both versions of the game are the same, the design of individual segments differ significantly. In the introduction segment of the older version, players were required to fill up their personal information using a virtual keyboard and adjust their height to a comfortable level before starting the game. They would then be guided around the modeled aircraft while a narrator explains the different parts of the aircraft, the hydraulic systems, as well as the dynamics of the fluid flow. After the introduction has been completed, players were left to look around the hydraulic system freely before external instructions, by the game designer, were given for them to proceed. However, if the players wanted to review the names of the components within the hydraulic system, they would have

Fig. 5 Comparison between the real filter and the model filter

to restart the introduction segment to do so. To better illustrate the flow of the gameplay sequence that takes place during the introduction stage, a flowchart diagram highlighting the introduction segment is provided below (Fig. 6).

In the current version of the hydraulic system game, the process of entering personal particulars was removed, and instead, it was simplified to a basic game title screen. The registration of student particulars would instead be done physically during the class session. The ergonomic feature such as the height adjustment menu was kept to ensure that the game can be played comfortably by players of varying heights (Figs. 7 and 8).

These changes were implemented firstly. The hydraulic system game was designed to be played during traditional class durations of an hour. The problem lies in that students who have not experienced VR before may take longer to familiarize themselves with the interface, resulting in them exceeding the allocated time for gameplay. This snowball effect could prevent other students from playing the game within the given time frame.

The other change that was made was the replacement of the introduction with a basic title screen. Firstly, for students who have little to no experience with VR, having to stay stationary while observing fast optical flow can result in virtual sickness [49], which may have adverse effects on gameplay for the other segments. Instead, a separate scene for the introduction of the hydraulic components was created. Secondly, a

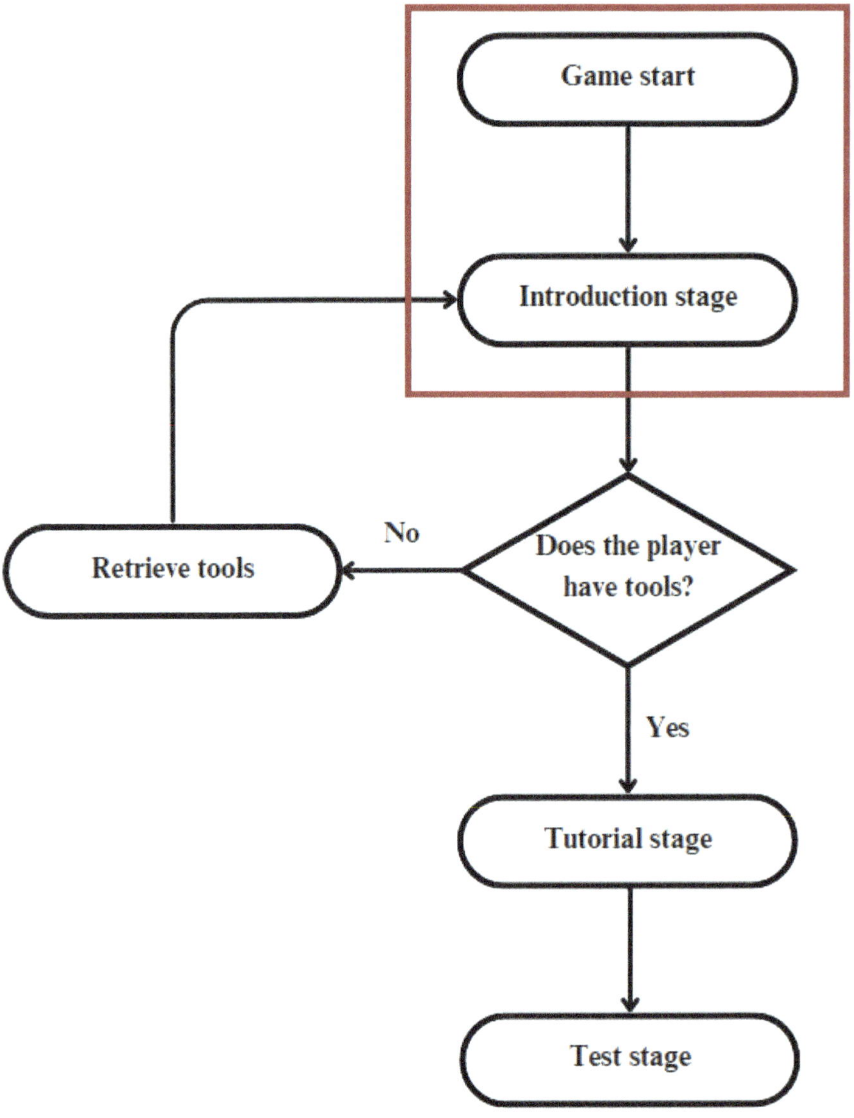

Fig. 6 Flowchart of the gameplay sequence during the introduction segment for the older version of the game

new title screen was created to provide a brief overview of the game and to focus on its theme. After the player is comfortable and ready to proceed, they can press on a start button that brings them to introduction scene. In this scene, the players are allowed to freely interact with and learn about the components that make up the hydraulic system. Within this same scene, players are also able to observe the fluid flow between the components. As this scene is a standalone and not a guided sequence, players

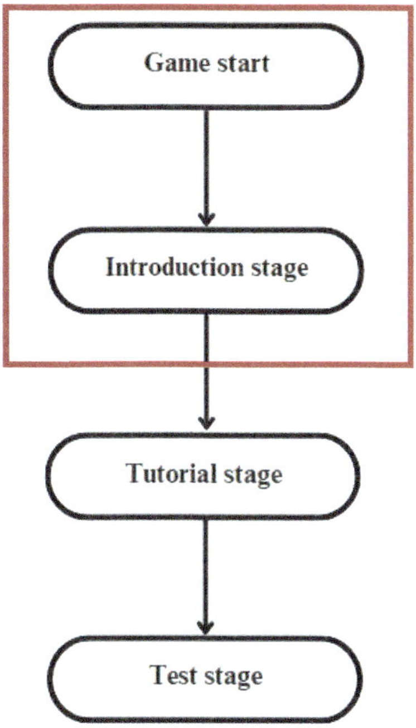

Fig. 7 Flowchart of the gameplay sequence during the introduction segment for the current version of the game

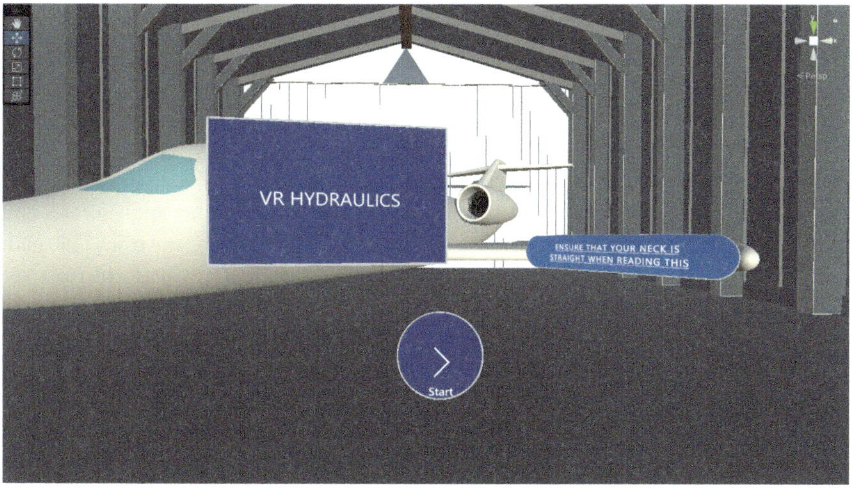

Fig. 8 Newly created title screen for the hydraulic system game

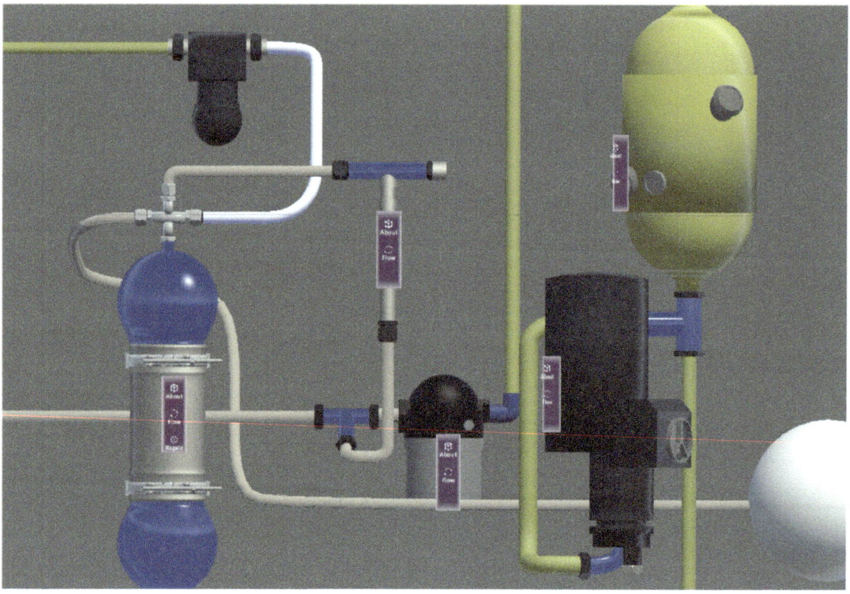

Fig. 9 Introduction segment for the current version of the hydraulic system game

are free to review and confirm if they missed any important information (Figs. 9 and 10).

The next gameplay segment in the hydraulic system game is the tutorial segment. The goal for the tutorial segment is the same for both versions of the game, however, the segment is designed differently for each version. The learning goals are for the player to understand how an accumulator is removed from the hydraulic system during repair. This involves familiarizing the players with the appropriate tools required, as well as the procedures a player must follow to remove it. The game mechanics for removing the accumulator from the hydraulic system is relatively straightforward. Players only needed two types of tools—a 7/8 wrench and a 3/8 wrench. The player must first use both wrenches to remove the three nuts at the cross-joint located at the top of the accumulator. By using the 3/8 wrench in the left hand to simulate stabilizing the joint in place, while the right hand simulates the loosening motion with the 7/8 wrench, players will be able to remove the nuts. The process is then repeated for the two clamp nuts that helps to secure the accumulator in place, this time using only the 3/8 wrench in the right hand. When all five nuts have been removed, the tutorial segment is considered completed, and the players can then be able to proceed to the test segment.

In the older version of the game, players had to exit the hydraulic system first to retrieve the necessary tools as they were not provided with any initially. Once the tools were obtained, players then have to place them onto a floating dock. This dock was designed to carry the tools so that when the player returns to the hydraulic system, they could proceed with the accumulator removing process. These movements of entering

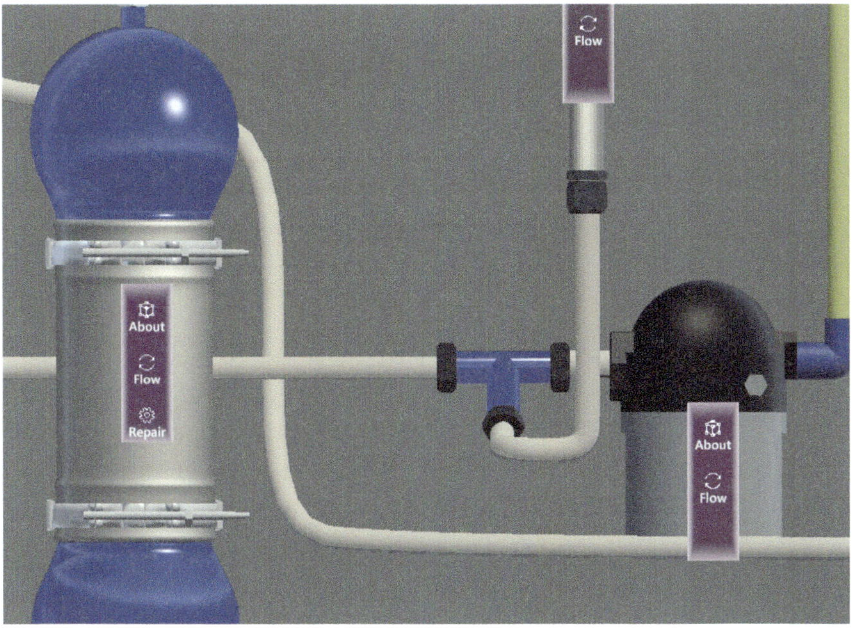

Fig. 10 Close up of the game options in the introduction segment

and leaving the hydraulic system would be done by interacting with a hand menu on the left hand. The instructions, however, are often guided by an external member, typically the game designer. Although this process was not entirely problematic, the procedures required the player to go back and forth to progress in the game, which could be messy and unclear, and often required a need for external help. The flowchart diagram below can help to visualize the gameplay sequence the players will have to go through (Fig. 11).

Therefore, to simplify the gameplay and reduce the dependency, the game design was modified to be more linear, with the tool interaction in the same scene as the tutorial. As such, the redesign of the current version of the hydraulic system game eliminates the need for additional movement required for the player, saving time, and reducing confusion. Having the area of interaction, tools, dock, as well as the instructions are all placed in the same scene allows the players to begin playing the tutorial segment immediately after the introduction segment (Fig. 12).

By redesigning the segments of the game, an improved and more linear gameplay sequence was produced. The intended outcome for both versions of the game also remained the same. Players are still able to identify components in the hydraulic system and understand the process of removing an accumulator during repair. The current version of the hydraulic system game offers a more seamless gameplay experience, making it easier for players to play the game on their own (Figs. 13 and 14).

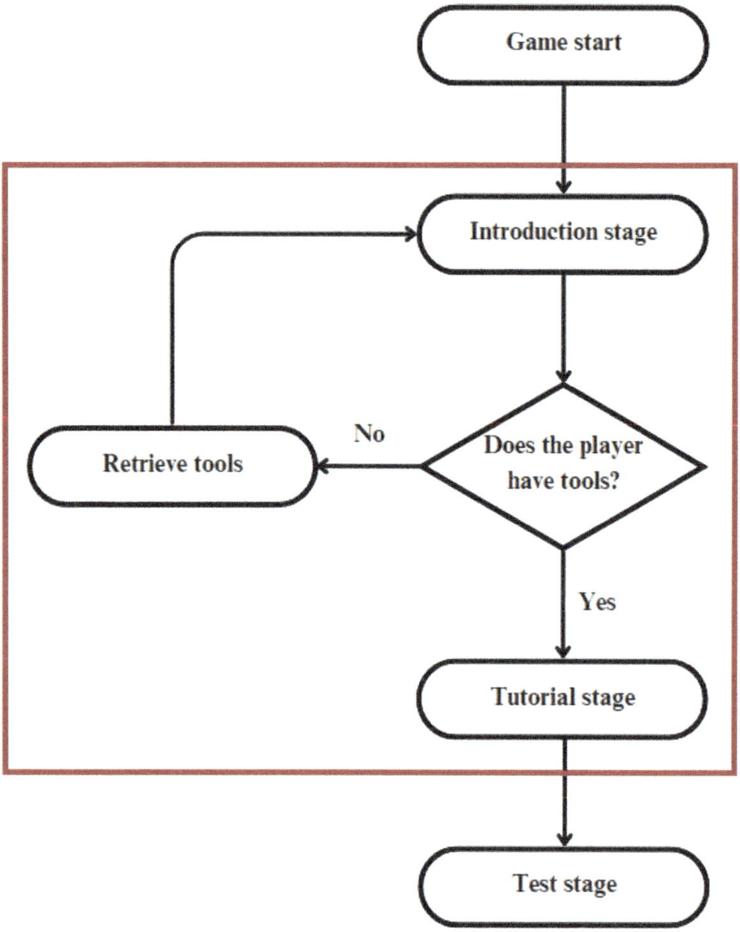

Fig. 11 Flowchart of the gameplay sequence during the tutorial segment for the older version of the game

The testing segments in both versions are the same. This segment aims to assess whether players are able to apply what they had learned from the introduction and tutorial segments. Players are once again tasked with removing the accumulator, a process that they should be familiar with after completing the tutorial segment, but now without the help of instructions. In the current version of the hydraulic game, when the required nuts have been removed, the player is then able to interact with the accumulator. The game is fully completed when the accumulator is placed onto its designated dock, and a pop-up notification appears, indicating that they have completed the task.

Some designs changes that were implemented in this segment were the creation of the dedicated dock for the accumulator as well as the task complete message.

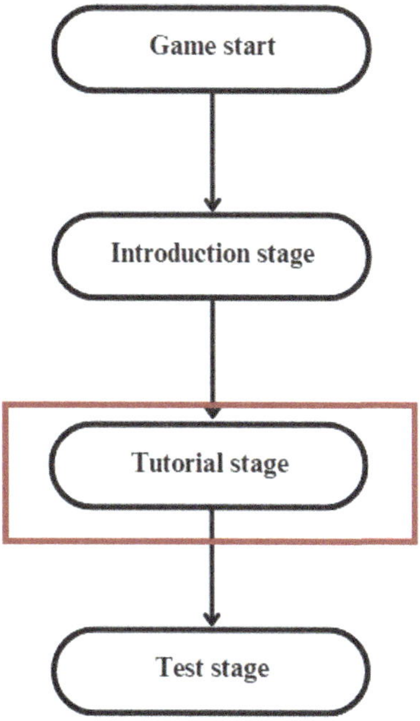

Fig. 12 More linear gameplay sequence after simplifying the gameplay in the tutorial segment

Fig. 13 Old design of the tutorial segment for the hydraulic system game

These changes were built upon the game mechanics of the previous version. In the older version of the game, players could interact with the accumulator once the task was completed. With the introduction of a dedicated dock for the accumulator to be placed, the game felt more completed and comprehensive. It also provided a more satisfying game experience for the player. The addition of a prominent pop-up game

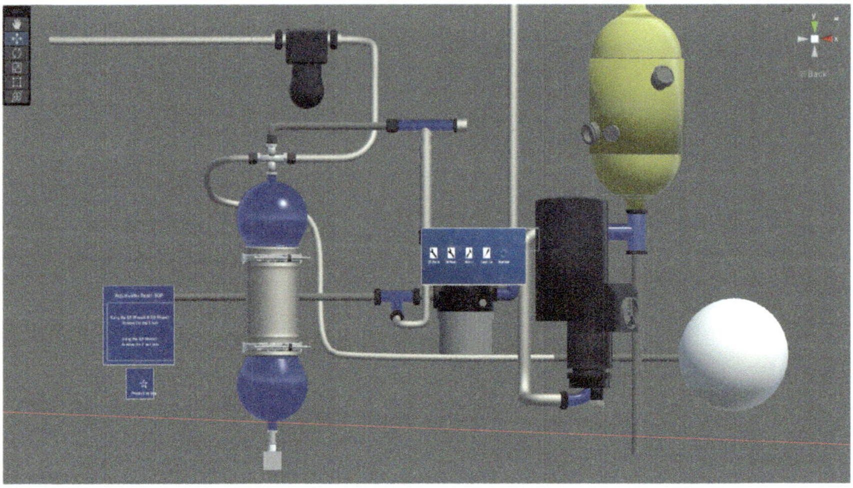

Fig. 14 New design of the tutorial segment for the hydraulic game

notification also provided a clearer indication to the player that they have completed the game and was free to pass it on to the next student to play.

Overall, the redesign of the individual segments from the older version of the hydraulic game to the current has resulted in improvements in the gameplay sequence. Features and game mechanics from the older version of the game were improved to reduce confusion, save time, and most importantly, reduce the likelihood of virtual sickness. The introduction segment was simplified and incorporated help from educators to register student particulars during the physical session, instead of solely relying on the game. The tutorial segment was made more linear by placing all the interactive tools and objects in the same scene, making the hydraulic system gameplay friendlier and more seamless. Creation of additional features were made in the test segment to promote satisfaction in gameplay and clarity. These improvements have been made in hopes to make the hydraulic system game more enjoyable and educational, as well as to provide a better learning experience for the students (Fig. 15).

3.3.4 Colors and Icons

Colors are crucial in video games as they help in information perception and a person's cognitive abilities [50]. From a research done by Tilburg University [51], it found that using the right kind of colors can invoke different types of emotions in a player. For instance, there was correlation between the color yellow and positive emotional responses, as the color yellow can be associated with the feeling of joy. Furthermore, when a game is well-designed, colors and proper lighting can help to enhance the level of immersion a player has with the environment [52].

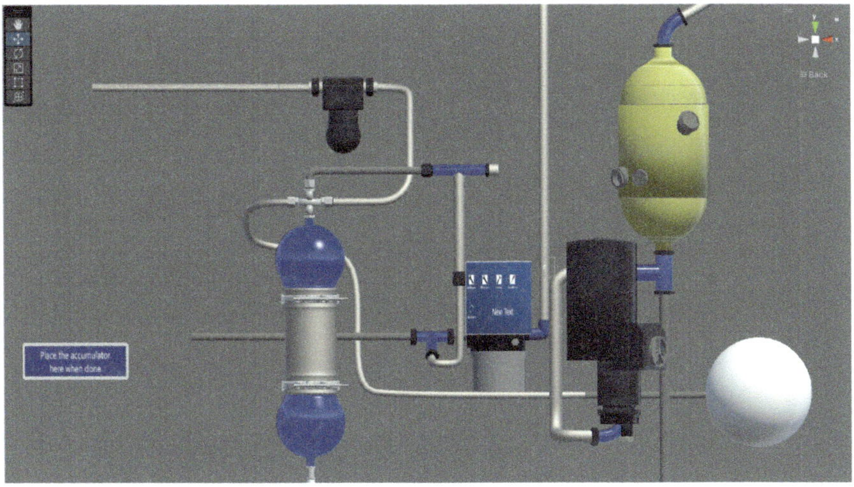

Fig. 15 Design of the test stage for the current version of the hydraulic system game

In the earlier version of the hydraulic game, while the intent of the game designer was to show the hydraulic system as accurately as possible, the representation of it proved to be a challenge due to the number of parts required to be modeled. An alternative approach was to project a low-rendered image of the hydraulic system onto a backboard and creating only the objects that required interaction. From a game design standpoint, the combination of low-quality and high-quality renders and objects could be distracting for the students. The amount of information presented to the player at once could also be overwhelming, resulting in a cognitive overload [53], the areas of focus of the game could also be lost in the process.

As such, for this project, while the number of components for interactions remained the same, the background however was reduced to a non-distracting one tone background, allowing the player to focus solely on the main objectives of learning about the hydraulic system components as well as fluid flow. Positive colors such as yellow was also used to highlight important points of focus during gameplay. Since the game consisted of similar objects of different sizes, colors were also used to differentiate them for a clearer gameplay, a red and green handle was used to differentiate the wrenches that were of similar sizes (Fig. 16).

Shaders were also used in the introduction segment of the game to show fluid dynamics. This was done by applying the pipes in a different material, and when the player interacts with the flow button, the pipes would light up in a different shade, indicating the fluid flow from one component to the other. To help with visualization and clarity, tool icons were also used in the tutorial and test segments of the game. These icons helped the players to visualize what tools would be spawned when they interacted with the respective buttons on the panel (Figs. 17 and 18).

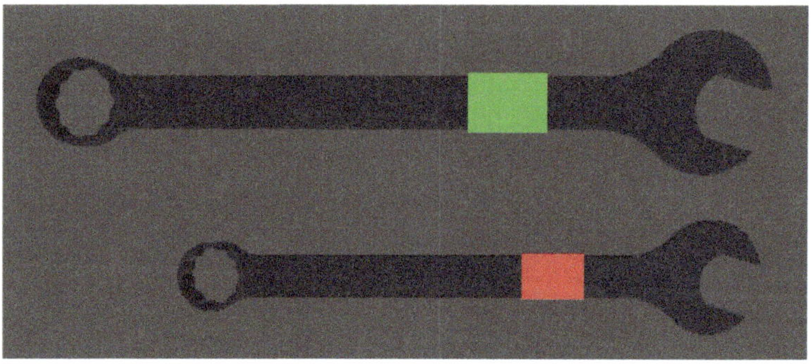

Fig. 16 Wrenches differentiated by color

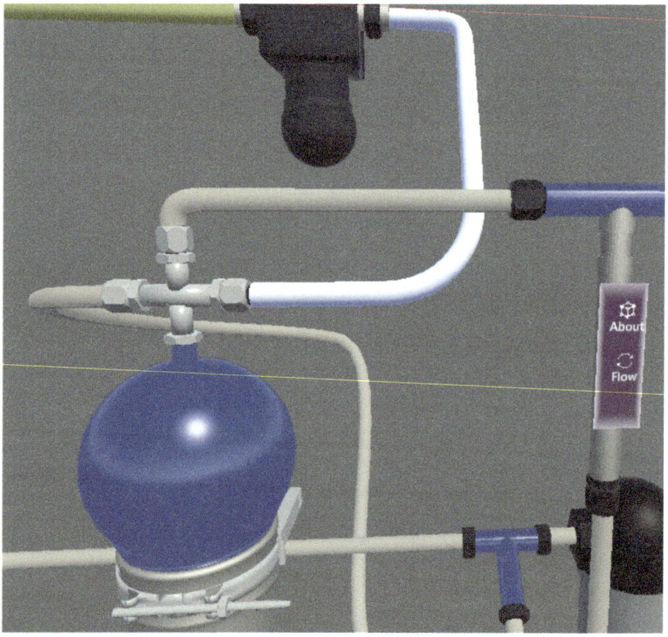

Fig. 17 Use of shaders to visualize fluid flow

3.3.5 Sounds and Audio Cues

Sounds and audio cues are typically incorporated as part of a game's mechanics, and they can play an important role in affecting a player's motivation [54], as well as providing the players with important information and feedback [55]. The selection of an appropriate background sound in video games is vital, as it sets the desired ambiance and experience for the player [56]. Examples of bad sound design would

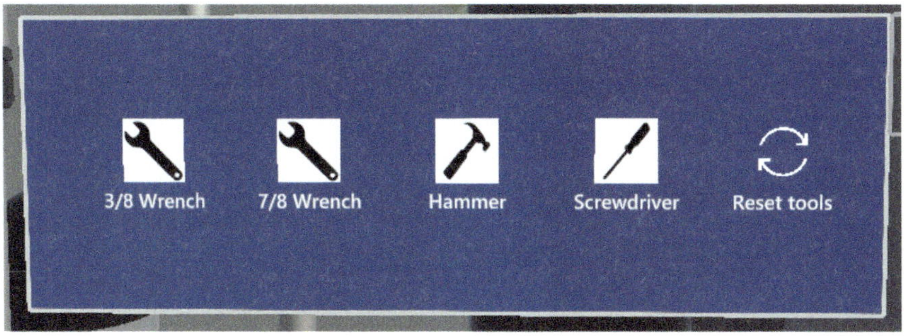

Fig. 18 Tool panel with respective tool icons

include having actions with no corresponding audio indications or choosing a background sound that is too loud or over-stimulating, causing the player to be distracted or lose focus from the task at hand in game [55].

The environment of the hydraulic system game is meant to take place in an aircraft hangar, where planes are usually parked for repair and maintenance. As such, the ambient sound used in this game was a white noise recording of airplane engines in the background. The audio is set to decrease gradually in volume as the player progressed through the game so that it is playing subconsciously in the background of the player's mind. These small changes in volume can help to influence the focus and attitude of players [57]. Audio cues can also help to enhance player engagement [55]. Thus, this project made used of Microsoft's Mixed Reality Toolkit buttons, as they are already pre-built to provide sound feedback when the button is pressed successfully by the player.

3.3.6 Cognitive Loads

Cognitive loads can play a big part in educational video games. When a game design is too overly complicated, students may have difficulty engaging with the learning tasks, leading to them experiencing a cognitive overload or a sense of feeling lost [58]. As the capacity of the human working memory is also limited, the amount of information delivery is crucial in preventing cognitive overload [59]. It is also important for game designers to take into consideration how much cognitive load is being received by the players, too little can result in players getting distracted, too much and the player might get overloaded. As such, the hydraulic system game is designed to introduce more information gradually as the game progresses. Each stage is also designed with the minimum required amount of game models to ensure that the player does not take in too much information at once.

3.3.7 Gestures in Virtual Reality Games

A benefit that VR games have over other types of games is its ability to replicate hand gestures from a player [60, 61]. VR games can facilitate a more natural experience, and also support students in learning through physical movement by enabling players to interact with objects in a three-dimensional environment [62, 63]. Researchers have also suggested that by incorporating gestures into the learning process, it may help to strengthen memory traces as well [64, 65]. Therefore, in designing the game mechanics, it was important to ensure that the gestures designed were accurate in replicating the motions required to remove the nuts from the accumulator, during the tutorial and test stages of the hydraulic game.

3.3.8 Game Mechanics

Upon entering the tutorial stage, the players will be introduced to the game mechanics of the hydraulic game. The players are first required to interact with the tool panel to the right of them. The tool panel comprises of four buttons with icons and labels, helping the players know what tools they will be spawning in.

Once they have interacted with the buttons on the tool panel and successfully spawned the required tools in, they can move on to the next stage, which is to remove the nuts from the accumulator. In order to remove the accumulator from the hydraulic system, the player first needs to loosen five nuts that secure the accumulator in place. Three at the top cross-joint of the accumulator and two are the clamp around its body (Fig. 19).

To remove the top three nuts, the player must hold the appropriate wrenches in each hand, sing a 3/8 wrench in the left, and a 7/8 wrench in the right. The 3/8 wrench is used to hold the joint in place while the 7/8 wrench will simulate the motion of loosening the nuts. The procedure for removing the two nuts located at the body of the accumulator remains the same, except the player only has to use a 3/8 wrench in their right hand (Figs. 20 and 21).

Tools and objects, such as the wrench and nut that require interaction from the player are fitted with the right tags and the box collider component in Unity. For example, the 3/8 wrenches and 7/8 wrenches will be tagged as "38wrench" and "78wrench." These tags and components will help to fulfill certain requirements later on as part of the game mechanics. C# scripts are also fitted onto the wrenches to ensure that the player is holding on to it when performing the repair process. This is mainly to prevent any form of cheating during gameplay. The game mechanic for the accumulator is operated by checking specific requirements through a C# script. In this script, it checks firstly whether if the player is using the right tool by checking the tag. If the player was to use a hammer, for example, the tag would be incorrect and thus not fulfilling the requirement for the game mechanic to activate. Next, it would check if the box collider of the wrenches is in collision with the box collider of the nut. This is an important requirement to check for as it ensures that the player knows where each tool is required to go, otherwise the player can simply perform the gesture

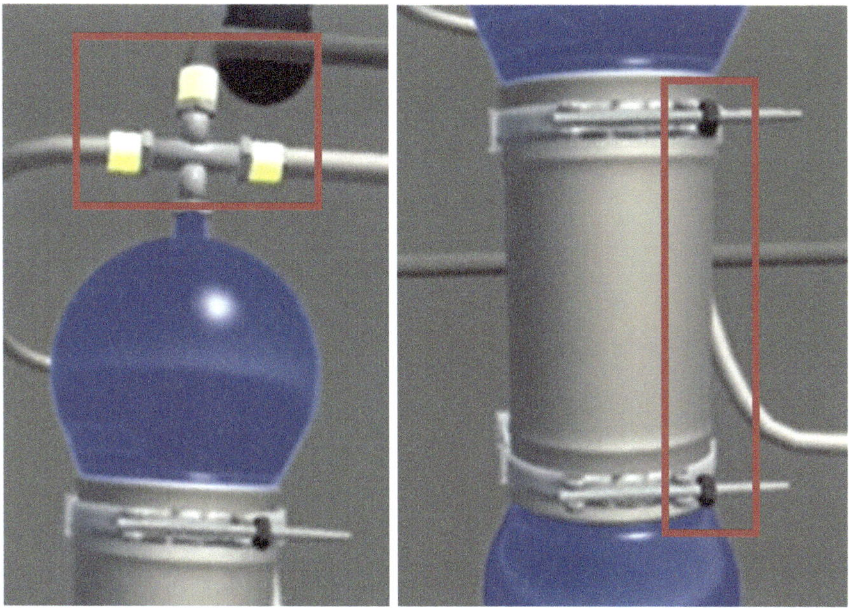

Fig. 19 Five nuts that hold the accumulator in place

anywhere else and achieve the same results. Lastly, while ensuring that the right tools are in the right hands and the box colliders are in contact, when the player performs the right gesture, only then will an animation of the nut being removed be played, therefore indicating that the player has accomplished the task successful. Once the player has removed all five nuts, they will then proceed on to the test stage. The C# scripts that were used in the development of this game can be found in appendix B.

4 Testing

Game testing is an important process in game development and is done throughout the life cycle of the hydraulic system game. It is essential to ensure that the game mechanics, model designs, and interfaces are functioning correctly as intended and free of any bugs or glitches that could detract from the learning experience. Testing also helps to ensure that the game is engaging and enjoyable for the players, which is vital in keeping them motivated and interested during the learning process.

This section of the report will talk about the different types of testing done in this project, mainly the functionality testing done by the game developers, user testing which is done by a group of residential students, and finally the integration of the hydraulic game into a traditional classroom setting. The team was also allocated six Oculus Quest headsets for this project.

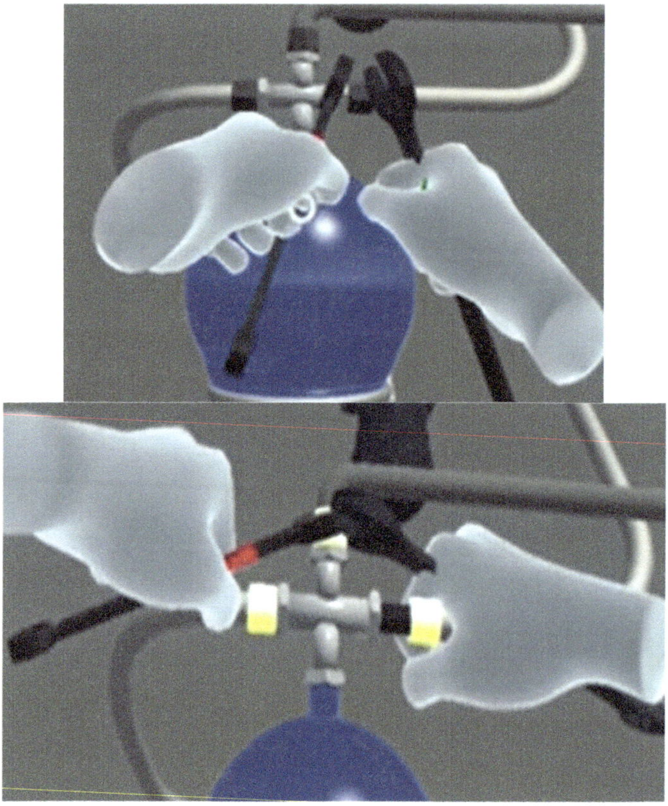

Fig. 20 Examples of removing of the three nuts at the top of the accumulator

4.1 Functionality Testing

During game development, functionality testing is an important stage that ensures the game's mechanics, features, and overall performance work as intended. This stage of testing is typically done during the alpha phase of the game, where the game developer spends most of their time testing each iteration of the game until they are satisfied with the results. This involves creating the game models, designing the game mechanics, setting up the right ambiance, placement of game models and scene setting, programming of interactions, and finally making improvements after each iteration of testing.

Another important aspect of functionality testing is ensuring that the game meets the goals and intended outcomes. This includes making sure that the hydraulic components are accurately designed to match their real-life counterparts, ensuring that the pipelines are connected correctly within the hydraulic system, fluid flow is shown correctly from one component to another. It also involves making sure that the operating procedures are accurate, so on and so forth.

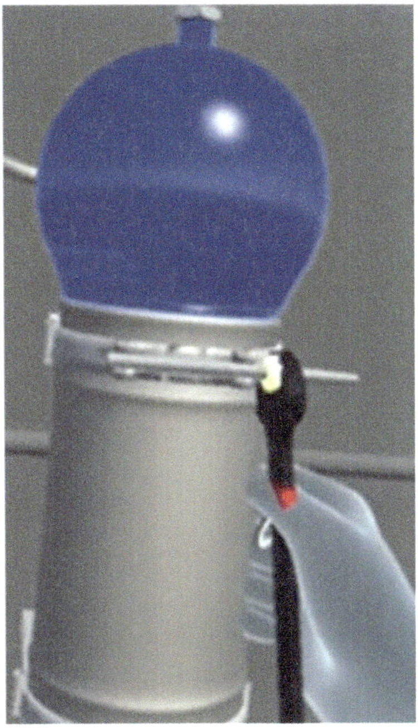

Fig. 21 Example of removing the nut located at the body of the accumulator

A critical process that also takes place during functionality testing is checking if the game is compatible with the intended platforms, in this case it would be the Oculus Quest. The older version of the game was designed to be used on the Oculus Rift, while the newer version is intended for use with the Oculus Quest. To ensure that the hydraulic game is fully compatible with the new VR headset, the game had to be updated from Unity version 2018 to Unity version 2021. Game packages that were used, such as MRTK, also had to be updated to version that worked with Unity 2021.

Overall, functionality testing is a key part of game development. It helps to ensure that the game developed is fun and of good design quality, and also meets the characterizing goals of set by the educator.

4.2 User Testing

User testing, also known as play testing, is the process of getting feedback from other users, outside of the team, in order to gather areas of improvements and identify issues that could have been missed during functionality testing. The goal in this testing phase is to ensure that the user playing the game is engaged, having fun, while checking if the learning goals have been achieved. Some questions that were asked in this testing phase were:

- "Is your neck comfortable?"
- "Are the game objects too high or low for you?" or "Are you at a comfortable height?"
- "Are you able to interact with the objects in the game?"
- "Are you feeling any motion sickness or headaches?"
- "Are you able to navigate through the game easily?"
- "Were they any problems you encountered in game?"

The game designer will try their best to observe the user interactions and gather feedback by interviewing the users when they have completed the play through. The users were interviewed to see if they were able to recognize the components in the hydraulic system, and if they had any difficulties during the tutorial and test segments.

In this project, the users involved are residential students from various academic backgrounds, which may prove to be valuable. These users are self-volunteered because of their interest in VR and gaming, and they will help by providing insightful feedback to further improve the game's quality, as well as allowing the game designer to address any issues before the game is released. Due to the nature of this users, each group would usually consist of five to ten students, which meant that each user had more time to interact and explore the game within the one-hour session. There was a total of five user testing sessions, conducted over the span of four weeks.

User testing can help to provide valuable insights into a player's preferences. The inclusion of the height menu for each individual stages was one suggestion that was made during the user testing process. Initially, players were only able to adjust their heights at the beginning of the game by using a height menu controlled by the right hand. The game would then take note of their height adjustments and carry it over to the next stages. This was a deliberate design choice to prevent the height menu from appearing each time the player has to use their right hand for an activity, and confusing the player in the process. However, due to some flaws in MRTK or the Oculus Quest, it is difficult to accurately ensure that the player starts at the same height each time they transit from one scene to another. While the height adjustments are saved in the player preference, the height of the stage itself may have changed. This was observed to be the case when the game was tested in two separate locations. Therefore, allowing users to have height control in each individual stages is more beneficial for players and will help to improve the player experience.

4.3 Integration of the Hydraulic Game

The final phase of this project involves the integrating of the hydraulic game into a traditional classroom setting. The target audience would be university students from the mechanical engineering field. The hydraulic game should have some sense of familiarity for these students which will help in engagement. Unlike the playtesting group, the classroom setting will have a significantly larger amount of audience of about 50 students or more. Given the length of a standard playthrough and taking into consideration the time each student needs to familiarize themselves with the controls of VR, it would be impossible for every student to play the game.

There were three solutions to overcome this problem. Firstly, students were divided into smaller groups and different time slots were allocated for each group. This allowed the team to better manage the students during the lesson. Secondly, the game was redesigned to only include the tutorial and test stages, shortening the duration of each playthrough. Lastly, the team developed a pre-lesson assignment in the form of a video recorded walkthrough, featuring gameplay from the tutorial segment of the hydraulic game. This walkthrough would also include step-by-step instructions for opening the hydraulic game from within the Oculus Quest, as well as how a player can exit the game when they are done. This pre-lesson assignment highlights all the important key points that a student should take note of when playing the game and is done in hopes that students can build familiarity with the game before playing it, and therefore reducing the time required for each student to complete the game. This walkthrough video will be sent to the students via email a day prior the class. During the class, the video will be continuously played on loop so that students can refer to it for assistance. Moreover, educators will also be moving around during the class period to assist and guide students who encounter any difficulties.

A pre-test and post-test survey were conducted before and after the gameplay session. The results were then compared at the end to see if the intended learning goals have been met by the students. As they were not able to playthrough the introduction segment of the hydraulic game, they were only asked to identify the accumulator, which is the component that was interacted with during their gameplay (Fig. 22).

The students were also asked on feedback for the tutorial and test segments, to see if they understood the procedure in removing the accumulator from the hydraulic system. Students were told to assume that the check gauge was already removed and asked to continue with the following steps instead. It was also to check if there were any difficulties during the repair process as well as difficulties performing the gesture required. Lastly, the survey also aims to see if the player had fun and was engaged during gameplay (Fig. 23).

5 Results

5.1 Survey Feedback

As the students in the target audience for this hydraulic game have a background in mechanical engineering, it is reasonable to assume that they would be more familiar with the hydraulic system, when compared with students from other fields of study. Therefore, it is expected that they would be able to recognize the accumulator component with ease. This hypothesis was supported by the pre-test survey results, which showed that more than half of the students already have a good sensing of what an

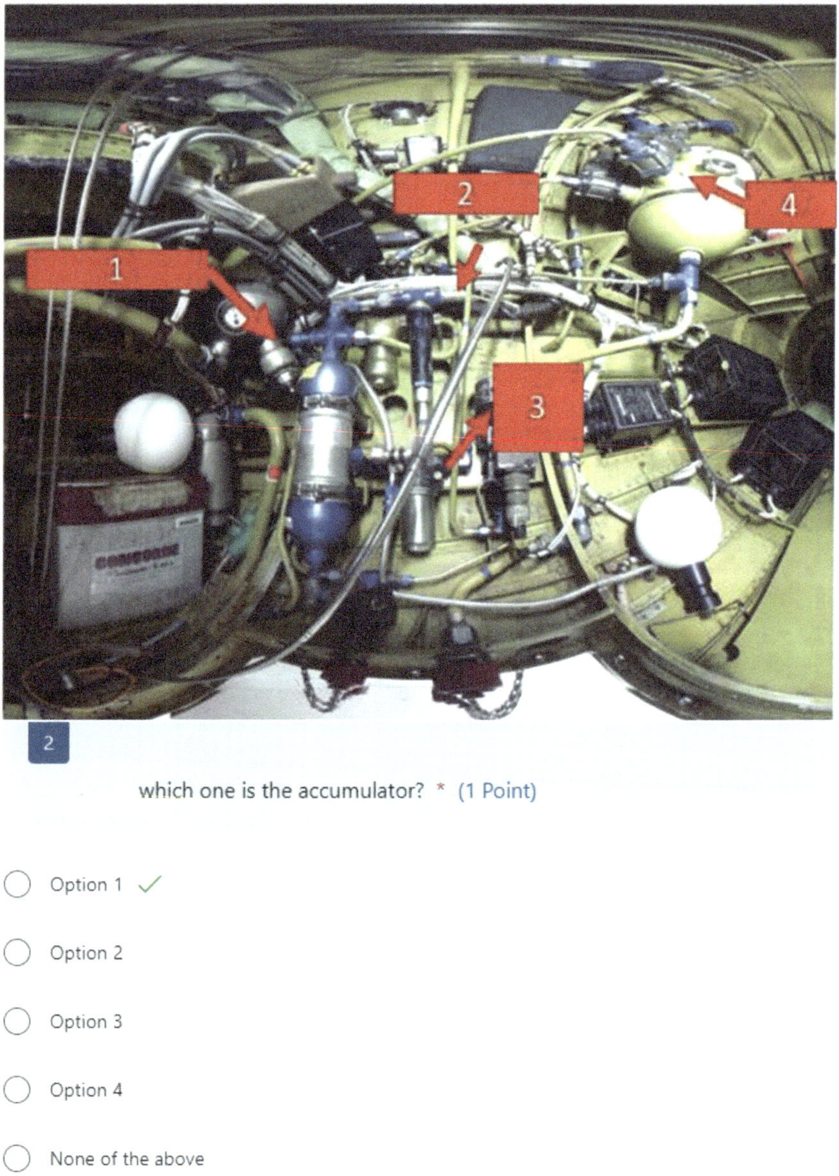

Fig. 22 Question from the survey on accumulator component recognition

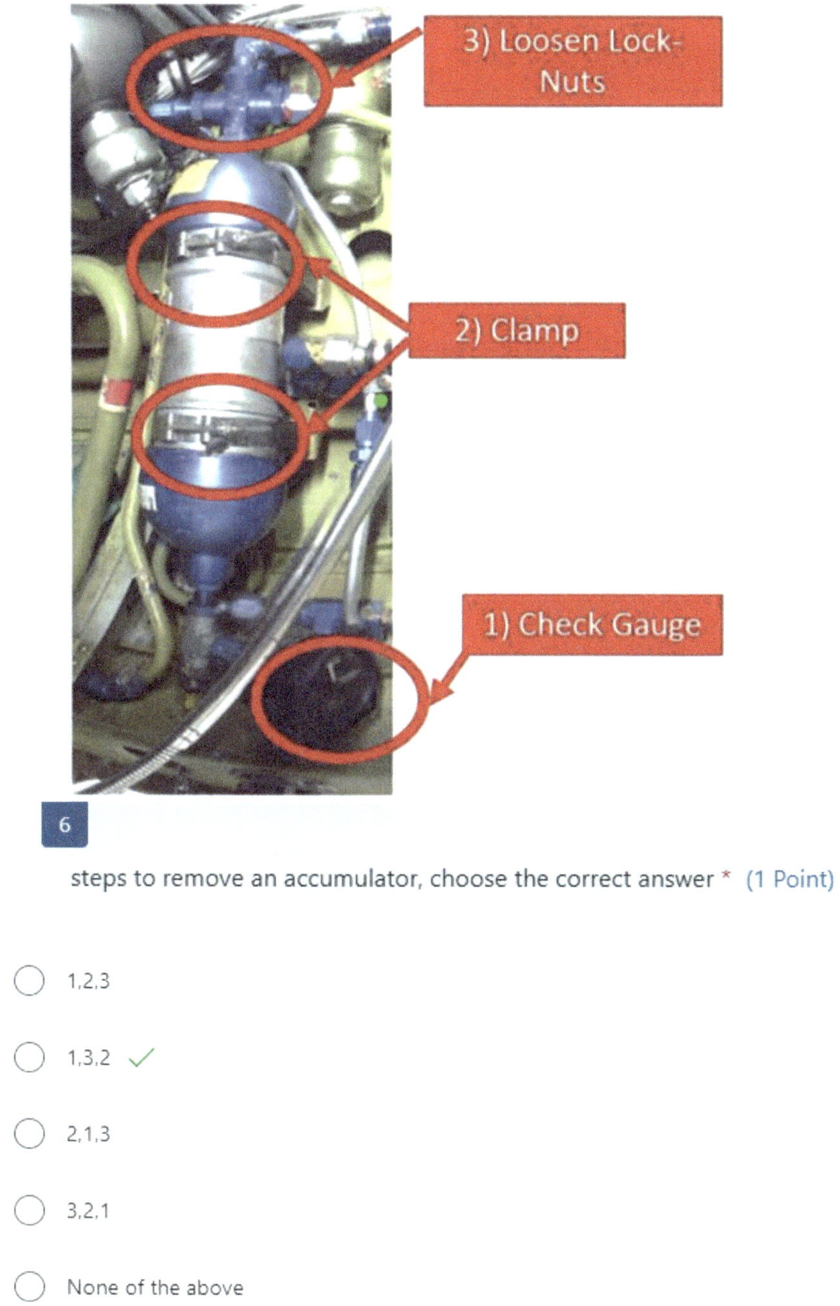

Fig. 23 Question from the survey on removal process of the accumulator from the hydraulic system

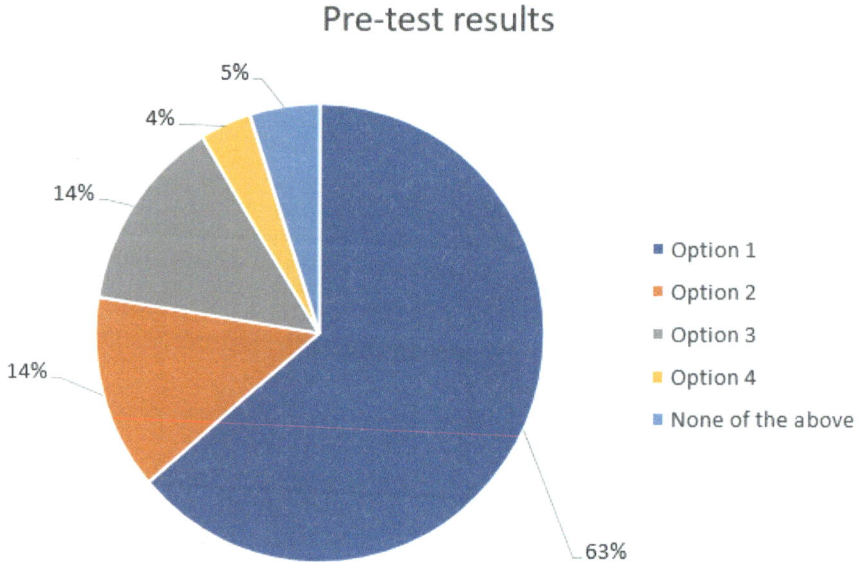

Fig. 24 Pre-test results for question on accumulator component recognition

accumulator looks like. After the gameplay session, majority of students, about 90%, were able to correctly identify the accumulator component in the hydraulic system. This was about a 27% increase in right answers from the pre-test survey. This could suggest that the hydraulic system game was effective in helping the students in either refreshing their memory on the hydraulic systems, or reinforcing their knowledge on the hydraulic system (Figs. 24 and 25).

Most students were also able to correctly identify the procedures in removing the accumulator from the hydraulic system. From the pre-test results, more than half of the students answered the procedures for removing an accumulator correctly, which could be once again linked to their background in mechanical engineering. Following the gameplay session, the post-test survey showed that majority of the students, about 82%, answered correctly when asked about the removal procedure for the accumulator which was about a 19% improvement in score from the pre-test survey. This indicates that the hydraulic system game was effective in helping the students learn the correct procedure for removing the accumulator from the hydraulic system. This high percentage of correct answers can also be attributed to the pre-class assignment walkthrough video. During the gameplay session, it was observed that many students would take off the VR headset, and recap on the steps with help from the walkthrough videos (Figs. 26 and 27).

From the results of both survey questions, it suggests that the students have a good grasp of identifying the accumulator in the hydraulic system of the Learjet 24. Most students have also gained familiarity and a good understanding of the accumulator removal process.

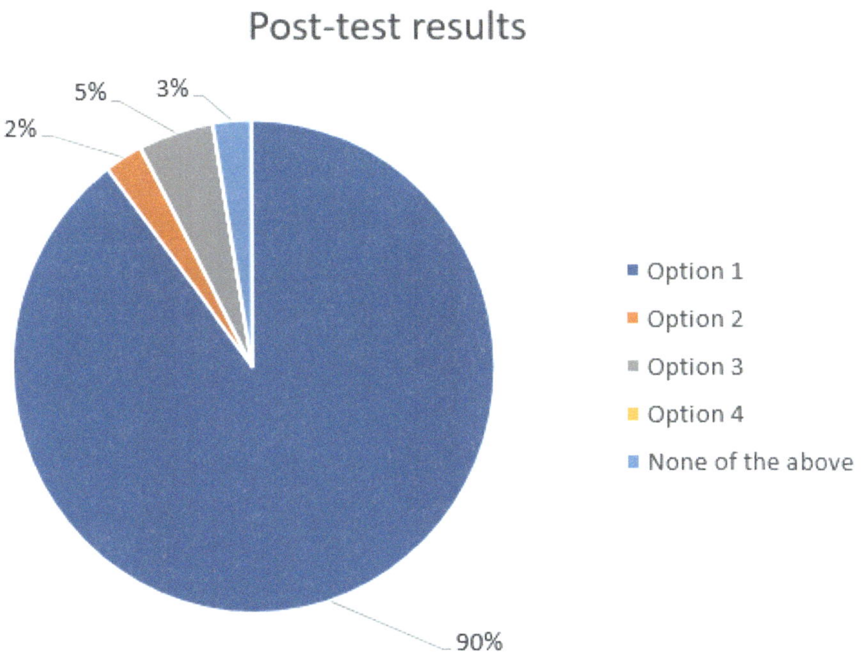

Fig. 25 Post-test results for question on accumulator component recognition

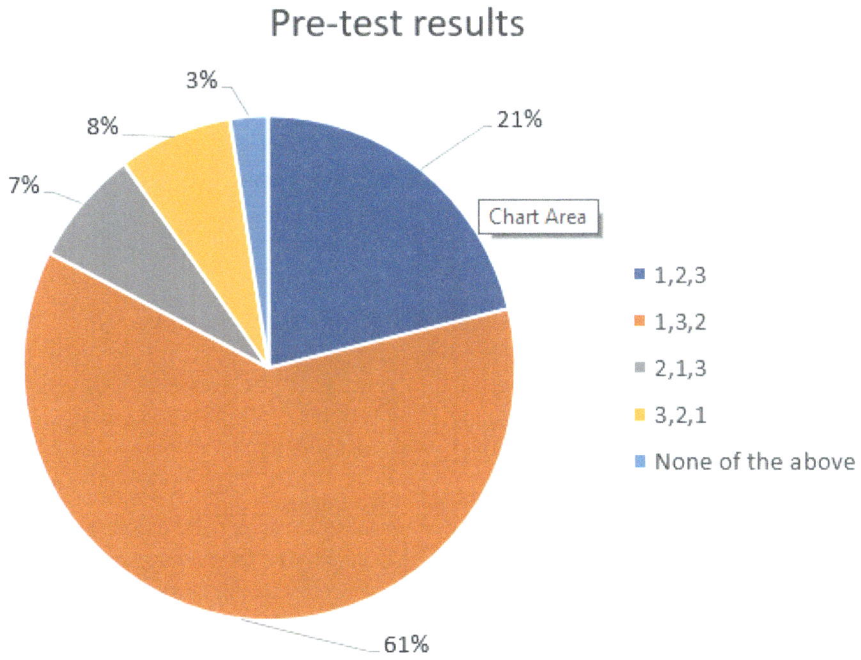

Fig. 26 Pre-test results for question on accumulator removal process

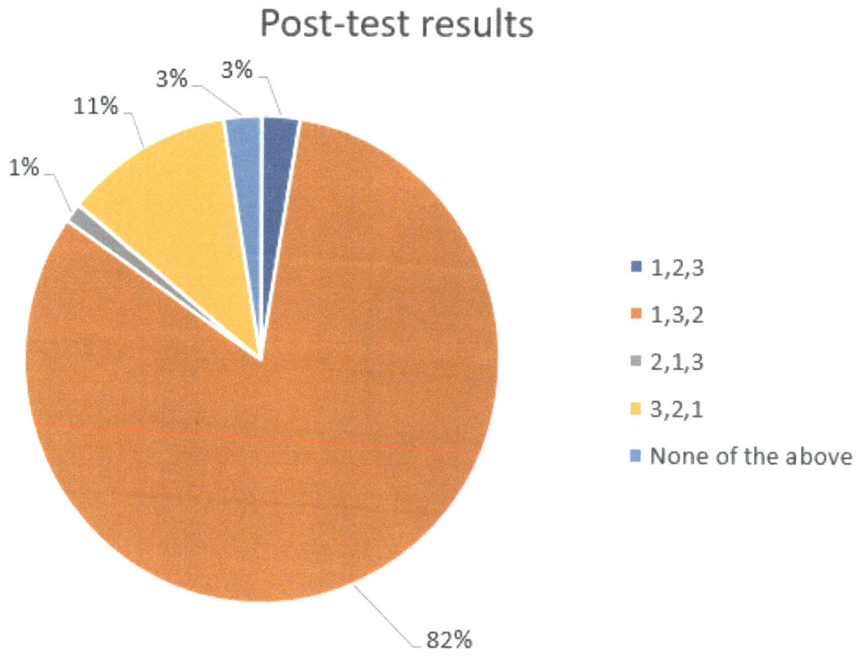

Fig. 27 Pre-test results for question on accumulator removal process

5.2 *Gameplay*

Other than asking about the serious aspect of the game, the survey was also used to measure the level of engagement and enjoyment. Questions asked in the post-test survey were related to a user's focus and interest in gameplay and the topic at hand.

The students were tasked to answer these questions using the Likert scale of 1–5, with 1 being strongly disagree and 5 being strongly agree. The questions that were asked in the post-test survey were as such:

- "I am fully focused on today's topic and activity; I am not distracted by other things."
- "I think today's topic and activity are interesting."
- "I will enjoy working on today's topic and activity."
- "Presently, I feel bored about today's topic and activity."

After collating the survey results, most of the students responded positively which is a good indication that their hydraulic system game has helped them to stay engaged and focused throughout the lesson period. This can also be further supported by their ability to complete the assigned task, as shown previously (Fig. 28).

According to feedback from residential students, the current version of the hydraulic system game is said to be more responsive and visually pleasing compared to the older version. This can be attributed to the redesign of the stages by reducing

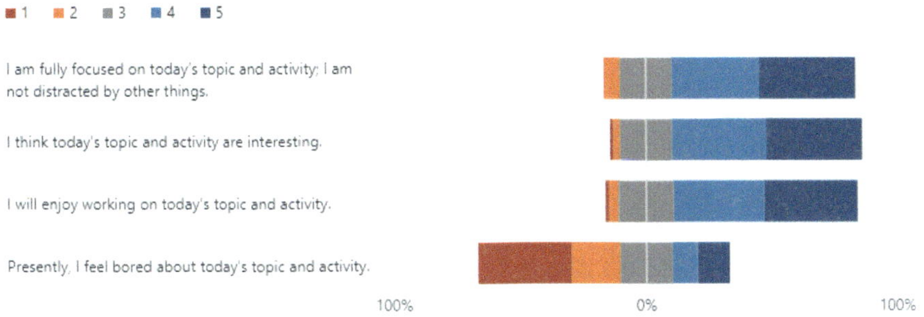

Fig. 28 Survey results of the overall student's enjoyment and engagement level

the amount of cognitive load a player has to handle upon starting the game, as well as to the technological advancements in Unity as well as the VR headsets. The Oculus Quest has a much better hand tracking camera when compared with the Oculus Rift with LEAP motion detector. MRTK also added new features that could help improve the player's experience. They also improved the handling of game objects, making interactions smoother and a better experience for the user. With these factors combined, it is possible to see why the residential students may prefer the current version of the hydraulic system game.

5.3 Discussion

The results of this study suggests that integrating the hydraulic system game as a form of interactive activity into the traditional classroom scene was effective in helping students improve their understanding of the hydraulic system, particularly in identifying the accumulator component and the removal process of it. The results from the post-test survey showed a higher percentage of correct answers than the pre-test survey, which could be an indication that the game was successful in helping the students refresh or reinforce their knowledge on hydraulic systems. It can also be observed that students were able to stay engaged and entertained throughout the duration of the lesson.

Positive feedbacks from the residential students about the redesign of the hydraulic game also suggests that the new design of the hydraulic game could be in the right direction, and that improvements in Unity, MRTK, and VR headsets have contributed to improving the overall experience of a user. Survey results from the post-test also suggests the importance of having accurate game models in educational games for learning purposes.

Other observations that can be made is that this activity has helped to build some teamwork within the groups. Some students were exceptionally quick on picking up the workings of VR games and were able to help instruct and guide other students, after playing through the game themselves. This helped to reduce the amount of time

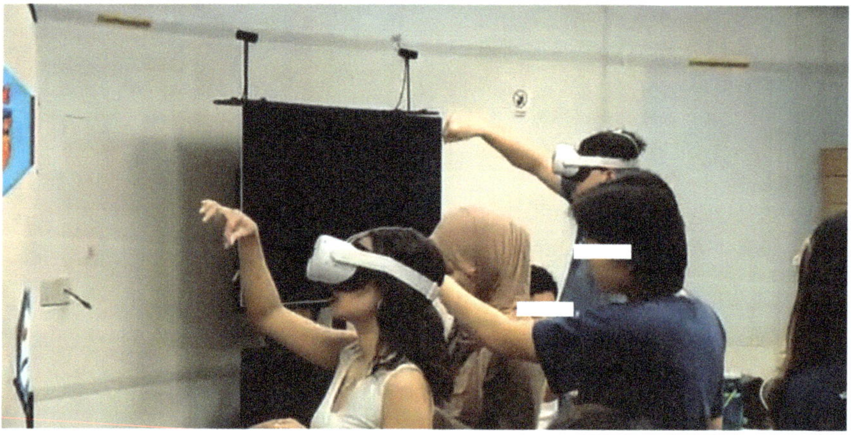

Fig. 29 Photo taken during the gameplay session, showing collaboration work among students

wasted when a student gets lost and has to refer back to the walkthrough video guide (Fig. 29).

Another notable finding in this study is the effectiveness of using video guides as a form of pre-class assignment, similar to the flipped classroom model. The video guide helps to provide students with a preview of the game's content, which allows them to gain a sense of familiarity with the gameplay and the game mechanics before they play the game. This may help in reducing the cognitive loads for students upon playing the game, allowing them to focus on the more challenging aspects of the game. The video guide also serves as a point of reference any time a student is lost and needs help to progress. The students have access to the video guide, and it is also being played on loop throughout the duration of the class. This learning model may also prove to be an effective method of enhancing a student's learning experience, as they are able to apply what they have learned immediately through the use of the VR hydraulic system game (Fig. 30).

5.4 Limitations and Future Works

One limitation is the use of self-report measures in the survey. While the post-test survey provides a valuable insight into the students' engagement and enjoyment for the hydraulic game, self-report measure is subjected to response biases and may not be accurate in reflecting the students' actual experiences. An approach that future studies could consider doing is the use of eye-tracking technology to measure the engagement and attention during gameplay.

This study was also conducted on university students from the mechanical engineering field. The students may already have a prior experience or knowledge with hydraulic systems which may affect the results of the study. Since this study was also

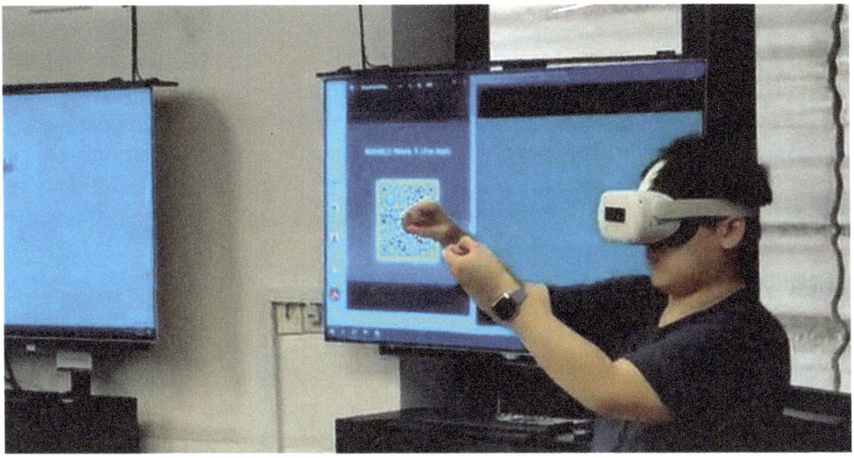

Fig. 30 Student trying out the hydraulic game with the video guide in the background

conducted on students from the mechanical engineering field, the results may not be the same if conducted for students from other educational fields. Future researchers should consider having a more diverse sample size of students to improve the external validity of the findings.

The study also only examined the immediate effects of the hydraulic system game for learning and engagement, it is difficult to say if the positive effects would be sustained over time, and whether if the game would be effective in improving long-term retention of the lesson learnt. Future studies could consider conducting a follow-up assessment to examine the durability of the learning outcomes.

Future considerations could be using physical interactable objects outside of the VR environment to help with player immersion, but that would require the construction of a real suitable environment to fit this need. Bug and glitches were also found outside of testing during the gameplay sessions, certain activities would result in one of the wrenches to be glitches in a fixed position. It is theorized that this occurs when a player leaves the VR boundary and could be a point to take note of for future developers of the game.

6 Conclusion

In conclusion, our research highlights the critical role that game design plays when creating an educational game. Game design is the component that helps to tie the serious aspects of educational games, as well as the fun and engaging aspect of gaming together. During the game design process, clear goals and learning outcomes should be set by the educator so that the game designer is able to tailor the game to fit the requirements, while ensuring a balance between education and enjoyment. In this sense, game design can serve as a bridge between education and gaming, becoming a useful tool for educators to enhance the students learning and engagement. This research has also explored the use of the flipped classroom model, by using video guides as a form of pre-class assessment that may be beneficial to students. It is important to note that educational games should not replace the traditional classroom teaching, but instead be integrated into it as a supplementary tool. Overall, this research has highlighted the potential for educational games, and the importance of understanding game design concepts in promoting effective learning and engagement.

6.1 Contributions

This project has laid the foundations for the game design of the hydraulic system game. While the gameplay and mechanics have only been completed for the accumulator component, future works can be done to expand on the game mechanics for other components, eventually completing the hydraulic system as a whole. The game may then be potentially used by aircraft technicians for simulation training in the future, or as a basic introduction to the hydraulic system for students who are new to mechanical engineering.

This project has also contributed to the understanding of the importance of game design and its purpose in educational games. Game design is a topic that should be researched on more as it is an important bridge that helps to connect education and gaming. As technology continues to advance at such a rapid rate, the limitations of game design continue to be pushed further, which may lead to even greater benefits for education in the future. From our findings, effective game design can prove to be highly beneficial for learning outcomes, while providing an enjoyable experience for the players. A clear and well-structured game design can also facilitate navigation for players, it may also promote collaboration among students as shown in the discussion above. These contributions highlight the significance of game design in educational games, and hopefully help to provide some insight for future research on this topic.

This project has also shown the potential benefits of integrating VR educational gams into a flipped classroom learning model. Rather than using traditional pen and paper assignments, this project has incorporated VR educational games to encourage students to apply what they have learned from the walkthrough videos physically. It also builds on the concept of Education 4.0, which talks about learning by doing. This approach offers an engaging and more modern way for students to learn and could be a point of exploration for other educators alike. This teaching approach could also prove to be the way forward in education of the future as technology.

References

1. Erin, C.: (2012). http://box2D.org/
2. Hsu, Y.S., Wang, C.Y., Zhang, W.X.: Supporting technology-enhanced inquiry through metacognitive and cognitive prompts: sequential analysis of metacognitive actions in response to mixed prompts. Comput. Hum. Behav. **72**, 701–712 (2017). https://doi.org/10.1016/j.chb.2016.10.004
3. Wong, J.Y., Azam, A.B., Cao, Q., Huang, L., Xie, Y., Winkler, I., Cai, Y.: Evaluations of virtual and augmented reality technology-enhanced learning for higher education. Electronics **13**(8), 1549 (2024). https://doi.org/10.3390/electronics13081549
4. Lee, T., Wen, Y., Chan, M.Y., Azam, A.B., Looi, C.K., Taib, S., Ooi, C.H., Huang, L.H., Xie, Y., Cai, Y.: Investigation of virtual & augmented reality classroom learning environments in University STEM education. Interact. Learn. Environ. 1–16 (2022). https://doi.org/10.1080/10494820.2022.2155838
5. Cai, Y., van Joolingen, W., Walker, Z.: VR, simulations and serious games for education. Springer (2019)
6. Ehmann, S., Lin, M.C.: Accurate and fast proximity queries between polyhedra using convex surface decomposition. Eurographics **20**(3), 500–510 (2021)
7. Clark, D.B., Tanner-Smith, E.E., Killingsworth, S.S.: A meta-analysis of digital games and learning gains. Human Behav. Emerg. Technol. **4**(1), 68–80 (2022)
8. Duty, L.L., Holden, K.N., Dompier, B.M.: A review of gamification and gamified applications in the context of higher education. IEEE Access **8**, 38478–38492 (2020)
9. Dicheva, D., Dichev, C., Agre, G., Angelova, G.: Gamification in education: a systematic mapping study. Educ. Technol. Soc. **18**(3), 75–88 (2015)
10. Koivisto, J., Hamari, J.: The rise of motivational information systems: a review of gamification research. Int. J. Inf. Manage. **45**, 191–210 (2019)
11. Toledo Palomino, P., Toda, A.M., Oliveira, W., Cristea, A.I., Isotani, S.: Narrative for gamification in education: why should you care? In: 2019 IEEE 19th International Conference on Advanced Learning Technologies (ICALT), Maceio, Brazil, pp. 97–99 (2019). https://doi.org/10.1109/ICALT.2019.00035
12. Akinoglu, M., Inan, A., Cigdem, H.: The impact of gamification on learning and instruction: a systematic review of empirical evidence. Educ. Inf. Technol. **25**(6), 5269–5298 (2020)
13. Alan, C., Salter, M., Longmuir, A., Benson, M., Adachi, C.: Transformation or evolution?: Education 4.0, teaching and learning in the digital age. Higher Educ. Pedagogies **5**(1), 223–246 (2020)
14. Halili, S.H.: Technological advancements in Education 4.0. Online J. Dist. Educ. e-Learn. **7**(1), 63–65 (2019)
15. Almeida, F.: The Role of Serious Games, Gamification and Industry 4.0 Tools in the Education 4.0 Paradigm, vol. 10, no. 2, pp. 120–136 (2019)
16. Krath, J., Schürmann, L., von Korflesch, H.F.O.: Revealing the theoretical basis of gamification: a systematic review and analysis of theory in research on gamification, serious games and game-based learning. Comput. Hum. Behav. **125**, 106963 (2021). ISSN: 0747-5632. https://doi.org/10.1016/j.chb.2021.106963
17. Nazry, N., Nazrina, M., Romano, D.M.: Mood and learning in navigation-based serious games. Comput. Hum. Behav. **73**, 596–604 (2017)
18. Watt, K., Smith, T., et al.: Research-Based Game Design for Serious Games, vol. 52, no. 5, p. 202. https://doi.org/10.1177/1046878121 1006758
19. Larson, K.: Serious Games and Gamification in the Corporate Training Environment: A Literature Review. TechTrends2020, 64, pp. 319–328 (2020). https://doi.org/10.1007/s11528-019-00446-7
20. Caserman, P., Hoffmann, K., Müller, P., Schaub, M., Straßburg, K., Wiemeyer, J., Bruder, R., Göbel, S.: Quality criteria for serious games: serious part, game part, and balance. JMIR Ser. Games **8**(3), e19037 (2020). https://doi.org/10.2196/19037

21. Dörner, R., Göbel, S., Effelsberg, W., Wiemeyer, J.: Serious Games: Foundations, Concepts and Practice. Springer International Publishing, Cham, Switzerland (2016)
22. Koster, R.: Theory of Fun for Game Design. O'Reilly Media Inc., Sepastopol, CA (2013)
23. Csikszentmihalyi, M.: Flow: The Psychology of Optimal Experience. Harper & Row, New York (1990)
24. Erhel, S., Jamet, E.: Improving instructions in educational computer games: exploring the relations between goal specificity, flow experience and learning outcomes. Comput. Hum. Behav. **91**, 106–114 (2019)
25. Ermi, L., Mäyrä, F.: Fundamental components of the gameplay experience: analysing immersion. Worlds Play Int. Persp. Digital Games Res. **37**(2), 37–53 (2005)
26. Becker, K.: What's the difference between gamification, serious games, educational games, and game-based learning? Acad. Lett. 209 (2021). https://doi.org/10.20935/AL209
27. Schmidt-Kraepelin, M., Thiebes, S., Tran, M.C., Sunyaev, A.: What's in the game? Developing a taxonomy of gamification concepts for health apps. In: 2018 51st Hawaii International Conference on System Sciences (HICSS), Waikoloa Village, HI, pp. 1217–1226 (2018). https://doi.org/10.24251/HICSS.2018.672
28. Eckardt, L., Grogorick, S., Robra-Bissantz, S.: Play to learn: conducting a playtest session for improving an educational game. In: Proceedings of the Americas Conference on Information Systems (AMCIS), pp. 1–10.
29. Fullerton, T.: Game Design Workshop: A Playcentric Approach to Creating Innovative Games, 4th edn.. A K Peters/CRC Press. https://doi.org/10.1201/b22309
30. Schöbel, S., Janson, A., Jahn, K., Kordyaka, B., Turetken, O., Djafarova, N., Saqr, M., Wu, D., Söllner, M., Adam, M., Gad, P.H., Wesseloh, H., Leimeister, J.M.: A research agenda for the why, what, and how of gamification designs. In: Outcomes of an ECIS 2019 Panel. Communications of the Association for Information Systems, p. 46 (2020). https://doi.org/10.17705/1CAIS.04630
31. Boller, S., Kapp, K.: Play to Learn. ATD Press, Alexandria, Virginia (2017)
32. Palomo-Duarte, M., Berns, A., Cejas, A., Dodero, J.M., Caballero, J.A., Ruiz-Rube, I.: Assessing foreign language learning through mobile game-based learning environments. Int. J. Hum. Capital Inf. Technol. Prof. (IJHCITP) **7**(2), 53–67 (2016). https://doi.org/10.4018/IJHCITP.201604010
33. Stenros, M., Paavilainen, J.: Game Definition: A Discussion About the Definition of Games. Tampere University, Tampere, Finland, Tech. Rep. (2015) [Online]. https://trepo.tuni.fi/bitstream/handle/10024/141136/GameDefinition_Stenros.pdf?sequence=1&isAllowed=y
34. Chang, C.Y., Hwang, G.J.: Trends in digital game-based learning in the mobile era: a systematic review of journal publications from 2007 to 2016. Int. J. Mobile Learn. Organis. **13**(1), 1–20 (2019)
35. Qian, M., Clark, K.R.: Game-based learning and 21st century skills: a review of recent research. Comput. Hum. Behav. **63**, 50–58 (2016). https://doi.org/10.1016/j.chb.2016.05.023
36. Cassola, F., Pinto, M., Mendes, D., Morgado, L., Coelho, A., Paredes, H.: A novel tool for immersive authoring of experiential learning in virtual reality. In: 2021 IEEE Conference on Virtual Reality and 3D User Interfaces Abstracts and Workshops (VRW), Lisbon, Portugal, pp. 44–49 (2021). https://doi.org/10.1109/VRW52623.2021.00014
37. Allcoat, D., Hatchard, T., Azmat, F., Stansfield, K., Watson, D., von Mühlenen, A.: Education in the digital age: learning experience in virtual and mixed realities. J. Educ. Comput. Res. **59**(5), 795–816 (2021)
38. Oyelere, S.S., Bouali, N., Kaliisa, R., Obaido, G., Yunusa, A.A., Jimoh, E.R.: Exploring the trends of educational virtual reality games: a systematic review of empirical studies. Smart Learn. Environ. **7**(1), 31 (2020). https://doi.org/10.1186/s40561-020-00142-7
39. Hu-Au, E., Lee, J.: Virtual reality in education: a tool for learning in the experience age. Int. J. Innov. Educ. **4** (2018). https://doi.org/10.1504/IJIIE.2017.10012691
40. Lau, K., Lee, P.Y.: The use of virtual reality for creating unusual environmental stimulation to motivate students to explore creative ideas. Interact. Learn. Environ. **23**, 3–18 (2012). https://doi.org/10.1080/10494820.2012.745426

41. Vélez, B., Verdugo, G., Mejía-Pesántez, M., Veintimilla-Reyes, J., Maldonado-Mahauad, J.: Playing in the classroom: a game proposal for the flipped classroom. In: 2020 XV Conferencia Latinoamericana de Tecnologias de Aprendizaje (LACLO), Loja, Ecuador, pp. 1–7 (2020). https://doi.org/10.1109/LACLO50806.2020.9381163
42. Mandasari, B., Wahyudin, A.Y.: Flipped classroom learning model: implementation and its impact on EFL learners' satisfaction on grammar class. Ethical Lingua **8**(1), 150–158 (2021)
43. Jarnac de Freitas, M., Mira da Silva, M.: Systematic literature review about gamification in MOOCs. Open Learn. J. Open Dist. e-Learn. **38**(1), 73–95 (2020)
44. Francis, R.: Can serious games contribute to developing and sustaining 21st century skills? Games Cult. **10**(2), 148–177 (2015)
45. Harris, C., Sun, B.: Assessing the effect of interactivity design in VR based second language learning tool. In: 2022 IEEE International Symposium on Mixed and Augmented Reality (ISMAR), Singapore, Singapore, pp. 18–25 (2022). https://doi.org/10.1109/ISMAR55827.2022.00015
46. Di Natale, A.F., Repetto, C., Riva, G., Villani, D.: Immersive virtual reality in K-12 and higher education: a 10-year systematic review of empirical research. Br. J. Edu. Technol. **51**(6), 2006–2033 (2020)
47. Rawas, H., Bano, N., Alaidarous, S.: Comparing the effects of individual versus group face-to-face class activities in flipped classroom on student's test performances. Health Prof. Educ. **6**(2), 153–161 (2020). https://doi.org/10.1016/j.hpe.2019.06.002
48. Mayer, R.E., Moreno, R.: Nine ways to reduce cognitive load in multimedia learning. Educ. Psychol. **38**(1), 43–52 (2010). https://doi.org/10.1207/s15326985ep3801_6
49. Li, Y., Ryan, C.T., Sheng, L.: Optimal Sequencing in Single-Player Games, Feb 2023
50. Joosten, E., Lankveld, G.V., Spronck, P.: Colors and emotions in video games. In: 11th International Conference on Intelligent Games and Simulation GAME-ON, 2010, pp. 61–65
51. Alamri, M.M.: Students' academic achievement performance and satisfaction in a flipped classroom in Saudi Arabia. Int. J. Technol. Enhanc. Learn. **11**(1), 103 (2019)
52. Microsoft.: Mixed Reality Toolkit for Unity, Microsoft Learn (2022) [Online]. https://learn.microsoft.com/en-us/windows/mixed-reality/mrtk-unity/mrtk2/?view=mrtkunity-2022-05. Accessed 2 Apr, 2023
53. Roohi, A., Forouzandeh, A.: Regarding color psychology principles in adventure games to enhance the sense of immersion. Entertain. Comput. **30**, 100298 (2019). https://doi.org/10.1016/j.entcom.2019.100298
54. Salen, K., Zimmerman, E.: Rules of Play: Game Design Fundamentals. MIT Press (2004)
55. Wongutai, K., Palee, P., Choosri, N.: The effect of sound in VR exergame to adult player: a primary investigation. In: 2021 Joint International Conference on Digital Arts, Media and Technology with ECTI Northern Section Conference on Electrical, Electronics, Computer and Telecommunication Engineering, Cha-am, Thailand, pp. 1–4 (2021). https://doi.org/10.1109/ECTIDAMTNCON51128.2021.9425698
56. Chang, E., Kim, H.T., Yoo, B.: Virtual reality sickness: a review of causes and measurements. Int. J. Hum. Comput. Interact. **36**(17), 1658–1682 (2020). https://doi.org/10.1080/10447318.2020.1778351
57. Thiparpakul, P., Mokekhaow, S., Supabanpot, K.: How can video game atmosphere affect audience emotion with sound. In: 2021 9th International Conference on Information and Education Technology (ICIET), Okayama, Japan, pp. 480–484 (2021). https://doi.org/10.1109/ICIET51873.2021.9419652
58. Kenwright, B.: There's more to sound than meets the ear: sound in interactive environments. IEEE Comput. Graphics Appl. **40**(4), 62–70 (2020). https://doi.org/10.1109/MCG.2020.2996371
59. Mott, R.L.: Sound effects: radio, television and film. Mcfarland (2014)
60. Petko, D., Schmid, R., Cantieni, A.: Pacing in serious games: exploring the effects of presentation speed on cognitive load, engagement and learning gains. Simul. Gaming **51**(2), 258–279 (2020). https://doi.org/10.1177/1046878120902502

61. Johnson-Glenberg, M.C.: Immersive VR and education: embodied design principles that include gesture and hand controls. Front. Robot. AI **5**, 81 (2018). https://doi.org/10.3389/frobt.2018.00081.PMID:33500960;PMCID:PMC7805662
62. Backlund, P., Hendrix, M.: Educational games—are they worth the effort? A literature survey of the effectiveness of serious games. In 2013 5th International Conference on Games and Virtual Worlds for Serious Applications (VS-GAMES). IEEE, pp. 1–8 (2013). https://doi.org/10.1109/VS-GAMES.2013.6624226
63. Jin, T., Zhou, S., Lang, X., He, J., Wang, W.: Combined effect of color and shape on cognitive performance. Math. Probl. Eng. 3284313, 1–12 (2022). https://doi.org/10.1155/2022/3284313
64. Li, Y., Huang, J., Tian, F., Wang, H.A., Dai, G.Z.: Gesture interaction in virtual reality. Virt Reality Intell. Hardw. **1**(1), 84–112 (2019). https://doi.org/10.3724/SP.J.2096-5796.2018.0006
65. Ibánez, J.D., Wang, A.I.: Learning recycling from playing a kinect game. Int. J. Game-Based Learn. (IJGBL) **5**(3), 25–44 (2015). https://doi.org/10.4018/IJGBL.2015070103

Stealth Assessment for VR-based Skill Training Utilizing Hand Motion Detection

Siang Jian Phang, Siyi Li, Yuan Xie, Lihui Huang, and Yiyu Cai

Abstract The ability of virtual reality (VR) games to provide immersive experiences is an extremely valuable property, especially in an educational setting for skill training, where players are quite often trained in scenarios that they are required to perform in real life. Assessing of training, including debriefing, however, remains largely a manual task in skill training. Video recording-based assessment is a common technique which is unfortunately data crunching particularly when a large number of students are involved in a lengthy training process. Also, automatic skill assessment for students' learning based on the recorded video is still a research topic. To address this issue, this research is keen to investigate stealth assessment or in-process assessment for VR-enabled skill training. A novel technique focusing on hand motion detection is developed for skill assessment and debriefing during the training process. In particular, tracking and detecting of specific and customizable hand motion are performed in real time for skill-based task training. Plugin scripts are designed to track and detect realistic hand motions into portions of the training game. The trajectory of hand motion for skill training can be used to reconstruct the virtual hand motion for debriefing as a replay of the player's hand movement step by step. This is significant because of the tremendous saving of the trajectory compared to video recording of the hand movement in terms of data size. Comparison can also be easily developed for automated assessment between the correct answers and the actual movement performed by the students during their skill training. An experiment was conducted with the students learning the aircraft hydraulic maintenance. They learn through playing a hand motion based game built on top of the plugin scripts developed. The script implemented in this game is set to detect a nut-loosening motion. The survey results demonstrate that most players agree that the hand motion tracking and detection enhance the interactivity and learning experiences of VR-based skill training. This technology has the potential to elevate the learning experience for various educational applications, such as medical, engineering, and entertainment.

S. J. Phang · S. Li · Y. Xie · L. Huang · Y. Cai (✉)
Nanyang Technological University, Singapore, Singapore
e-mail: myycai@ntu.edu.sg

Y. Cai et al. (eds.), *Virtual and Augmented Reality Technology-Enhanced Learning*, Gaming Media and Social Effects, https://doi.org/10.1007/978-981-96-2332-7_4

1 Introduction

1.1 Background

Virtual reality (VR) is a type of computer-generated environment that is commonly experienced using a specialized headset, usually coupled with supplementary devices such as controllers and sensors (Fig. 1). The concept of VR has been around since the 1960s but has only recently garnered mainstream popularity due to the COVID-19 pandemic [1] and advancements in computer graphical hardware to a point where consumer grade product is now capable of running VR environments with high fidelity and performance. The cumulative number of VR headsets sold by 2024 is 34 million, which is almost triple the amount of 2019 [2], showing a clear increase in demand.

The development of VR marks a significant milestone in the evolution of training techniques, by empowering educators to provide realistic and flexible training environments to learners with significant advantages over traditional methods. In contrast to traditional training methods, such as those employed in aircraft maintenance training, where the use of physical equipment can result in high costs and space requirements [4], VR-based teaching, on the other hand, provides an affordable and space-efficient alternative. Furthermore, VR-based training also allows for training of scenarios in hazardous or inaccessible environments, such as nuclear-related training or interplanetary exploration [5], that would otherwise be difficult and costly to implement.

VR-based education [6–8] has a wide variety of applications in many different fields, such as (i) Informal or continual education, where students can obtain knowledge through VR lessons which increases motivation [9], enjoyment [10] and long-term learning retention [11] as compared to traditional learning; (ii) medical field, where patients can utilize VR experiences as an alternative form of treatment for chronic pain instead of using opioids [12, 13], pain management, eating disorders, and cognitive and motor rehabilitation for inpatients [14], non-technical healthcare skills [15], training for minimally vascular surgery [16] and first responders training [17].

Fig. 1 Meta quest Pro VR headset and controller [3]

1.2 Motivation

Even though VR-based training systems have been adopted by many different industries, there is still room substantial for improvement. One key such area is the detection of motion-based gestures that players are required to perform in the virtual environment to simulate real-life scenarios in VR training. By implementing an algorithm that detects simple yet important movements of the player's hands, the realism and immersion of performing specific actions in a VR game can be meaningful for training purpose; hence the learning retention and effectiveness can be enhanced and carried over to real-life scenarios. This is significant in skill-based training where hands movement plays a critical role.

1.3 Objective and Scope

The objective of this project is to develop plugins that track and detect the player's virtual hand movements with support for flexibility and easy implementation into upcoming or existing VR games. This can allow training of more nuanced hand motions such as using wrenches to loosen bolts on a mechanical system, and a replay system that reads the recorded motion data and translates them into a representation of a virtual hand for debriefing purposes. It is different to the conventional video approach which may end up with a significant amount of video data recorded, particularly when a large number of players are involved. In contrast, hand motion trajectory is substantially concise compared to video data. By reconstructing the virtual hand movement from trajectory data, learners can do debriefing on their own after their virtual hands-on experiments.

The developed plugins are designed using the Unity game engine, one of the most widely used game engines in the world and would ensure that the plugins would be compatible with the largest number of existing games. The plugins are also based off Microsoft's Mixed Reality Toolkit (MRTK), a free and well-developed Mixed Reality development package.

2 Literature Review

The most distinct advantage of VR as compared to traditionally 2D-rendered environments is immersion and interaction with the aid of motion tracking, stereoscopic rendering, and spatial audio. These technologies allow the player's movement to be translated accurately into the virtual environment, making interactions feel natural and intuitive, plus an increased sense of presence and interactivity [18]. This feature is particularly beneficial to education and training [19].

2.1 Haptic Feedback

Haptic feedback pertains to the utilization of tactile sensations like tremors or pressure to mimic touch and furnish sensory feedback to users. In VR games, haptic feedback may heighten immersion in various ways, such as enriching realism by mimicking the tactile sensations found in the real world [20]. For instance, haptic feedback may imitate the sensation of touching surfaces or objects, or the perception of impact or force. It can also better the feedback received by players, helping them comprehend and interact with the virtual environment. Haptic feedback can be employed to specify the location or proximity of objects or to provide feedback on actions during gameplay. A study done in 2017 showed that the use of haptic communication in virtual environments for collaborative problem-solving can facilitate communication and collaboration between players, making navigation more efficient, reducing the need for complex verbal directions, and allowing for more goal-focused discussions [21]. Downsides of incorporating haptic feedback are the high cost of equipment and setup time.

2.2 Gesture Detection

Gesture recognition is the capability of a VR system to identify and detect hand movements and gestures executed by users. In VR games, gesture recognition can amplify immersion by enabling users to interact with the virtual environment in a more intuitive and realistic manner [22], resembling actions performed in the real world. This enhances the feeling of being present and immersed in the virtual environment. Gesture recognition can empower users with greater control over their virtual avatar or environment, allowing them to manipulate objects, execute actions, and interact with the environment in a more fluid and precise manner. This can result in a more gratifying and immersive experience for users.

Gesture detection can be divided into two main categories: glove-based and vision-based. Glove-based detection requires players to wear specialized gloves equipped with sensors that detect the current state of the player's hands and relay that information to the computer [23], while vision-based detection utilizes cameras to identify players' hand and its current gesture [24].

The benefit of glove-based detection is that it is superior in terms of accuracy compared to vision-based detection, as it utilizes more sensors and data processing. The drawback of wearing gloves is that it affects the naturality of the interaction to a certain extent [25] and is usually more costly to setup, implement, and maintain compared to vision-based detection. Vision-based detection accuracy was severely limited by the computational power of older hardware during its early stages but has advanced to a stage where its accuracy is much greater than before [26]. As hand tracking gains wider adoption in VR, there has been an increase in research examining its impact on VR experiences [27, 28]. Some interesting examples of recent utilization of vision-based detection are in flight simulators as an alternate form of input over pedals and sticks [29], real-time automated hand gesture detection of elderly in home or clinical setting [30], and real-time sign language recognition [31]. These examples utilize machine learning to detect the gestures shown in a video format.

2.3 *Hand Controller*

Controllers for VR are handheld devices developed to allow users to interact with the virtual environment. These devices typically feature buttons, triggers, joysticks, and/or touchpads, as well as sensors for tracking their orientation and position in three-dimensional space. Some VR controllers also provide haptic feedback, which can produce tactile sensations to enhance the sense of presence in the virtual world. Various VR platforms utilize distinct types of controllers, such as the Oculus Touch for the Oculus Rift, the Valve Index Controllers, or the PlayStation Move Controllers for the PlayStation VR (Fig. 2). VR controllers facilitate a broad range of interactions in VR, from simple clicking and pointing to intricate gestures and manual movements.

Using controllers versus hand tracking each has its own benefits and drawbacks. One of the advantages of controller-based interactions is the ability to provide users with tactile feedback, which aids in executing precise actions like aiming a weapon or manipulating small objects. Controllers are also equipped with onboard sensors that increase the accuracy of motion tracking, leading to an increase in precision and immersion [33, 34]. Furthermore, controllers can deliver a more authentic and immersive feel for actions that require a tangible object, such as swinging a sword or throwing a ball. Conversely, hand-tracking interactions possess the advantage of being more instinctive and natural since users don't need to learn how to operate a separate device to interact with the virtual world. Hand tracking can also allow for more subtle and complicated interactions, such as using hand gestures to convey emotions or perform intricate manipulations. Research published in 2021 concluded that when using glove-based hand tracking as an input modality compared

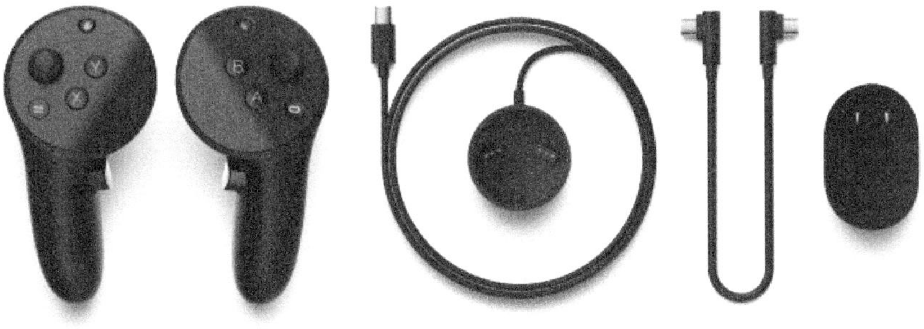

Fig. 2 Quest Touch Pro Controller [32]

to controller, ownership, realism, enjoyment, and presence is substantially increased [35].

Ultimately, both controller-based and hand-tracking interactions come with their own set of advantages and disadvantages in VR. While controllers may provide more precise and tactile feedback, hand tracking can furnish a more instinctive and natural interaction. The decision on which interaction method to use may depend on the specific VR application and the desired user experience.

2.4 Debriefing

In the realm of education, debriefing is the post-test or post-examination process of reflection and discussion that takes place after the learning has ended. This process provides learners with the opportunity to dissect their experiences, evaluate their performance, and assess the lessons learned from the test or examination results. [36, 37] reported their interesting research on debriefing with the aid of games. The debriefing process is typically led by an instructor conducting reflection and discussion either individual or group based.

The goal of a game-based debriefing is to help players draw connections between the game and real-life situations, pinpoint the skills and knowledge they have acquired, and strategize how they can use these skills in other settings. Additionally, it can enable players to recognize any preconceptions or biases they may have had and cultivate a more profound comprehension of the subject matter.

Debriefing is a culmination of multiple concepts and processes working together, such as data collection, sensitivity, internal validity, generalizability, ecological validity, and robustness. A debriefing framework allows a gaming simulation to adequately address the unpredictable context of innovation processes. By integrating debriefing as a fundamental aspect of the design aspects of a research game, the game designer can better manage this unpredictability [38].

A proven tool that is effective for debriefing is a video replay or re-enactment system. A study showed that re-enactment recordings can be a valuable addition to the debriefing procedure, providing an exclusive opportunity to enhance its overall quality. The participants showed a favorable attitude toward the videos and even voluntarily watched them, despite it not being mandatory [39]. Yet the data size of the video recorded can be exponentially increasing especially when a large number of learners are involved. This work is interested to record learners' trajectory data of virtual hand motion during their hands-on experiment. The trajectory data of each learner can be used to reconstruct the virtual hand movement for debriefing purpose.

3 System Setup

3.1 Unity Game Engine

The Unity game engine is a widely used [40] and popular platform for developing 2D, 3D, VR, and AR games across a variety of devices, including mobile phones, consoles, PCs, and the web. A major strength of Unity is its user-friendliness and accessibility, which has made it a popular choice for both professionals and hobbyists alike. As a result, Unity has developed a large and active community of developers and has been applied in various fields such as gaming, simulations, education, and more, making it a highly versatile and sought-after engine. The scripts in this project were written using Unity version 2021.3.18f1 compatible with all the other components of the project.

3.2 Mixed Reality Toolkit

The Mixed Reality Toolkit (MRTK) is an open-source development framework designed to help developers create mixed reality applications for various platforms such as HoloLens, Oculus, and Windows Mixed Reality [41]. It provides a collection of components and features that make it easier to develop mixed reality applications (Fig. 3), including input management, spatial mapping, spatial audio, and more. It also supports a range of development tools and platforms, including Unity and Visual Studio. MRTK is maintained by Microsoft and is available for free on GitHub and is widely used for games, training simulations, and enterprise applications.

During the project, MRTK version 2.7.3.0 was used while coding the scripts, version 2.8 was used. No compatibility issues were raised during the version change.

3.2.1 Hand Joints

In MRTK, the hand of the player is rendered in the virtual environment as shown in Fig. 4. This hand is represented using 26 different joints, 25 of which are shown in Fig. 5, with the last joint being the palm joint located at the center of the palm. In-game design, the term "Transform" refers to the position, rotation, and scale of an object. The transform of the individual joints can be obtained at any given frame using the MRTK "RequestJointTransform" method. Using this method, the rotation and position of each joint can be obtained, allowing the tracking of movement of joints over a period of time, and can also be recorded for later use.

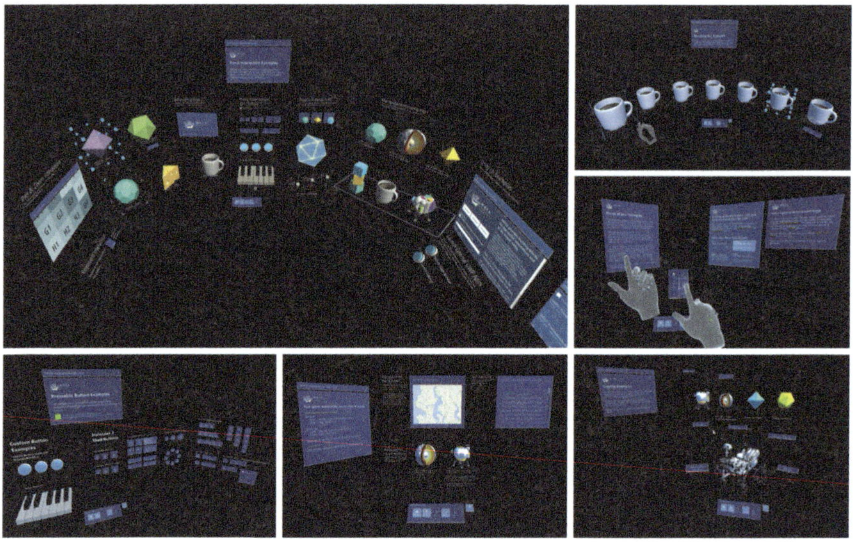

Fig. 3 MRTK feature showcase [41]

3.3 *Leap Motion Controller*

The Leap Motion Sensor (Fig. 6) is a compact gadget that employs motion-tracking technology to identify and trace the movement of hands and fingers [43]. It comprises a small rectangular block that can be positioned on a surface or attached to a VR headset and is equipped with two cameras and three infrared LEDs. When stationed in front of a user, the Leap Motion Sensor captures the motion of the user's hands and fingers in 3D space with high precision and minimal delay. This enables users to interact with virtual environments and applications using natural hand movements and gestures, without the need for a physical controller or keyboard (Fig. 7).

The Leap Motion Sensor is engineered to function with various applications and platforms. It also supports a range of programming languages and frameworks, enabling developers to create customized applications and integrations. Overall, the Leap Motion Sensor is a potent instrument for intuitive and organic interaction in virtual and augmented reality environments. The version of Leap Motion software used was Orion version 4.1.0 + 52,211.

Fig. 4 A rendered hand in virtual environment [42]

4 Hand Motion Detection

4.1 Approach for Detecting Motion

There are different approaches that can be taken to detect hand motions in a VR environment, such as a machine learning approach to train a model using a dataset, or a more simplistic approach of comparing the player's motion with a pre-recorded defined motion.

A frame-by-frame comparison method was selected as the method of detection, due to several reasons. Firstly, the motions that the scripts developed will be detecting are simple motions; hence, using a machine learning approach is excessive for this purpose. Secondly, the scripts designed are intended to be a simple, lightweight, and flexible add-on for developers of all skill levels to implement. Thirdly, a machine

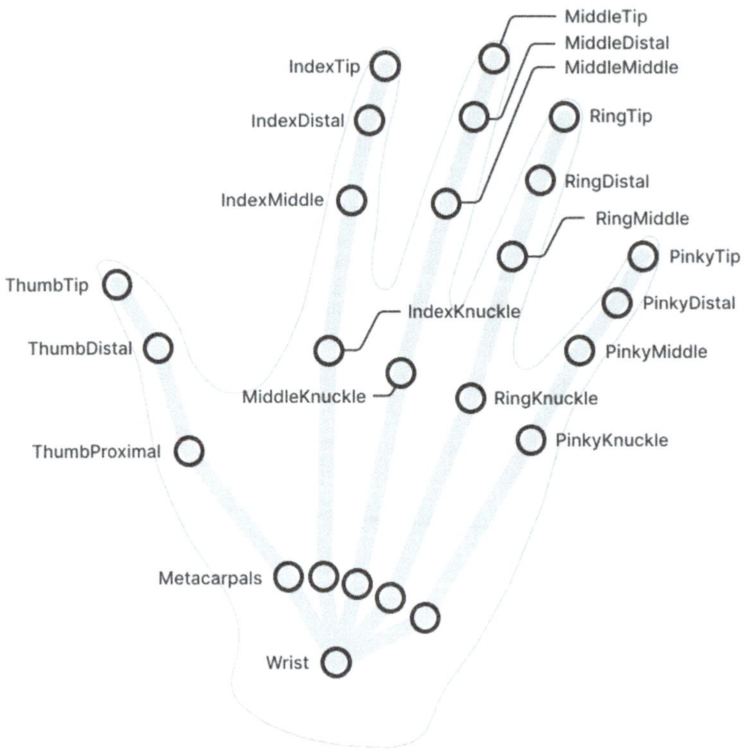

Fig. 5 MRTK hand joints with labels [42]

Fig. 6 Leap Motion Sensor [43]

learning trained model can be quite large, and can be more performance heavy resulting in higher hardware requirements and longer response time.

4.2 Design

The frame-by-frame comparison method works by comparing the hand's current frame position and rotation with a pre-recorded set of position and rotation, checking if the current position and rotation is within a set threshold from the pre-recorded

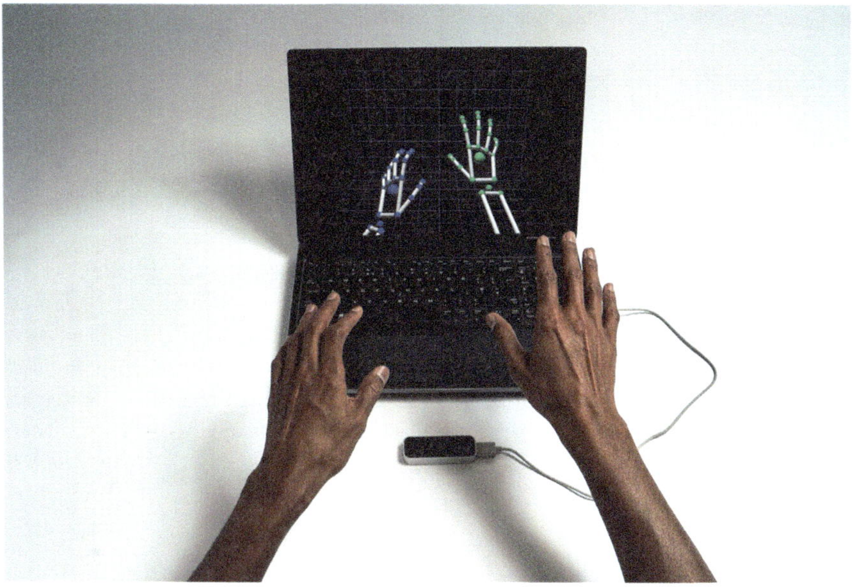

Fig. 7 A leap motion controller in desk mode [43]

data. The threshold is a variable that can be manually set by the developer in the Unity editor, allowing them to fine tune the sensitivity of the detection. Since a hand motion is performed across multiple frames, and the order in which the frames are compared needs to be sequential, a checkpoint system is created as a means for the script to keep track of the progress of the hand motion, with each checkpoint representing one frame in the recorded pattern. Once a checkpoint is reached, it will not compare the data with previous checkpoints. The core logic behind the motion detection scripts is presented in the flowchart shown in Fig. 8 representing what happens in every frame.

4.2.1 Deviation

Deviation refers to the difference between the current frame transform compared to the current checkpoint frame of the pre-recorded motion and is calculated using quaternion math. In the context of rotational comparison, the deviation would be the angle between the two rotations. In Unity, the angle in degrees between two rotations can be obtained using the "Quaternion.Angle" function. In the context of positional comparison, the deviation would be the distance between two points in space, which can be obtained using simple vector geometry.

The deviation obtained is used to determine whether a player performed motion is deemed to be too different from the expected motion. When the deviation is above the prescribed threshold, the motion is regarded as a failed attempt and is discarded.

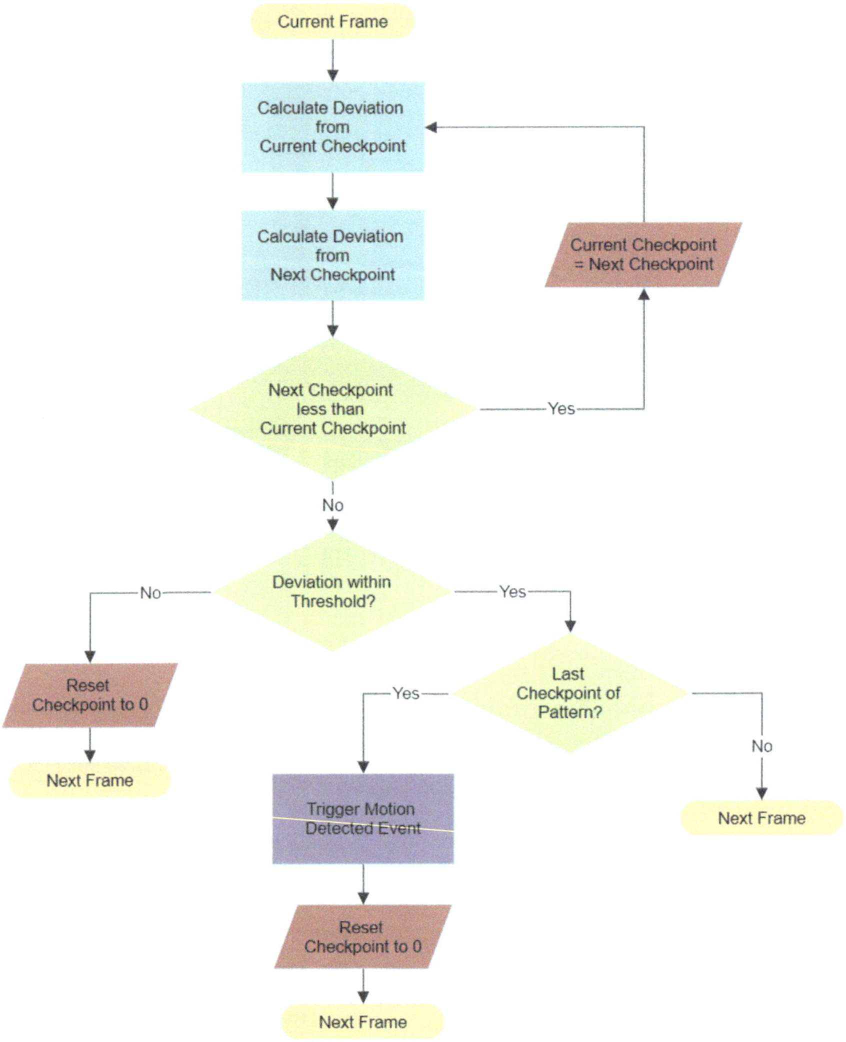

Fig. 8 Flowchart of detection logic

When the deviation is within the threshold throughout the entire motion, the motion is deemed successful, which triggers the motion detected event.

4.2.2 Pre-recorded Motion Pattern

There are two ways of creating a pre-recorded motion pattern, manual input of each frame's rotation, or using a recording function to record the movement of the hand.

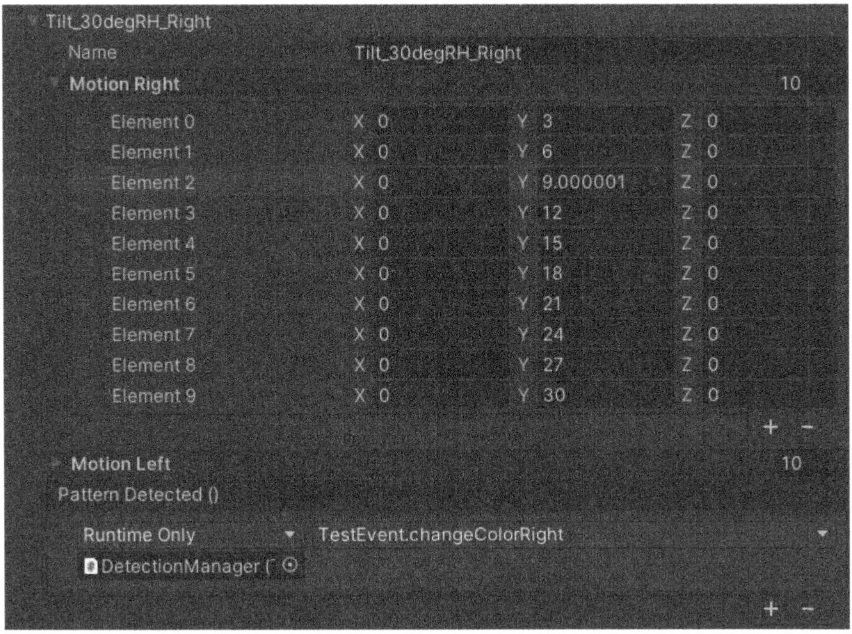

Fig. 9 Recorded pattern in Unity editor

For simple motions such as rotation in one axis, it is easier to just manually input the rotation angles in the Unity editor. A recording functionality was designed into the scripts, allowing developers to easily record and implement their own motions into their games. This feature is intended to be used when more complex motions are required where movements in more than one axis are required. Figure 9 shows a manual input set of rotations that detect a 30-degree rotation about the Y-axis spanning over 10 checkpoints/frames. The pre-recorded set of data is represented in the form of Euler angles, which is automatically converted internally by Unity into quaternions for calculation purposes. This is beneficial for developers as Euler angles as significantly more readable compared to quaternions; hence, it is easier for developers to visualize the rotation of the motion.

4.2.3 Base Transform

When the checkpoint is zero, which happens either at the start of a new detection, when the deviation exceeds the threshold or when a motion is successfully detected, the current transform will be set as the base transform. The base transform is then used as a reference point for future frame comparisons. This is done to allow detection regardless of the direction of the hands, as the net change is always with respect to the base transform. The calculation of the rotational shift required to compare is done using quaternion math as follows:

$$\text{Shift} = a^{*}\text{Inverse}(b)$$

where the variable “Shift” is the transforming quaternion, and the variable “aa” is the base rotation quaternion, and the variable “bb” is the rotation quaternion of the first checkpoint/frame of the pre-recorded motion. It can then be multiplied by the incoming rotation quaternion to obtain a shifted quaternion using this formula:

$$\text{Shifted} = \text{Shift}^{*}c$$

where the variable “ShiftedShifted” is the quaternion of the corresponding checkpoint/frame after shifting, and the variable “cc” is the checkpoint/frame before shifting. This variable “cc” can now be used to compare against the current frame transform of the player’s hand using the “Quaternion.Angle” function as described earlier, obtaining the deviation of the motion.

4.3 Tracked Joint

For the scripts designed, it is decided that only one joint will be tracked for comparison. The chosen joint is the palm joint, as it is the central joint of the hand, making it a good candidate to represent the movement of the entire hand. Compared to tracking all 26 joints, this brings several advantages and disadvantages, which will be discussed in the next section.

As a proof of concept, only the rotation of the palm joint will be tracked. This means that positional-based gestures will not be supported in the current iteration of the scripts. This is adequate for simple gestures, such as twisting and turning motions, where the position of the hand is fixed throughout the motion. For the purpose of the experiment was conducted, rotational tracking is sufficient.

The main advantage of tracking and comparing only one joint is performance. Naturally, if the game engine only needs to compare the transform data of one joint instead of 26, the performance of the script would be significantly greater. This is in line with the goal of the scripts by lowering the hardware requirement to run and increases the compatibility meantime.

Another advantage is flexibility. The same hand motion pattern can be recycled to be used to detect different sets of actions with different finger positions with minimal effort. For example, a waving hand motion can be recycled into a wrench turning motion, since in both cases, the palm joint goes through the same motion of rotation in the same axis, with the differences being the finger positions, as shown in Figs. 10 and 11.

The third advantage is more forgiving detection during player usage. At the present moment, the hand-tracking precision of VR devices is not perfect due to technical limitations, resulting in possible glitches that may cause the virtual fingers and hands

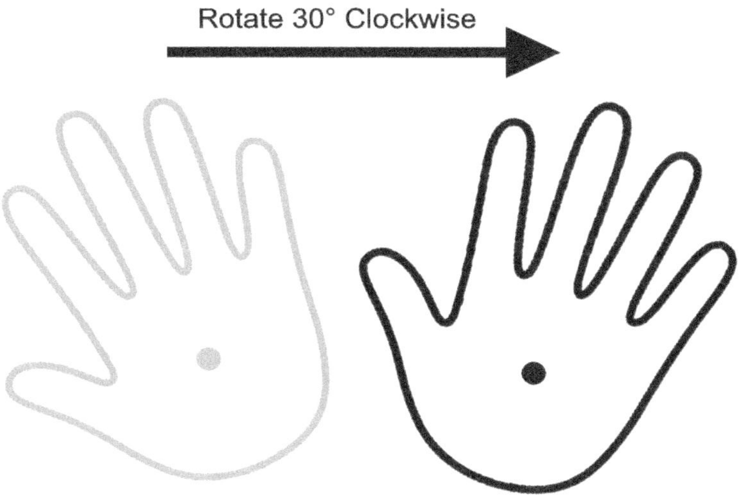

Fig. 10 Wave hand motion

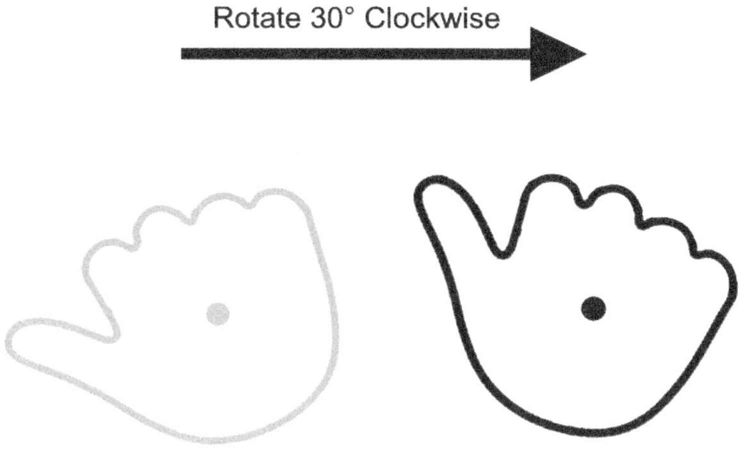

Fig. 11 Wrench hand motion

to be inaccurately tracked, such as the fingers being opened when it should actually be closed. This is typically caused by several reasons such as hands or fingers being occluded, insufficient lighting in the play environment, smudges on the lenses, hand moving too fast, or technical glitches. By only tracking the palm joint, it becomes significantly easier and more forgiving for the player to perform the required hand motions.

The main disadvantage of only tracking one joint is that the script is limited to a smaller set of simple hand motions. Complex hand motion that combines specific finger positions at specific timings cannot be tracked and detected.

4.4 Python Scripts

A Python version of the same scripts was developed for the purpose of analyzing past recorded data. The scripts take in two sets of data, the motion pattern that the users wish to detect, and the dataset that the user wishes to analyze. The working principle behind the scripts is the same as the Unity version, except for the pre-recorded motion input, where it is changed into a comma-separated values (CSV) format and uses Quaternions instead of Euler angles. The same set of pre-recorded motion as shown in Fig. 9 is now converted into the CSV version as shown in Fig. 12. Columns A, B, C, and D represent the x, y, z, and w components of a Quaternion, respectively, and each row represents one checkpoint/frame of the pre-recorded motion. The dataset consists of rows of rotational data in Quaternion form representing the rotation of the hand in each frame in the same configuration as the pre-recorded motion earlier, as shown in Fig. 13.

The main purpose of the Python version is to allow analysis of data recorded earlier. For example, if the goal is to detect how many times a student performs a 30-degree counterclockwise rotation about the Y-axis, the Python scripts can be ran using inputs as shown in Fig. 14, which returns a result of two detections.

By analyzing motion data, valuable insights into the learning process of the student can be obtained. For example, it can be used to detect if the student is doing well or frustrated during the play session by checking if any corresponding motion is detected.

Fig. 12 CSV version of pre-recorded motion

	A	B	C	D
1	0	-0.02618	0	0.999657
2	0	-0.05234	0	0.99863
3	0	-0.07846	0	0.996917
4	0	-0.10453	0	0.994522
5	0	-0.13053	0	0.991445
6	0	-0.15643	0	0.987688
7	0	-0.18224	0	0.983255
8	0	-0.20791	0	0.978148
9	0	-0.23345	0	0.97237
10	0	-0.25882	0	0.965926

Fig. 13 Recorded rotation data of past gameplay

	A	B	C	D
3938	0.759408	0.586348	0.161473	0.231132
3939	0.759271	0.587326	0.160331	0.229891
3940	0.758699	0.588803	0.158694	0.229137
3941	0.758246	0.589588	0.15728	0.229592
3942	0.758129	0.589273	0.155484	0.231999
3943	0.758064	0.588612	0.154934	0.234243
3944	0.757956	0.588214	0.155022	0.235534
3945	0.757642	0.588733	0.154973	0.235279
3946	0.757196	0.589619	0.155524	0.23413
3947	0.756366	0.591717	0.155306	0.231651
3948	0.755562	0.593927	0.155089	0.228746
3949	0.754998	0.593752	0.157017	0.229744
3950	0.754135	0.594894	0.158188	0.22882

5 Replay Scripts

5.1 Purpose

A replay system that shows the player a visual representation of their actions performed during gameplay has several benefits in educational-based games. It enhances the learning experience by providing immediate feedback to the player, showing what they did well or where they can improve, which motivates them to continue playing and learning. It can also be used as a means of tracking the progression of the player over multiple sessions, further reinforcing the learning outcome of the game. The replay can be shown during the debriefing portion of the game.

5.2 Design

The scripts are designed for modularity and ease of implementation. The scripts take in a specifically formatted set of data in a CSV file and uses cubes and line renderers to draw a visual representation of a hand performing the motion as stored in the CSV file. The rendered representation of the hand is shown in Fig. 15.

```
Patterns:
0  -  Tilt_30deg_Down
1  -  Tilt_30deg_Left
2  -  Tilt_30deg_Right
3  -  Tilt_30deg_Up
Enter pattern number: 1
------------------------------------------
Data sets:
0  -  RecordedData2
1  -  rtn0
2  -  rtn1
3  -  rtn2
4  -  rtn3
5  -  rtn4
6  -  rtn5
7  -  rtn6
8  -  testdata1
9  -  testdata2
Enter data number: 2
------------------------------------------
Pattern detected! [6106, 6111]
Pattern detected! [6643, 6803]
=====================================================
Data ended. Total Patterns Detected:  2
Detected frames list [Starting Frame, Ending frame]:
 [[6106, 6111], [6643, 6803]]
=====================================================
```

Fig. 14 Python script execution for detecting rotation

5.3 *CSV Configuration*

The scripts designed take in a CSV file configured in a specific format (Fig. 16). The first 4 columns represent the rotation of the palm joint in Quaternion form (x, y, z, w), which can be directly obtained via the transform of the palm joint, the next 3 columns represents the position of the palm joint relative to the first frame of the recording, which is why the first row if this column will always be zeros, since it is the reference position. The subsequent groups of 3 columns represent the relative positions of the remaining 25 joints with respect to the reference position that was mentioned earlier. The order in which the joints are arranged can be found in Appendix A. The last column is an optional column. In the case of the motion detection script, it was used to store the deviation of that frame. Each row represents one frame of the motion. The last row is also an optional row. In the case of the motion detection script generated CSV, it was used to store the average deviation of the entire motion. In total, each

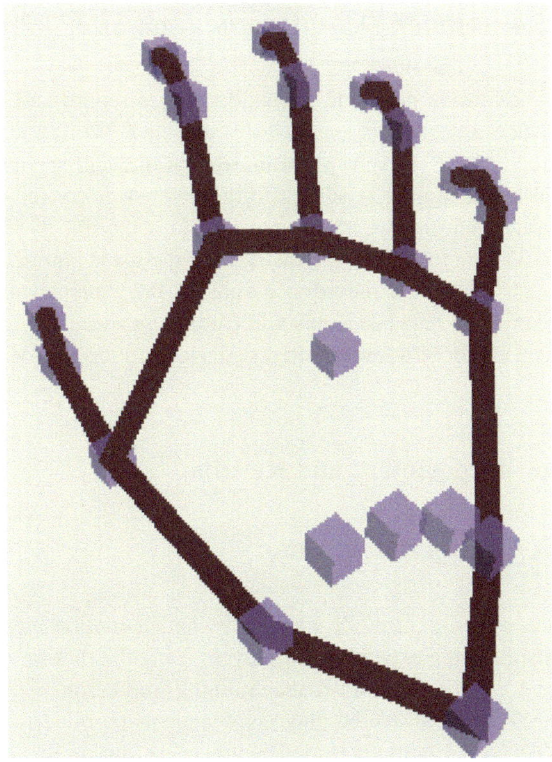

Fig. 15 Rendered visual representation of hand

row will have 82 mandatory columns, consisting of 4 columns for the quaternions of the palm joint, and three times 26 columns for the positions of the 26 joints.

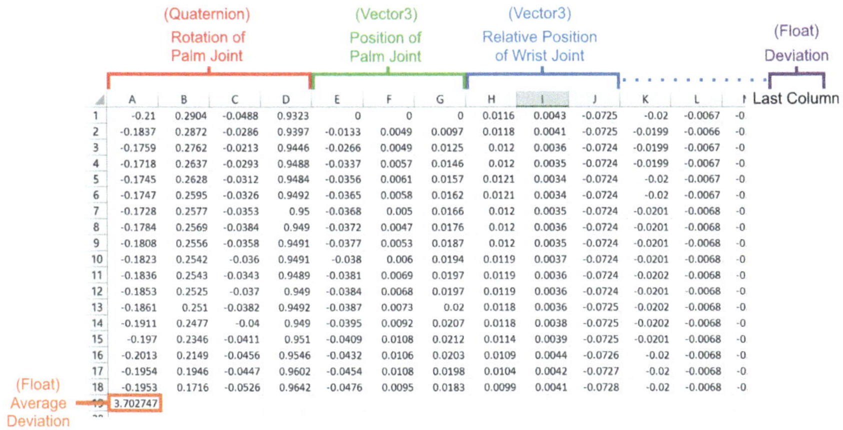

	A	B	C	D	E	F	G	H	I	J	K	L	…
1	-0.21	0.2904	-0.0488	0.9323	0	0	0	0.0116	0.0043	-0.0725	-0.02	-0.0067	-0
2	-0.1837	0.2872	-0.0286	0.9397	-0.0133	0.0049	0.0097	0.0118	0.0041	-0.0725	-0.0199	-0.0066	-0
3	-0.1759	0.2762	-0.0213	0.9446	-0.0266	0.0049	0.0125	0.012	0.0036	-0.0724	-0.0199	-0.0067	-0
4	-0.1718	0.2637	-0.0293	0.9488	-0.0337	0.0057	0.0146	0.012	0.0035	-0.0724	-0.0199	-0.0067	-0
5	-0.1745	0.2628	-0.0312	0.9484	-0.0356	0.0061	0.0157	0.0121	0.0034	-0.0724	-0.02	-0.0067	-0
6	-0.1747	0.2595	-0.0326	0.9492	-0.0365	0.0058	0.0162	0.0121	0.0034	-0.0724	-0.02	-0.0067	-0
7	-0.1728	0.2577	-0.0353	0.95	-0.0368	0.005	0.0166	0.012	0.0035	-0.0724	-0.0201	-0.0068	-0
8	-0.1784	0.2569	-0.0384	0.949	-0.0372	0.0047	0.0176	0.012	0.0036	-0.0724	-0.0201	-0.0068	-0
9	-0.1808	0.2556	-0.0358	0.9491	-0.0377	0.0053	0.0187	0.012	0.0035	-0.0724	-0.0201	-0.0068	-0
10	-0.1823	0.2542	-0.036	0.9491	-0.038	0.006	0.0194	0.0119	0.0037	-0.0724	-0.0201	-0.0068	-0
11	-0.1836	0.2543	-0.0343	0.9489	-0.0381	0.0069	0.0197	0.0119	0.0036	-0.0724	-0.0202	-0.0068	-0
12	-0.1853	0.2525	-0.037	0.949	-0.0384	0.0068	0.0197	0.0119	0.0036	-0.0724	-0.0201	-0.0068	-0
13	-0.1861	0.251	-0.0382	0.9492	-0.0387	0.0073	0.02	0.0118	0.0036	-0.0725	-0.0202	-0.0068	-0
14	-0.1911	0.2477	-0.04	0.949	-0.0395	0.0092	0.0207	0.0118	0.0038	-0.0725	-0.0202	-0.0068	-0
15	-0.197	0.2346	-0.0411	0.951	-0.0409	0.0108	0.0212	0.0114	0.0039	-0.0725	-0.0201	-0.0068	-0
16	-0.2013	0.2149	-0.0456	0.9546	-0.0432	0.0106	0.0203	0.0109	0.0044	-0.0726	-0.02	-0.0068	-0
17	-0.1954	0.1946	-0.0447	0.9602	-0.0454	0.0108	0.0198	0.0104	0.0042	-0.0727	-0.02	-0.0068	-0
18	-0.1953	0.1716	-0.0526	0.9642	-0.0476	0.0095	0.0183	0.0099	0.0041	-0.0728	-0.02	-0.0068	-0
19	3.702747												

Fig. 16 Replay script CSV configuration

5.4 Motion Detection and Data Size

The motion detection scripts developed record and save the player's hand motion when a motion is successfully detected. While the motion detection is active, it records the relevant joints in every frame and appends it into a list of float values. When a motion is successfully detected, it converts the list and saves it into the aforementioned CSV format. When a motion is discarded due to exceeding the deviation threshold or deactivation, the list is emptied.

Compared to recording a video of the gameplay, using this technique for storing gameplay data takes up significantly less space. A video file of an equivalent length can easily be a few hundred times larger than a CSV file.

6 Experiment and Results

6.1 Experiment Setup

An experiment is set up to test whether motion-based gesture detection improves the learning experience of players using the motion detection scripts developed for the VR aircraft maintenance training simulation.

The game will be played by students enrolled in the major prescribed elective, MA4831 Computer-Aided Engineering, during the lab session on the 10th week of the semester.

Before playing the game, they were encouraged to watch a demonstration video showing the entire process of playing the game. This is to get them familiar with the task they are about to perform, and speed up the process of the session.

After playing the game, they are required to take a post-test survey with questions regarding the motion detection system of the game. A total of 77 students participated in the lab session.

6.2 Aircraft Maintenance Game

The VR game for aircraft maintenance training is designed in house for students to learn hydraulic system reparing by assuming the role of an aircraft maintenance trainee to virtually disassemble the hydraulic components of an aircraft. The player's task is to use virtual wrenches to loosen five nuts and remove the aircraft's accumulator (Figs. 17 and 18). The hand motion detection is utilized in this game to detect the twisting motion of the player's right hand as they attempt to loosen the required nuts while holding on to the virtual wrenches.

In order for the game to trigger that the nuts are loosened, the player performed action must satisfy three conditions. (i) The required virtual wrenches must be held

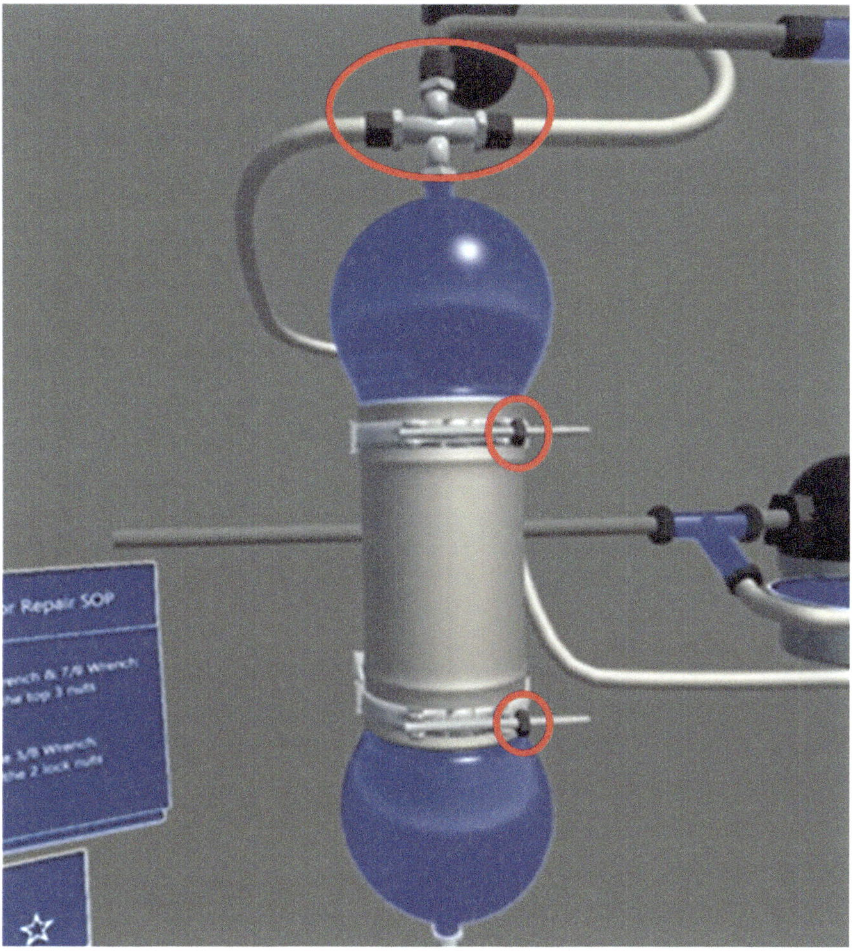

Fig. 17 Accumulator and the nuts to be loosened

by the hands; (ii) The head of the wrenches must be within a boundary of the nut throughout the entirety of the twisting motion; and (iii) A counterclockwise twisting motion traveling nine degrees must be detected on the right hand. When all three of these conditions are met, the nut is successfully loosened, and the player can proceed to the next step.

There are five nuts that the player must loosen. The first three nuts require both types of wrenches, one on each hand, while the last two only require the right hand to be holding the red wrench.

The motion that the game is detecting is a nine-degree counterclockwise rotation about the Y-axis as shown in Fig. 19, as it is the same motion that the hand goes through while performing the same action in real life. Figure 20 shows an illustration of the hand motion that the players are required to perform to loosen the nut.

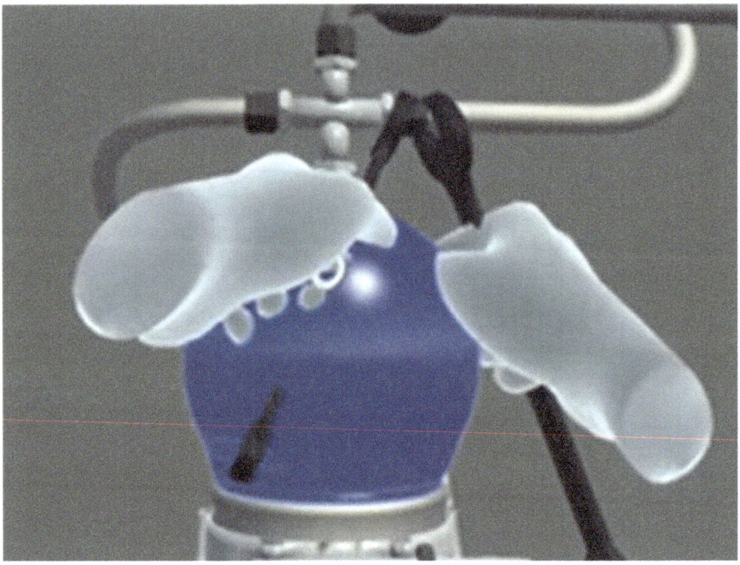

Fig. 18 Loosening the nuts

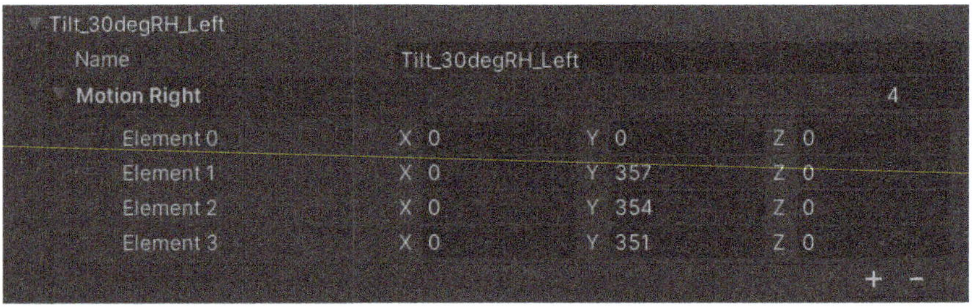

Fig. 19 Motion detection setup

The nine-degree rotation was used during internal testing in keeping the wrench head within the collider of the nut. This is due to the fact that the action is being performed on intangible virtual objects, which means there is no resistive force that is keeping the head of the wrench in place as it would in real life. The players are essentially trying to pivot the virtual wrenches along the head of the wrench without any forces supporting the rotation. Hence, a nine-degree rotation was chosen after testing multiple values of rotation. The threshold deviation is set to 10 degrees, which worked well during testing.

The scripts implemented in the game will only start detecting for motion if the wrench head collider is within the nut collider, so that unnecessary detections are eliminated, improving the overall performance of the game.

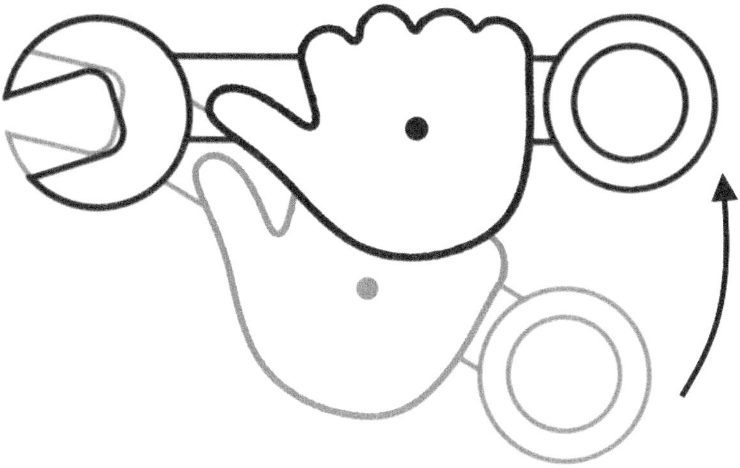

Fig. 20 In-game hand motion

6.3 *Survey and Discussion*

A total of five Likert scale questions and two open-ended questions (Table 1). The questions were designed to obtain insight into the player's experience with the motion detection part of the game.

For the Likert scale questions, the first question (Fig. 21) is intended to test if it was intuitive and natural for the players to perform the motion. The second question (Fig. 22) is intended to test if the action itself was simple to perform. The third question (Fig. 23) is intended to test if the accuracy of the motion detection is good, that is, if the motion was repeatably detected throughout the different bolts. The fourth question is intended to check if the players think that the motion performed during the test is realistic. The fifth question is to get the players' opinion about

Table 1 Survey questions

Question	Answer
It was easy to understand the hand motion that I am required to perform	Likert scale
The action of loosening of the bolts was easy to perform	
The action of loosening of the bolts was accurately detected	
The action of loosening of the bolts was accurate as compared to loosening of bolts in real life	
The hand motion helped me learn the educational content effectively	
Did you encounter any difficulties or issues while performing the action of loosening of the bolts?	Open ended
What can be improved regarding the motion detection?	

whether they think incorporating motion-based actions in educational type games is an effective way of learning content.

For the open-ended questions, the first question is intended to gather any feedback regarding any specific difficulties that were faced during the task, and the second question is to gather any points of improvements of the system.

The Likert scale questions are asked as follows:

- 1 = Strongly disagree.
- 2 = Disagree.
- 3 = Neutral.
- 4 = Agree.
- 5 = Strongly agree.

Question 1

Median: 4—Agree

Most students agree that the hand motion was simple to understand. This can be due to the fact that a demonstration video was shown to them prior to playing the game, which familiarized them with the type of motion they needed to perform. It can also be due to the fact that it is a common motion, especially for mechanical engineers; hence, they were already familiar with how to loosen a nut based on prior experiences.

Question 2

Median: 3—Neutral

Most students are neutral toward this question, with more students agreeing that it was easy than disagreeing. During the experiment, it was noted that some students

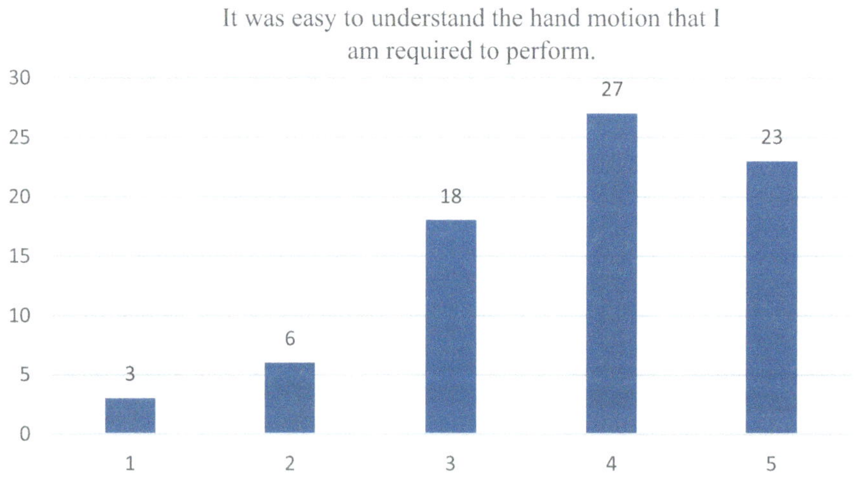

Fig. 21 Question 1 responses

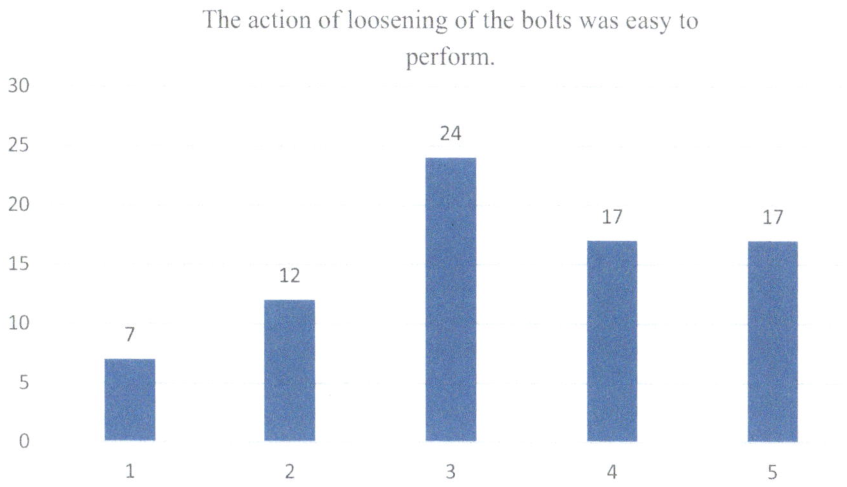

Fig. 22 Question 2 responses

found the task to be simple to perform, while others struggled a little bit. This is designed to let students doing some initial exercise before they can get familiar with the operation using the controls of games and the VR games.

Question 3

Median: 3—Neutral

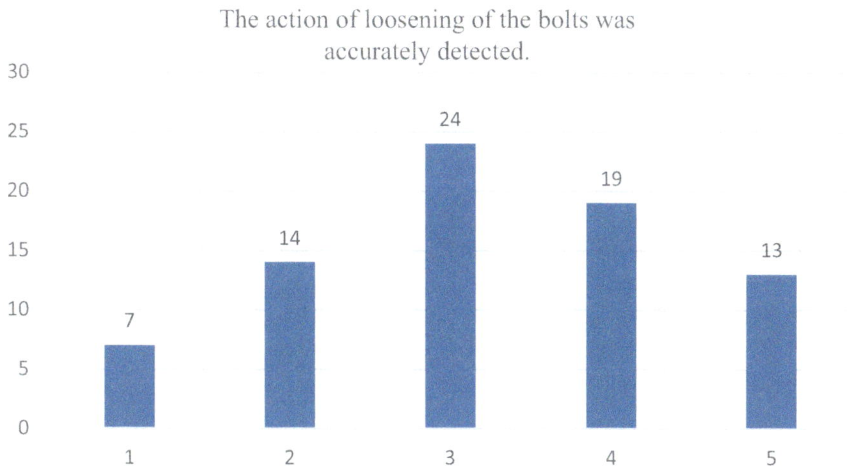

Fig. 23 Question 3 responses

Like the previous question, most students are neutral toward this question, with more students agreeing that it was easy than disagreeing. This could be due to the fact that the three conditions stated earlier were a little bit strict, leading to players needing to retry the motion multiple times.

It was also observed that some players were performing the rotation in the wrong axis; most notably, some students were performing an up-to-down motion instead of a right-to-left motion. This can be caused by the fact that the players are unfamiliar with how a wrench is held, leading to them holding the virtual wrench wrongly, as shown in Fig. 24, and the players thought that the rotation needed is with respect to the axis of the wrench instead of the axis of the hands.

This can be improved by changing the game's implementation of the wrench interaction, by having the wrench be at a fixed orientation relating to the hand, so the play always holds the wrench the right way.

Question 4

Median: 3—Neutral

The response was distributed relatively evenly (Fig. 25). As mentioned in the previous question's discussion, the wrench interaction implementation in the game can be improved to be more in line with how a wrench is properly held in real life.

Question 5

Median: 4—Agree

The majority of participants agreed that hand motion helped them learn the educational content effectively (Fig. 26). A total of 45 participants either agreed or strongly

Fig. 24 Correct versus wrong wrench position

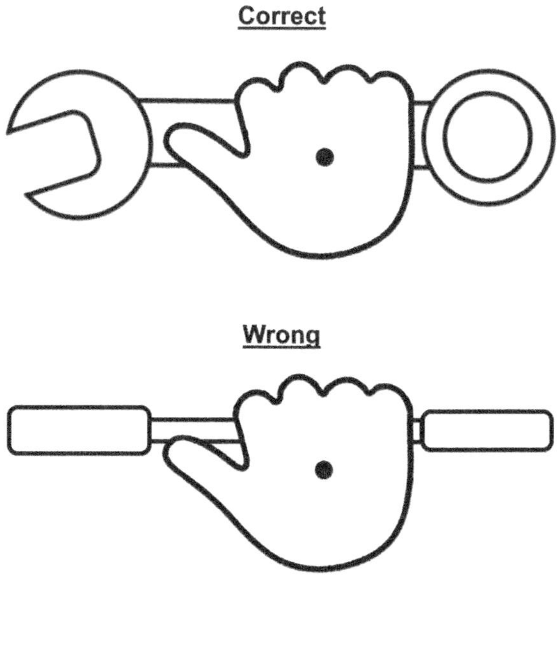

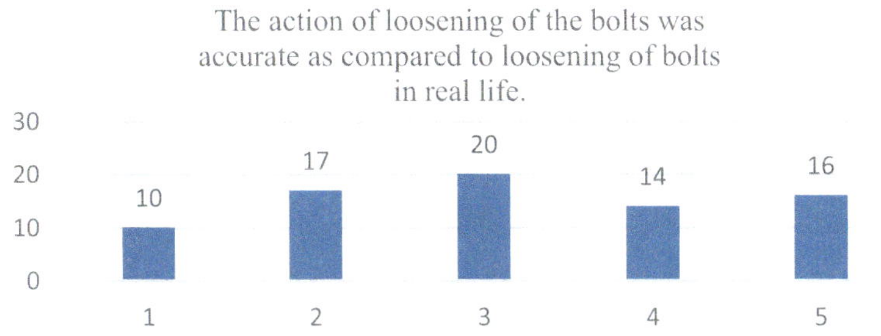

Fig. 25 Question 4 responses

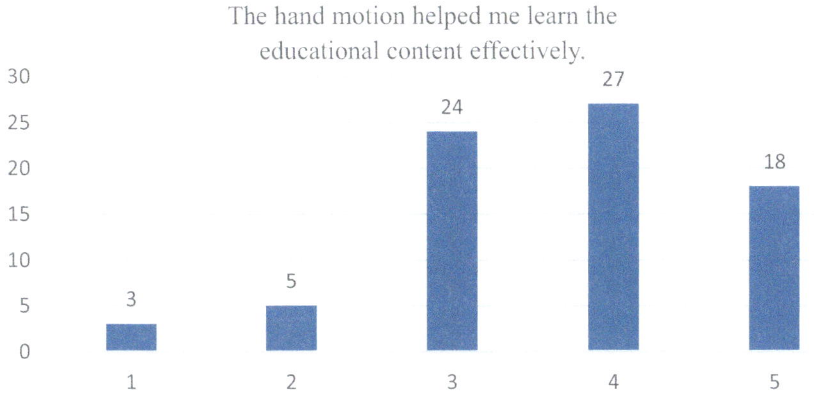

Fig. 26 Question 5 responses

agreed, while only 8 participants disagreed. The high proportion of agreement with the statement suggests that hand motion may be an effective tool for enhancing learning in some individuals. However, it is important to note that the results of this survey may not be generalizable to all learners, and further research is needed to confirm the effectiveness of hand motion as a learning aid.

Question 6

Since it is an open-ended question, the answers will vary from person to person, but can be split into three main categories. (1) No difficulties faced. (2) The motion was difficult. (3) Non-related answers, usually referring to other game mechanics. Appendix B shows the full response of the survey.

The result of this question is mixed (Fig. 27), which is in line with the results gathered in questions 2 and 3, where the difficulty in completing the tasks had mixed responses. Again, this could be due to the fact that some players are more familiar with VR systems and hence found it easier to perform the task. As developers of the script and game, the skill level of players needs to be taken into account and balanced. As an example, the team members that developed the game were able to

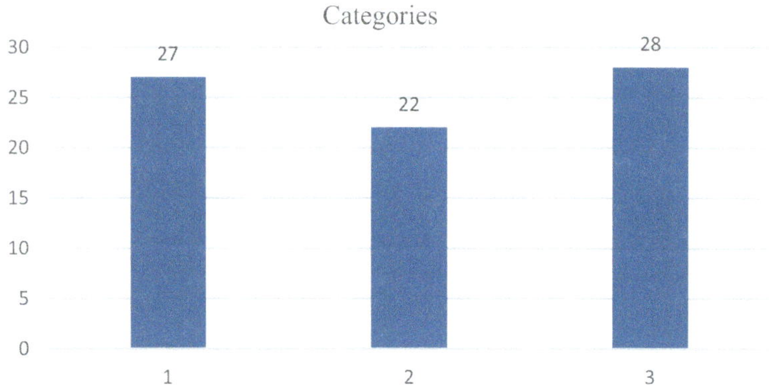

Fig. 27 Question 6 response categories

consistently complete the task due to their familiarity with the system and how to manipulate objects in VR, while first-time players could struggle to perform the task.

As for responses in category 3, most of them are feedback in difficulty in manipulating the wrench, such as properly aligning the wrench with the hand, or the aforementioned glitch causing the wrench to be locked in a certain rotation, or difficulties in spawning the wrenches.

Question 7

Since it is also an open-ended question, the answers also vary from person to person, but can be split into four main categories. (1) No improvements suggested or feedback. (2) Related to motion detection. (3) Non-related answers, usually referring to other game mechanics. Appendix C shows the full response of the survey.

The majority of the feedback was either not related to the motion detection system or no improvements suggested (Fig. 28). Only 17 responses were related to the motion

Fig. 28 Question 7 response categories

detection system. Most of those responses stated that the accuracy and tuning of the detection system need to be improved.

It is also notable that there were several responses in category 3 that stated that having real-time feedback mechanisms such as tactile or haptic feedback would improve the system. Some suggested that the physical tools and equipment should be implemented and held by the players to give them a greater sense of realism and immersion.

7 Conclusion

This research has developed three scripts that explore the possibilities and benefits of motion detection in VR environments, specifically in VR-based skill training. The scripts are highly compatible with cross-platform applications using Windows, Android, and Meta Quest 2.

Through an implementation into an aircraft hydraulics maintenance training game, an experiment is conducted detailing the student's experiences while playing the game and gather feedback through a post-test survey regarding the potential of motion detection in educational VR games.

The incorporation of hand motion tracking and detection in VR-based training games proves to be promising in revolutionizing the approach of skill acquisition and paving the way for more immersive and engaging learning experiences. Yet the implementation has limitations.

Full left-hand support

As the script was mostly developed with the implementation into the aircraft maintenance game in mind, some of the functionality was only coded for right-hand support.

Mirroring motion patterns

A first recommendation with the system is to allow easy mirroring of motion patterns for ease of changing handedness in game, to increase players support by not alienating left-handers.

Replay script

The second recommendation is to create an independent script that records the required joints and saves them in the right format for usage in the replay script.

Explore alternatives to CSV format

The third recommendation is seeking and experimenting with alternate save file formats for both the replay script and the Python script such as JSON format. Potential performance benefits may be untapped.

Future improvement includes finding a different approach to detecting motion and storing data, such as curve functions, as there could be potential optimization and accuracy improvements. To test the full capability of motion detection in educational games, more rigorous tests need to be conducted for the purpose of identifying weaknesses of the system with complex motions introduced.

References

1. Albany, N.Y.: Rise in popularity of VR games and digitalized classrooms during COVID-19 pandemic offers prodigious sales opportunities to virtual and augmented reality market. Gale General OneFile 1286 (2021)
2. Thomas, A.: Virtual reality (VR) headset unit sales worldwide from 2019 to 2024 (in million units) (2021)
3. Meta. https://www.meta.com/quest/quest-pro/ (2023)
4. Zhang, S., Chen, M., Wen, G., Wan, J.: Research on virtual maintenance training for airborne electronic equipment. In: IEEE International Conference on Information Technology, Big Data and Artificial Intelligence (ICIBA), pp. 132–136 (2020)
5. Andaluz, V.H., Quevedo, W.X., Chicaiza, F.A., Varela, J., Gallardo, C., Sánchez, J.S., Arteaga, O.: Transparency of a bilateral tele-operation scheme of a mobile manipulator robot. In: Augmented Reality, Virtual Reality, and Computer Graphics, pp. 228–245 (2016)
6. Wong, J.Y., Azam, A.B., Cao, Q., Huang, L., Xie, Y., Winkler, I., Cai, Y.: Evaluations of virtual and augmented reality technology-enhanced learning for higher education. Electronics **13**(8), 1549 (2024). https://doi.org/10.3390/electronics13081549
7. Lee, T., Wen, Y., Chan, M.Y., Azam, A.B., Looi, C.K., Taib, S., Ooi, C.H., Huang, L.H., Xie, Y., Cai, Y.: Investigation of virtual and augmented reality classroom learning environments in university STEM education. Interact. Learn. Environ. 1–16 (2022). https://doi.org/10.1080/10494820.2022.2155838
8. Gao, S., Lu, Y., Ooi, C.H., Cai, Y., Gunawan, P.: Designing interactive augmented reality application for studentas directed learning of continuous distillation process. Comput. Chem. Eng. 169 (2023). https://doi.org/10.1016/j.compchemeng.2022.108086
9. Cheung, S.K.S., Fong, J., Fong, W., Wang, F.L., Kwok, L.F.: Hybrid learning and continuing education. In: 6th International Conference, ICHL, p. 8038 (2013)
10. Apostolellis, P., Bowman, D.: Evaluating the effects of orchestrated, game-based learning in virtual environments for informal education. In: Proceedings of the 11th Conference on Advances in Computer Entertainment Technology, pp. 1–10 (2014)
11. Huang, H.M., Rauch, U., Liaw, S.S.: Investigating learners' attitudes toward virtual reality learning environments: based on a constructivist approach. Comput. Educ. **55**(3), 1171–1182 (2010)
12. Jones, T., Moore, T., Choo, J.: The impact of virtual reality on chronic pain. PLoS ONE **11**(12), 1–10 (2016)
13. Federation of State Medical Boards.: Model policy on the use of opioid analgesics in the treatment of chronic pain (2013)
14. Dascal, J., et al.: Virtual reality and medical inpatients: a systematic review of randomized. Controlled Trials. Innov. Clin. Neurosci. **14**(1–2), 14–21 (2017)
15. Bracq, M.S., Michinov, E., Jannin, P.: Virtual reality simulation in nontechnical skills training for healthcare professionals. Simul. Healthcare **14**(3), 188–194 (2019)
16. Cai, Y., Chui, C.K., Ye, X., Fan, Z., Anderson, J.H.: Tactile VR for hand–eye coordination in simulated PTCA, Computers in Biology and Medicine 36(2), 167–180, ISSN 0010-4825 (2006) https://doi.org/10.1016/j.compbiomed.2004.10.002

17. George, K., Scott, S., Grayson, L.: Performance evaluation of AR/VR training technologies for EMS first responders. Virtual Reality, p. 25 (2021)
18. Fang, X.: Hermeneutic relations in VR: immersion, embodiment, presence and HCI in VR gaming. In: HCI in Games, vol. 12211, pp. 23–38. Springer International Publishing AG, Switzerland (2020). https://doi.org/10.1007/978-3-030-50164-8_2
19. Yan, Y., Zhang, L., Chen, M.: AGRMTS: a virtual aircraft maintenance training system using gesture recognition based on PSO-BPNN model. Comput. Animation Virt. Worlds **33**(1) (2022)
20. Schölkopf, L., et al.: Haptic feedback is more important than VR experience for the user experience assessment of in-car human machine interfaces. Procedia CIRP **100**, 601–606 (2021)
21. Moll, J., Pysander, E.L.S.: Haptic communicative functions and their effects on communication in collaborative multimodal virtual environments. In: Proceedings of the Swecog Conference (2017)
22. Zhao, H., et al.: A virtual surgical prototype system based on gesture recognition for virtual surgical training in maxillofacial surgery. Int. J. Comput. Assisted Radiol. Surgery (2022)
23. Antillon, D.W.O., Walker, C.R., Rosset, S., Anderson, I.A.: Glove-Based hand gesture recognition for diver communication. IEEE Trans. Neural Netw. Learn. Syst. (2022)
24. Wario, R., Nyaga, C.: A survey of the constraints encountered in dynamic vision-based sign language hand gesture recognition. In: Universal Access in Human-Computer Interaction. Multimodality and Assistive Environments, pp. 373–382 (2019)
25. Tubaiz, N., Shanableh, T., Assaleh, K.: Glove-Based continuous Arabic sign language recognition in user-dependent mode. IEEE Trans. Human-Machine Syst. **45**(4), 526–533 (2015)
26. Singha, J., Roy, A., Laskar, R.H.: Dynamic hand gesture recognition using vision-based approach for human–computer interaction. Neural Comput. Appl. **29**(4), 1129–2114 (2018)
27. Dewez, D., Hoyet, L., Lécuyer, A., Sanz, F.A.: Towards "Avatar-Friendly" 3D manipulation techniques: bridging the gap between sense of embodiment and interaction in virtual reality. In: Proceedings of the 2021 CHI Conference on Human Factors in Computing Systems (2021)
28. Seinfeld, S., Feuchtner, T., Pinzek, J., Muller, J.: Impact of information placement and user representations in VR on performance and embodiment. IEEE Trans. Visual Comput. Graphics **28**(3), 1545–1556 (2022)
29. Qianzheng, Z., Xiaodong, L., Jie, R., Yuanyuan, Q.: Real time hand gesture recognition applied for flight simulator controls. In: IEEE 7th International Conference on Virtual Reality (ICVR), pp. 407–411 (2021)
30. Huang, G., Tran, S.N., Bai, Q., Alty, J.: Real-time automated detection of older adults' hand gestures in home and clinical settings. Neural Comput. Appl. **35**(11), 8143–8156 (2022)
31. Zhao, K., Zhang, K., Zhai, Y., Wang, D., Su, J.: Real-Time sign language recognition based on video stream. In: 39th Chinese Control Conference (CCC), pp. 7469–7474 (2020)
32. Meta. https://www.meta.com/quest/accessories/quest-touch-pro-controllers-and-charging-dock/ (2023)
33. Argelaguet, F., Hoyet, L., Trico, M., Lecuyer, A.: The role of interaction in virtual embodiment: Effects of the virtual hand representation. In: IEEE Virtual Reality (VR), Greenville, SC, USA, pp. 3–10 (2016)
34. Pyasik, M., Tieri, G., Pia, L.: Visual appearance of the virtual hand affects embodiment in the virtual hand illusion **10**(1), 5412 (2020)
35. Adkins, A., Lin, L., Normoyle, A., Canales, R., Ye, Y., Jörg, S.: Evaluating grasping visualizations and control modes in a VR game. ACM Trans. Appl. Percept. **18**(4), 1–14 (2021)
36. Roungas, B., Wijse, M., Meijer, S., Verbraeck, A.: Pitfalls for debriefing games and simulations: theory and practice. Intersections in Simulation and Gaming, 101–115 (2018)
37. Kim, Y.J., Pavlov, O.: Game-based structural debriefing: How can teachers design game-based curricula for systems thinking? Inf. Learn. Sci. **120**(9–10), 567–588 (2019)
38. Hoogen, J., Lo, J., Meijer, S.: Debriefing research games: context, substance and method. Simul. Gaming **47**(3), 368–388 (2016)
39. Dudas, K., Wheeler, J.: Faculty re-enactment videos as a tool in simulation debriefing. Clin. Simul. Nurs. **40**, 25–30 (2020)

40. Toftedahl, M.: Which are the most commonly used game engines? Game Developer (2019)
41. Microsoft. https://learn.microsoft.com/en-us/windows/mixed-reality/mrtkunity/mrtk2 (2023)
42. Microsoft Learn. https://learn.microsoft.com/en-us/windows/mixed-reality/mrtk-unity/mrtk2/features/input/hand-tracking?view=mrtkunity-2022-05 (2023)
43. Ultraleap. https://www.ultraleap.com/product/leap-motion-controller/ (2023)

VARTeL for STEAM Education: Qualitative and Quantitative Analysis

Junhong Song, Abu Bakr Azam, Lihui Huang, Yuan Xie, Siyi Li, and Yiyu Cai

Abstract Virtual and Augmented Reality Technology Enhanced Learning (VARTeL) classroom is an innovation from Nanyang Technological University, Singapore. VARTeL introduces a novel learning experience. This research is to examine the effectiveness of VARTeL from different aspects including situational interest, self-efficacy, affordances and continuous learning intention in Science, Technology, Engineering and Mathematics (STEM) education. Experiments are designed for a group of engineering students to see the benefits in their learning with VARTeL. Data collected are analyzed and interpreted to find out VARTeL's impact on affordances, continuous learning intention and academic achievement for individual students. This research methodologies are classified into two: quantitative and qualitative. The quantitative approach is administrated by collecting questionnaire responses from students. The qualitative approach is based on in-class observation and post class focus-group interviews. By considering both quantitative and qualitative approaches, we conclude that VARTeL does trigger and maintain individuals' situational interest, help to increase the self-efficacy, continuous learning intention and eventually the academic achievement for individual students. It is also founded that VARTeL has pedagogical affordances of immersive learning, collaborative learning, and interactive learning in educational aspects.

1 Introduction

The Sword of Damocles, the world first Head-mounted Display (HMD) was invented by Dr Ivan Sutherland and his students in the 1960s [1]. Reality Virtuality Continuum was proposed by Paul Milgram et al. in 1994 [2]. The first ACM SIGGRAPH Virtual-Reality Continuum and Application in Industry (VRCAI) conference was organized in 2004 [3]. Virtual-Reality Continuum spans the entire spectrum from Physical to Virtual Worlds including Virtual Reality (VR), Augmented Reality (AR), Mixed Reality (MR), etc. VR for education is one of the fast-growing fields in the recent

J. Song · A. B. Azam · L. Huang · Y. Xie · S. Li · Y. Cai (✉)
Nanyang Technological University, Singapore, Singapore
e-mail: myycai@ntu.edu.sg

Y. Cai et al. (eds.), *Virtual and Augmented Reality Technology-Enhanced Learning*, Gaming Media and Social Effects, https://doi.org/10.1007/978-981-96-2332-7_5

years with a huge interest from the education community consisting of Tertiary, Vocational, K12, Preschool and Special Needs Education [4–12].

Virtual and Augmented Reality Technology-enhanced Learning (VARTeL) classroom is a 'flipped' classroom invented at Nanyang Technological University (NTU). The HIVE (Fig. 1a) is a student-centric learning hub in NTU. A circular room in HIVE is equipped with 6 units of HD 3D capable TVs and VR headsets associated with high-end PCs (Fig. 1b). The classroom is designed for students, after video-based lectures, to do virtual hands-on to learn various topics such as Jet Engine Inspection, Formula 1 Assembly, Turning and Milling, Building Information Modeling and Crane Lifting. The classroom has 6 tables each hosting 5 or less students for team-based learning. While one student (master player) in each team is doing experience-based learning with the VR headset, the rest of teammates will also be able to view the scene in 3D through polarizing lenses, except the scene is under the control of the master player. Students in a team will take turn to serve as the master player. The VARTeL classroom has been used by NTU students for curriculum-based learning. This Chapter is interested to investigate its effectiveness of VARTeL for learning which is yet systematically and scientifically studied in either qualitative or quantitative ways with respect to situational interest (SI), self-efficacy (SE), continuous learning intention (CLI), affordances and academic achievement. In this work, a group of engineering students attending the module of Computer-Aided Engineering are invited to participate in an experiment study.

This research is interested in the impact of the pedagogical intervention on different aspects including SI, SE, affordances, and CLI using the VARTeL flipped classroom. Efforts are made to study the correlations between the above-mentioned aspects and students' academic achievement (AA).

(a) (b)

Fig. 1 (**a**) The HIVE at NTU, and (**b**) VARTeL flipped classroom in the Hive

2 Literature Review

2.1 Situational Interest

Research interest in educational psychology is increasing in recent years [13–17]. Personal interests are described as the individual's intention to achieve goals over time and situationism is a temporary phenomenon triggered by situational stimuli in a learning context [17, 18]. SI consists of attention and reaction response to stimuli, in two forms: triggered SI and sustained SI [15, 19]. Triggered SI involves enhancing an individual's emotional experience with the environment. Triggered-SI is similar to the "catch" conceptualization brought up by Mitchell [20]. The SI's configuration involves arousing or capturing individual attention. However, triggered SI implies particularly in early stage of interest development [21–23].

In contrast, maintained SI, also known as "hold", is more complex. It is a deeper form of situational concern in which the individual begins to shape personal interest which could last longer. It is linked to the content of the material to realize a deeper learning [19, 20]. SI can mean the interest that appears in a variety of settings, but it focuses specifically on the situational interest that appears in an educational setting. The ability to promote the triggered SI within these settings is the student's lasting on the focus. This provides an emotional experience for students to be actively involved in the material [20, 24, 25]. However, the triggered SI can have a short lifespan if the environment does not continue to support it [16, 19, 20, 22, 23, 26].

Unlike situational interests, individual interests are relative [16, 27]. Conceptualizes an individual's interest as deepening an individual's long-term and personal connection to the domain. At this point, interest in the individual deepens. Gaining knowledge of the domain and assigning it value that inspires curiosity and further navigation of the domain is considered as part of deepening the personal interest. Although personal and situational interest concerns are conceptually different, some models show that SI can grow into individual interest [23, 28]. SI triggered students are interested in the material, but it does not imply that the interest will maintain as SI is generally short-term. Through the continuous triggered SI transformation to maintained SI, it is believed to develop as individual interest. If a maintained SI can develop into individual interest, then a maintained SI can be constructed similarly to that of an individual.

2.2 Self-Efficacy, Affordance, and Continuous Learning Intention

Apart from SI, this research is also interested in the change of SE, CLI, and AA as well as the affordances of VARTeL classroom throughout the entire experiment. Moreover, the correlations between the above-mentioned measurements are part of the research.

SE is students' belief in their own efficacy in learning and completing tasks. Students tends to be more motivated during learning and will achieve success more easily [29]. Faculty-focused research reveals that engineering students have a strong sense of academic self-efficacy resulting in achievement [30] and persistence [31]. For example, engineering students were greatly influenced by their SE instead of self-report effort, seeking of help or critical thinking [32]. Students' beliefs about competence in engineering-related fields are also influencing career choices after graduation [33, 34].

Affordance is defined as "The affordances of the environment are what it offers the animal, what it provides or furnishes, either for good or ill" [35]. Affordance theories have been used to explain the interaction between material technology and human subjective use [36] and provide an explanation for the results use of IT artifacts [37]. Individuals first form perceptual affordances before interacting with technology [38] and then actors begin to take advantage of the realized affordance of perceived affordance [37].

CLI is individual's intention to learn an object in long-term standpoint. This is examined by asking individual's willingness of using a specific learning system, VARTeL systemin this case.

3 Methodology

3.1 Research Approach

To study the arousing of SI, SE, affordance and CLI during the VARTeL class, it was decided to conduct the research in both qualitative and quantitative ways. The quantitative research will be conducted through questionnaire that will be distributed to students before, during, and after the class. While interviews and class observations will also be considered as the qualitative approach for this research. More specifically, the quantitative questionnaire will be classified into 4 different categories to measure situational interest, self-efficacy, continuous learning intention and affordances. To track the academic achievement of individuals, pre and post-tests are implemented before and after each class. Hands-on tests were considered as measurement of academic achievement on VAR knowledge, but it was decided not to include it due to the time constraint and uncertainties of quantifying the result. Qualitative research will be conducted though post-lesson focused group interview and in-class observation.

The SI questionnaire consists of 4 questions which are adopted from existing research thesis [39]. The questionnaires aim to examine the SI from different aspects, including triggered SI and maintained SI. Due to the limitation of class size and schedule, the development of individual interest from situational interest will not be measured in this research. SI generally refers to "focused attention and the affective reaction that is triggered in the moment by environmental stimuli, which may or may not last over time" [23].

SI_1 and SI_2 are designed to measure focused attention while SI_3 and SI_4 are designed to measure affective reaction (Table 1). All questions are subject to the 5-point Likert scale: Strongly disagree (1); Disagree (2); Neither agree nor disagree (3); Agree (4); Strongly Agree (5).

The SE questionnaire consists of 5 questions adapted from [40] as in Table 2.

The Continuous Learning Intention questionnaire consists of 3 questions as shown in Table 3.

The affordance questionnaire (Table 4) is designed to measure from three different aspects including Social Presence (SP_1 to SP_3), Peer Influence (PI_1 to PI_4) and Satisfactory (SF_1 to SF_3).

For qualitative analysis, 6 questions (Table 5) were designed to be asked during focus-group interviews. The questions aim to get specific response from students on SI, SE and affordances. This is different from the questionnaire setup which is a passive approach where students can only select the options provided.

The interview questions also aim to collect feedback from students on the shortcomings of VARTeL class and potential improvements in the future.

Table 1 SI questionnaire

S/N	Questions
SI_1	I am fully focused on today's topic and activity; I am not distracted by other things
SI_2	I think today's topic and activity are interesting
SI_3	I will enjoy working on today's topic and activity
SI_4	Presently, I feel bored about today's topic and activity (reversed)

Table 2 SE questionnaire

S/N	Questions
SE_1	I can master the content in this module
SE_2	I can master the content in even the most challenging engineering course
SE_3	I can do a good job on almost all my engineering coursework
SE_4	I can learn the content taught in my engineering-related courses
SE_5	I can earn a good grade in my engineering-related courses

Table 3 CLI questionnaire

S/N	Questions
CLI_1	I would continuously choose VARTeL system to improve my learning abilities
CLI_2	I would continuously choose using VARTeL as much as possible in the future
CLI_3	I would recommend others to take this course through VARTeL

Table 4 Affordances questionnaire

S/N	Questions
SP_ 1	In this learning environment, it feels as if I see 'real' objects through the headset
SP_ 2	In this learning environment, it feels as if I interact with 'real' objects
SP_ 3	In this learning environment, I can form distinct impressions of the object that I deal with
PI_ 1	The group discussions inspired me to improve my critical thinking
PI_ 2	The group discussions inspired me to improve my communication skills
PI_ 3	I was able to build positive relationships with my group members
PI_ 4	My group members and I were able to help each other
SF_ 1	VAR's visual and interactive learning experience creates relevance to the topics taught in class
SF_ 2	I can achieve a deeper level of learning, i.e., I will not forget the material as easily with the passage of time
SF_ 3	When learning something new, I would rather learn it visually and interactively

Table 5 Interview questions

S/ N	Questions	Measurement
1	What is the most interesting part throughout the VARTEL class? Why?	SI & AFFORDANCE
2	What is the least interesting part throughout the VARTEL class? Why?	SI & AFFORDANCE
3	What is the biggest difference between VRL (video recorded lecture) and VARTEL?	AFFORDANCE
4	Does VARTEL help you participate more in group discussions?	SE & AFFORDANCE
5	Does VARTEL make you more confident in learning the subject?	SE & AFFORDANCE
6	Is the class easier to understand with VARTEL?	SE & AFFORDANCE

Pre & post-tests are to measure the academic achievement of individual students before and after the class. The questionnaires are specifically designed to match the content taught for the particular class. The pre & post-test questionnaire will be the same, while pre-test is conducted before the class, post-test is conducted after the class. Comparison of the test result will shed light on the academic performance change for individual students.

The initial plan was to use a Concept Recognition test to design the questions, however after studying the content of the class, it was decided to design content knowledge-based question as the class is not aiming to tell a specific concept, instead it is aiming to introduce the immersive learning environment through different engineering approach. The pre and post-tests will not be attached in this report, but the result will be discussed shortly.

3.2 Research Schedule

Table 6 shows the detail activities for 5-week class.

VARTeL class is conducted in 5 weeks with 2 sessions each week for 2 different group of students. The activities and content taught in two sessions will be exactly the same. To increase the sample size, both classes will be considered, and research responses will be consolidated for analysis.

Starting with the Introduction in week 1, the following week 2's topic is Aircraft Engineering Experiment on Jet Engine and Hydraulic System. Week 3 is for Milling and turning machine simulation, and week 4 is building and construction on crane lifting. 5th week will be make-up class and hydraulic class (cont.). There is a class in week 6 for debate activity, but it is not part of our research interest, hence it is not considered here.

As the content is only taught in week 2, 3 and 4, the pre & post-tests will be administrated in week 2–4 (Table 7).

Since SI has been defined as an immediate affective response to certain conditions and/or stimuli in the learning environment that focuses one's attention on the task, which may or may not last over time, we decided to measure it repeatedly. 'Triggered SI' is measured immediately after each class, while 'maintained SI' will be investigated during continuous measuring of SI through 3 weeks. It will provide more accurate results as 'maintained SI' is likely to last over time. The SE survey will only be conducted on week 1 and week 5. This is to compare student's confidence level on learning with VARTeL between the start and end of this VAR enabled class. The

Table 6 VARTeL class activities

Week	Topics
1	Class introduction
2	Aircraft engineering experiment on jet engine
3	Manufacturing engineering exp on milling & turning
4	Building & construction engineering exp on crane lifting
5	Hydraulic system and make-up class
6	Groups debate

Table 7 Weekly schedule

Instrument	Week No					
	1	2	3	4	5	6
Pre & Post-test		✔	✔	✔		
Questionnaire on SI		✔	✔	✔		
Questionnaire on SE	✔				✔	
Continuous learning intention + affordances					✔	
Post-lesson focus-group discussion	✔	✔	✔	✔	✔	
Class observation	✔	✔	✔	✔	✔	

CLI and affordances questionnaires are administrated at the end of teaching plan. This is to have an overview on student's intention on learning with VARTeL system and affordance level.

Post lesson focus-group interview will be audio recorded; transcript will be attached at the end of this report. Class observation will be recorded for each session, this includes student's behaviour under certain stimuli during the class and student's performance, attitude and personal focus while experiencing the VR headset. This will give us a qualitative reference of the effectiveness of VARTeL system.

3.3 Hypothesis and Conceptual Model

One of the research objectives is to find the Correlations between the elements of SI, SE, CLI and AA. It was expected that the sample size is not sufficient to run statistics analysis for interrelation finding. However, the hypothesis and model shall be created for reference and future use in the second phase of the research.

Below, 5 hypotheses are designed based on the research setup. It suggests the positive impact of one intervention on one another. Details are as below:

- H1: VAR helps to maintain students' situational interest.
- H2: VAR helps to improve students' self-efficacy.
- H3: VAR helps to improve students' academic achievement.

Based on the hypotheses above, 2 structure models are designed as Fig. 2.

Model 1 is designed to measure the SI impact on academic achievement through the teaching week 2 to week 4. This is to verify the hypothesis "Increasing of SI has positive effect on student's academic performance". The prior knowledge that is governed by the pre-test is to actuate SI in week 2, and the change in weeks 3 and 4 of SI will have continuous impact on academic achievement (Fig. 3).

The second model is to investigate the relationships between SI, SE, affordances and CLI. The model suggests that affordances will have direct impact on SI and SE, and SI triggers SE. Both SI and SE will have impact on continuous learning intention.

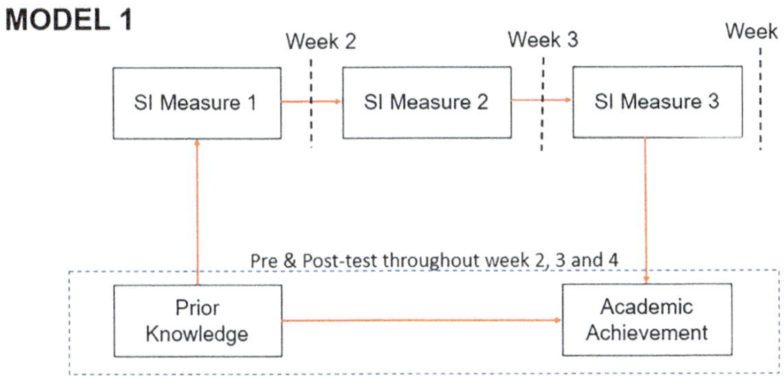

Fig. 2 SI model

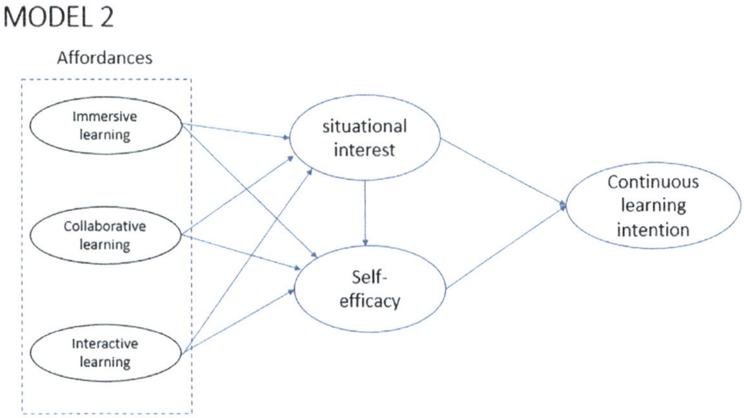

Fig. 3 General model

3.4 Data Collection and Analysis Methodology

The questionnaire will be administrated through google forms; students will need to scan the designated QR-code to access the questionnaire. To get the most accurate and realistic result, a short briefing will be conducted at the start of class to encourage them to fill up the form by their first impression, and they shall honestly provide their responses as it is not graded and will only be used for research purpose. Responses will be collected and consolidated to Excel sheets and can be further imported into SPSS for statistical analysis.

4 Results Discussion

4.1 Individual Findings and Result Overview

The quantitative data has revealed that VAR helps to maintain students' situational interest, improve their self-efficacy and academic achievement.

The quantitative data also revealed that VARTeL has pedagogical affordances of (a) immersive learning; (b) collaborative learning; (c) interactive learning.

Table 8 and Fig. 4 reveal that there is positive improvement on students' academic achievement for each week. However, the percentage of improvement is not proportional to the number of lessons that students have attended even though the last two weeks' results tend to be better than first 2 weeks.

Table 8 Pre and post-test mean value for each week.

PRE-Post test			
Week	Pre	Post	Improvement, %
2	1.172414	1.563636	33%
3	0.95082	1.135593	19%
4	2.428571	4.385965	81%
5	3.403509	4.396552	29%

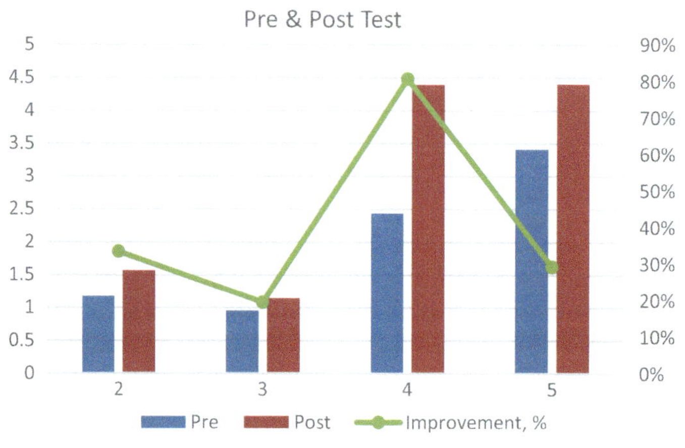

Fig. 4 Pre and post-test mean value figure

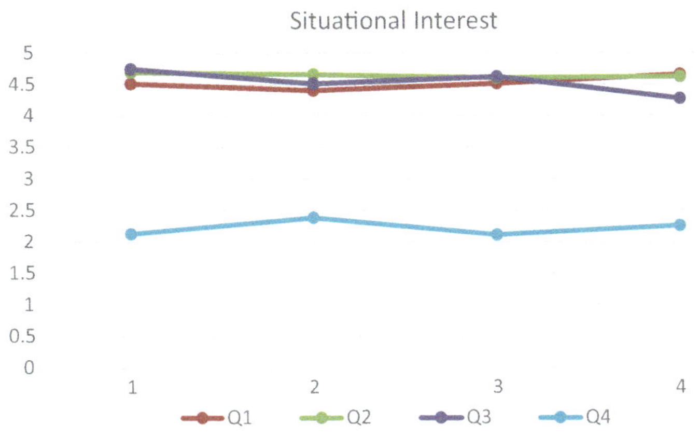

Fig. 5 SI value for each week

The variance of different improvement level might be due to different topics and is correlated to student's interest in the content. The variance can also be observed from the mean value—the mean values become higher at the last 2 weeks. The difficulty of the content remains the same for the 4 weeks, so the incremental change is due to the change in situational interest.

Figure 5 is the SI mean value for each week. Note that week 5 is not in the figure, this is because the week's SI questionnaire is different from week 2, 3 and 4. It can be used for reference but cannot be used for analysis. The SI value remains high throughout the entire experiment. There are some visible correlations with the pre and post-test results: the tendency of academic achievement is following the SI value. It reveals that SI value helps to increase academic achievement to the standpoint of mean value. Note that this is not considered as a statistics assumption, instead, it is more on qualitative conclusion by looking at mean value.

Correlations shall only be statistically defined by statistic approach, but as mentioned earlier, due to the small sample size, it is not possible to run any correlation analysis and thus quantitative approach will be considered as a relatively reliable source to conclude this experiment. It is also observed that the highest SI value happens in Week 2 even though there are some fluctuations in terms of SI value in next few weeks. Week 2 SI is considered as 'triggered' SI, and the entire VARTEL system is considered as stimuli. The value goes lower in week 2 but it is within 5% of deviation from the week 1 value. The SI is gradually transferring to 'maintained' SI as the SI values remain at the high side. This will be further explained in the next section. Q4 is a flipped question, the responses have large variance with standard deviation of 1.45. This is because students did not read the question carefully. This point shall be highlighted in the second phase of this project.

Table 9 and Fig. 6 show that there is positive improvement on students' SE for week 1 and week 5. However, there are some variations when it comes to specific aspects (by referring to different question numbers).

Table 9 SE mean value for each week

SE					
	Q1	Q2	Q3	Q4	Q5
Pre_Wk 1	3.603773585	3.188679245	3.566037736	3.962264151	3.528301887
Post_Wk 5	4.1875	3.953125	4.140625	4.296875	4.09375
Improvement	16%	24%	16%	8%	16%

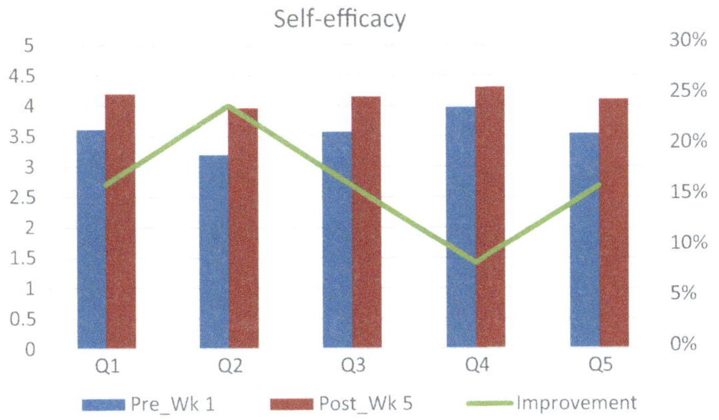

Fig. 6 SE responses

The highest percentage of improvement goes to Q2: I can master the content in even the most challenging engineering course. This reveals that students' confidence level in terms of learning the content is significantly increased. The lowest percentage of improvement goes to Q4: I can learn the content taught in my engineering-related courses. However, the mean value of Q4 is the highest among 5 SE questions. This is the preferred case and statistics analysis shall be interpreted here for further study.

Cronbach's alpha values are interpreted as below. All the values are larger than acceptance criteria value 0.70, meaning the results concluded above are relatively reliable.

- Self-efficacy (SE): Cronbach's alpha = 0.94
- Affordance of Immersive learning Cronbach's alpha = 0.84
- Affordance of collaborative learning Cronbach's alpha = 0.87
- Affordance of interactive learning Cronbach's alpha = 0.85
- Continuous learning (ContL) Cronbach's alpha = 0.93
- Situational Interest (SI) Cronbach's alpha = 0.72

4.2 Correlations

4.2.1 Overall Assumptions

The mean value and standard deviation are calculated as above. The mean values are extremely high. This supports the assumptions in the above section (Fig. 7).

The coefficient of variation (CV) is defined as standard deviation divided by mean value. Taking interactive learning under affordance, the CV is 0.158. This is considered as a very low variation value and shows high consistency for the results. Similar to other values in the model, it suggests that each measurement has high consistency and is reliable of what is concluded above.

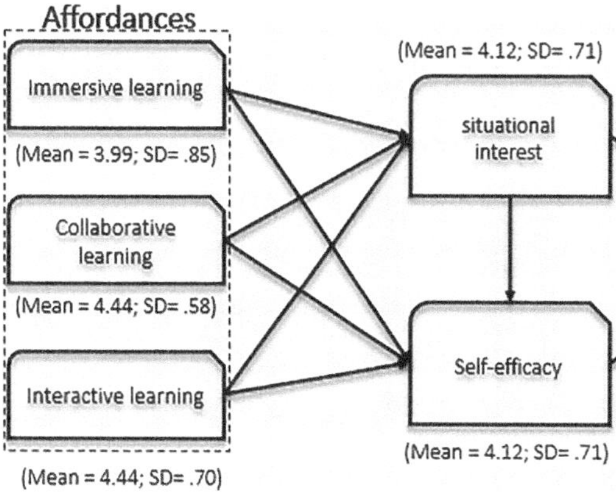

Fig. 7 Model correlations verification

Table 10 R value and R square value and SI and SE

Model summary[b]					
Model	R	R square	Adjusted R square	Std. Error of the estimate	Durbin-Watson
1	0.468[a]	0.219	0.139	0.62840	1.875

[a]Predictors: (Constant), SI_W11, SI_W10, SI_W9
[b]Dependent Variable: SE_Post

Note that due to that low sample size (valid sample size for analysis is 43), it is unable to run any correlations analysis based on the responses collected. To examine the correlations, qualitative approach is preferred here. Refer to above sect 4.1 for details.

4.2.2 SI and SE Correlations

A scatterplot was plotted as above; SE-post as dependent, SI among 3 weeks set as independent; no obvious correlation can be observed from the graph (Table 10) (Fig. 8).

The adjusted R square is much smaller than the R Square value. It means the SI value can explain 13.9% of SE values (median effect) (Table 11).

$F = 2.716$, $Sig(P) > 0.05$, it reveals that the dependent and independent values have no linear regression relations.

4.2.3 SI and Academic Achievement Correlations

See the Table 12.

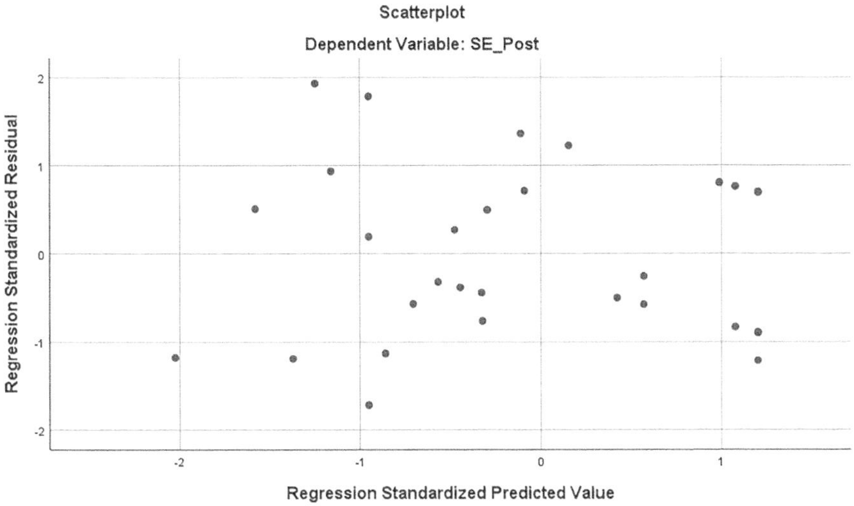

Fig. 8 Scatterplot of SI against SE

Table 11 ANOVA test of SI and SE

ANOVA[a]						
Model		Sum of Squares	df	Mean square	F	Sig
1	Regression	3.218	3	1.073	2.716	0.063[b]
	Residual	11.452	29	0.395		
	Total	14.669	32			

[a]Dependent Variable: SE_Post
[b]Predictors: (Constant), SI_W11, SI_W10, SI_W9

Table 12 R value and R square value of SI and academic achievement

Model summary[b]					
Model	R	R square	Adjusted R square	Std. error of the estimate	Durbin-Watson
1	0.356[a]	0.126	0.107	1.87794	1.731

[a]Predictors: (Constant), SI_W11
[b]Dependent Variable: Gain_W11

Adjusted R square is smaller than R Square value. It means the SI value can explain 10.7% of SE values (median effect) (Tables 13 and 14).

$F = 6.510$, $Sig(P) < 0.05$, it reveals that the dependent and independent values have linear regression relations. The formulae of this correlation can be expressed as:

Table 13 ANOVA test of SI and academic achievement

ANOVA[a]						
Model		Sum of squares	df	Mean square	F	Sig
1	Regression	22.959	1	22.959	6.510	0.014[b]
	Residual	158.700	45	3.527		
	Total	181.660	46			

[a]Dependent Variable: Gain_W11
[b]Predictors: (Constant), SI_W11

Table 14 Coefficients

Coefficients[a]								
Model		Unstandardized coefficients		Standardized coefficients Beta	t	Sig	95.0% confidence interval for B	
		B	Std. error				Lower bound	Upper bound
1	(Constant)	−2.649	1.809		−1.464	0.150	−6.293	0.996
	SI_W11	1.107	0.434	0.356	2.552	0.014	0.233	1.981

[a]Dependent Variable: Gain_W11

$$\text{Academic Achievement} = -2.649 + (0.037 \times \text{Situational Interest})$$

The 95% of confidence interval for the coefficient falls between 0.233 and 1.981.

4.3 Hypothesis

Repeated ANOVA analysis is conducted to verify the hypothesis. The major advantage of running a repeated ANOVA over a Single ANOVA is that the test is generally much more accurate and powerful. That said, it is preferred to run repeated ANOVA when the responses are received from different time points.

Details of statistics results and assumptions will be given below:

H1: VAR helps to maintain students' situational interest (Fig. 9).

Repeated ANOVA is conducted for SI, and no statistical difference is found. Mean values remain high and standard deviation remains low for 3 weeks. The $F(2,31) = 1.897$ and $p = 0.167$ reveals that the null-hypothesis will be accepted as F values are close to 1 and p value is small.

H2: VAR helps to improve students' SE (Fig. 10).

The paired t-test is a statistical approach to determine whether the variation of two sets of observations under analysis is zero. The p value is extremely small and t value

H1: No Statistical Difference of SI [Repeated ANOVA]

Descriptive Statistics

	Mean	Std. Deviation	N
SI_W9	4.2121	.49656	33
SI_W10	4.0758	.48608	33
SI_W11	4.1818	.55647	33

Multivariate Tests[a]

Effect		Value	F	Hypothesis df	Error df	Sig.	Partial Eta Squared	Noncent. Parameter	Observed Power[c]
Week	Pillai's Trace	.109	1.897[b]	2.000	31.000	.167	.109	3.794	.364
	Wilks' Lambda	.891	1.897[b]	2.000	31.000	.167	.109	3.794	.364
	Hotelling's Trace	.122	1.897[b]	2.000	31.000	.167	.109	3.794	.364
	Roy's Largest Root	.122	1.897[b]	2.000	31.000	.167	.109	3.794	.364

a. Design: Intercept
Within Subjects Design: Week

b. Exact statistic

c. Computed using alpha = .05

- $F(2, 31) = 1.897, p = .167$

Fig. 9 Hypothesis H1 analysis

H2: Significant difference of SE between Pre and Post [Paired *t*-test]

Paired Samples Statistics

		Mean	N	Std. Deviation	Std. Error Mean
Pair 1	SE_Pre	3.5560	50	.73322	.10369
	SE_Post	4.1240	50	.71130	.10059

Paired Samples Test

		Paired Differences					t	df	Sig. (2-tailed)
					95% Confidence Interval of the Difference				
		Mean	Std. Deviation	Std. Error Mean	Lower	Upper			
Pair 1	SE_Pre - SE_Post	-.56800	.59021	.08347	-.73573	-.40027	-6.805	49	.000

- $t(1, 49) = 6.805, p < .001$

Fig. 10 Hypothesis H2 analysis

is on the high side, this reveals that the 'null' hypothesis—high chance of having same result for SE pre and post-tests are invalid. This means the H2 is accepted.

H3: VAR helps to improve students' academic achievement (Figs. 11 and 12).

No significant difference between W9 and W10, but W9 and W11, W10 and W11 are observed to be significantly different. In sect 4.1. (Qualitative Result Analysis),

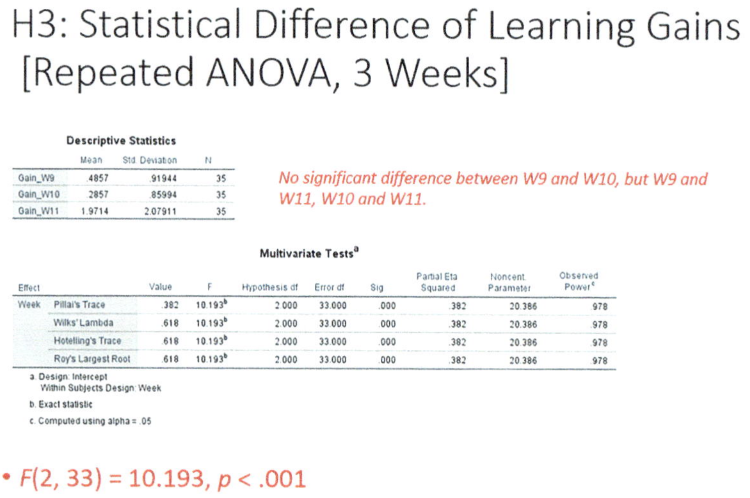

Descriptive Statistics

	Mean	Std. Deviation	N
Gain_W9	.4857	.91944	35
Gain_W10	.2857	.85994	35
Gain_W11	1.9714	2.07911	35

Multivariate Tests[a]

Effect		Value	F	Hypothesis df	Error df	Sig.	Partial Eta Squared	Noncent. Parameter	Observed Power[c]
Week	Pillai's Trace	.382	10.193[b]	2.000	33.000	.000	.382	20.386	.978
	Wilks' Lambda	.618	10.193[b]	2.000	33.000	.000	.382	20.386	.978
	Hotelling's Trace	.618	10.193[b]	2.000	33.000	.000	.382	20.386	.978
	Roy's Largest Root	.618	10.193[b]	2.000	33.000	.000	.382	20.386	.978

a. Design: Intercept
Within Subjects Design: Week

b. Exact statistic

c. Computed using alpha = .05

Fig. 11 Hypothesis H3 analysis_3 weeks

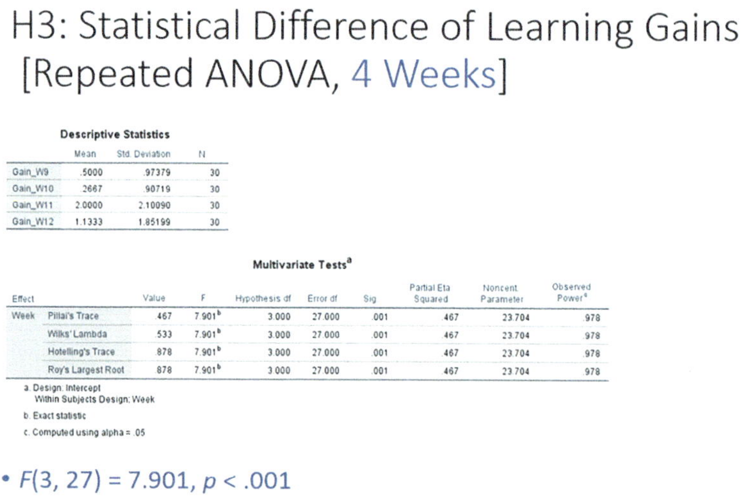

Descriptive Statistics

	Mean	Std. Deviation	N
Gain_W9	.5000	.97379	30
Gain_W10	.2667	.90719	30
Gain_W11	2.0000	2.10090	30
Gain_W12	1.1333	1.85199	30

Multivariate Tests[a]

Effect		Value	F	Hypothesis df	Error df	Sig.	Partial Eta Squared	Noncent. Parameter	Observed Power[c]
Week	Pillai's Trace	.467	7.901[b]	3.000	27.000	.001	.467	23.704	.978
	Wilks' Lambda	.533	7.901[b]	3.000	27.000	.001	.467	23.704	.978
	Hotelling's Trace	.878	7.901[b]	3.000	27.000	.001	.467	23.704	.978
	Roy's Largest Root	.878	7.901[b]	3.000	27.000	.001	.467	23.704	.978

a. Design: Intercept
Within Subjects Design: Week

b. Exact statistic

c. Computed using alpha = .05

Fig. 12 Hypothesis H3_4 weeks

it was observed that for each week, there is significant change before and after the class in terms of academic achievement.

When it comes to parallel comparison between weeks, no tendency of significant change between W9 (W2) and W10 (W3) can be observed. But there is a tendency of significant change between W9 and W11, W10 and W11. This can be explained by the qualitative approach, that on W9 there was a technical issue on the devices. That increased the waiting time for students to experience and familiarize with the devices.

Hence only part of the students got the chance to experience the VAR devices. This explained why there is no tendency of improvement in the academic achievement domain between W9 and W10.

Once students familiarized with the devices, they could master the content more easily, meaning that the SI, SE and academic achievement domain will be improved as well.

5 Conclusions and Future Work

5.1 Contribution

This research evaluated the VARTeL class in both qualitative and quantitative approaches. It studied the details of the triggered and maintained situational interest during 5 weeks of VARTeL classes. Meanwhile, it measured the students' self-efficacy level and continuous learning intention of VAR related context in the future. VARTeL's affordances were also measured and confirmed in this research.

This research enhanced developers' confidence to improve the class continuously. It also increases tutors' and instructors' confidence level in teaching and guiding during VARTEL class. This research studies the shortcomings of the class and provides constructive feedback to the class in terms of SI, SE, CLI and affordances aspects. The feedback will eventually benefit the class to become better in the next phase.

The quantitative and qualitative data has revealed that VARTeL helps to trigger and maintain students' situational interest, improve their self-efficacy and academic achievement. The quantitative and qualitative data also revealed that VARTeL has pedagogical affordances of (a) immersive learning; (b) collaborative learning; (c) interactive learning.

5.2 Limitations

The hypothesis model was not verified statistically in this research. The model cannot be run in SPSS due to its small sample size. SI was not measured in detail (situational triggering and maintained situation under different circumstances and stimulus in one single class.) There is no control group in this experiment. This is due to many reasons, such as limited resources, and not being able to set-up a class as a control group without VAR devices as the class is designed to be VAR technology embedded.

Waiting and setting-up time lowers students' focus and interest in class, hence lowering the self-efficacy, continuous learning intention and academic achievement eventually.

The gesture identification system does not function 100% as real life, which sometimes depressed students' interest and expectancy. For instance, some students complained that it is difficult to use gestures to control in the first place. They managed to get used to it after a while. Some students reported that the machine is not 100% close to real life such as only one-hand is allowed when operating of lathe machine while reality differs.

5.3 Future Improvement

There are some recommendations for the second phase of research, this is interpreted from qualitative responses (interview).

- Sample size for research shall be increased to enable the running of correlations analysis for hypothesis models.
- Simplify the model if increasing of sample size is impossible.

Situational Interest can be measured broadly in terms of weeks with a design of the research schedule to enable multiple measurements of SI in one single class. Specially a control group can be arranged to gain more confidence on the effectiveness of VARTeL class. Waiting and setting-up time in class can be shortened by re-arranging the schedule to have multiple activities within one group. The VR game can be refined to avoid un-natural interaction gestures with device by optimizing gesture recognition system as well as the learning system to enable more realistic interaction.

References

1. Sutherland, I.E.: A head-mounted three dimensional display. Proc. AFIPS **68**, 757–764 (1968)
2. Milgram, P., Takemura, H., Utsumi, A., Kishino, F.: Augmented reality: a class of displays on the reality-virtuality continuum. Proc. Telemanipulator Telepresence Technol. **2351**, 282–292 (1994)
3. Brown, J., Cai, Y.: Proceedings of the 2004 ACM SIGGRAPH International Conference on Virtual Reality Continuum and its Applications in Industry. Association for Computing Machinery, New York, USA (2004)
4. Cai, Y.: 3D Immersive and Interactive Learning. Springer (2013)
5. Cai, Y., Goei, S.L., Trooster, W.: Simulations and Serious Games for Education. Springer (2017)
6. Cai, Y., van Joolingen, W., Walker, Z.: VR, Simulations and Serious Games for Education. Springer (2019)
7. Cai, Y., van Joolingen, W., Veermans, K.: Virtual & Augmented Reality, Simulations and Serious Games for Education. Springer (2021)
8. Liu, D., Bhagat, K.K., Gao, Y., Chang, T.-W., Huang, R.: The potentials and trends of virtual reality in education. In: Virtual, Augmented, and Mixed Realities in Education, Springer, pp. 105–130 (2017)
9. Freina, L., Ott, M.: A literature review on immersive virtual reality in education: state of the art and perspectives. Int. Sci. Conf. eLearn. Softw. Educ. **1**, 133–141 (2015)

10. Nooriafshar, M., Williams, R., Maraseni, T.N.: The use of virtual reality in education. In: Proceedings of the 7th American Society of Business and Behavioral Sciences International Conference (ASBBS 2004) (2004)
11. Martín-Gutiérrez, J., Mora, C.E., Añorbe-Díaz, B., González-Marrero, A.: Virtual technologies trends in education. Eurasia J. Mathem. Sci. Technol. Educ. **13**(2), 469–486 (2017)
12. Zhan, C., Yu, X., Wu, Y., Zheng, J.: The Application of "Virtual Reality+ Education" in College Practical Teaching (2019)
13. Linnenbrink-Garcia, L., et al.: Measuring situational interest in academic domains. Educ. Psychol. Measur. **70**(4), 647–671 (2010)
14. Ainley, M.: Connecting with learning: motivation, affect and cognition in interest processes. Educ. Psychol. Rev. **18**(4), 391–405 (2006)
15. Hidi, S.: Interest: a unique motivational variable. Educ. Res. Rev. **1**(2), 69–82 (2006)
16. Krapp, A.: Interest, motivation and learning: an educational-psychological perspective. Eur. J. Psychol. Educ. **14**(1), 23–40 (1999)
17. Schraw, G., Flowerday, T., Lehman, S.: Increasing situational interest in the classroom. Educ. Psychol. Rev. **13**(3), 211–224 (2001)
18. Bergin, D.A.: Influences on classroom interest. Educ. Psychol. **34**(2), 87–98 (1999)
19. Dewey, J.: Interest and Effort in Education. Houghton Mifflin (1913)
20. Mitchell, M.: Situational interest: its multifaceted structure in the secondary school mathematics classroom. J. Educ. Psychol. **85**(3), 424 (1993)
21. Hidi, S.: Interest, reading, and learning: theoretical and practical considerations. Educ. Psychol. Rev. **13**(3), 191–209 (2001)
22. Harackiewicz, J.M., Barron, K.E., Tauer, J.M., Carter, S.M., Elliot, A.J.: Short-term and long-term consequences of achievement goals: predicting interest and performance over time. J. Educ. Psychol. **92**(2), 316 (2000)
23. Hidi, S., Renninger, K.A.: The four-phase model of interest development. Educ. Psychol. **41**(2), 111–127 (2006)
24. Deci, E.L., Koestner, R., Ryan, R.M.: The undermining effect is a reality after all—extrinsic rewards, task interest, and self-determination: reply to eisenberger, pierce, and cameron and Lepper, Henderlong, and Gingras (1999)
25. Schiefele, U., Rheinberg, F.: Motivation and Knowledge Acquisition: Searching for Mediating Processes (1997)
26. Rathunde, K., Csikszentmihalyi, M.: Undivided interest and the growth of talent: a longitudinal study of adolescents. J. Youth Adolesc. **22**(4), 385–405 (1993)
27. Renninger, K.A.: Individual interest and its implications for understanding intrinsic motivation. In: Intrinsic and Extrinsic Motivation. Elsevier, pp. 373–404 (2000)
28. Krapp, A.: Structural and dynamic aspects of interest development: theoretical considerations from an ontogenetic perspective. Learn. Instr. **12**(4), 383–409 (2002)
29. Bandura, A.: Self-efficacy. Corsini Encyclopedia Psychology 1–3 (2010)
30. Hsieh, P.-H., Sullivan, J.R., Sass, D.A., Guerra, N.S.: Undergraduate engineering students' beliefs, coping strategies, and academic performance: an evaluation of theoretical models. J. Exp. Educ. **80**(2), 196–218 (2012)
31. Concannon, J.P., Barrow, L.H.: Men's and women's intentions to persist in undergraduate engineering degree programs. J. Sci. Educ. Technol. **19**(2), 133–145 (2010)
32. Vogt, C.M., Hocevar, D., Hagedorn, L.S.: A social cognitive construct validation: determining women's and men's success in engineering programs. J. Higher Educ. **78**(3), 337–364 (2007)
33. Lent, R.W., Sheu, H.-B., Singley, D., Schmidt, J.A., Schmidt, L.C., Gloster, C.S.: Longitudinal relations of self-efficacy to outcome expectations, interests, and major choice goals in engineering students. J. Vocat. Behav. **73**(2), 328–335 (2008)
34. Zeldin, A.L., Britner, S.L., Pajares, F.: A comparative study of the self-efficacy beliefs of successful men and women in mathematics, science, and technology careers. J. Res. Sci. Teach. Official J. Nat. Assoc. Res. Sci. Teach. **45**(9), 1036–1058 (2008)
35. Gibson, J.J.: The theory of affordances. Hilldale, USA **1**(2), 67–82 (1977)

36. Parchoma, G.: The contested ontology of affordances: implications for researching technological affordances for collaborative knowledge production. Comput. Hum. Behav. **37**, 360–368 (2014)
37. Leonardi, P.M.: When flexible routines meet flexible technologies: affordance, constraint, and the imbrication of human and material agencies. MIS Quarterly, 147–167 (2011)
38. Mettler, T., Wulf, J.: Physiolytics at the workplace: affordances and constraints of wearables use from an employee's perspective. Inf. Syst. J. **29**(1), 245–273 (2019)
39. Rotgans, J.I., Schmidt, H.G.: Situational interest and academic achievement in the active-learning classroom. Learn. Instr. **21**(1), 58–67 (2011)
40. Mamaril, N.A., Usher, E.L., Li, C.R., Economy, D.R., Kennedy, M.S.: Measuring undergraduate students' engineering self-efficacy: a validation study. J. Eng. Educ. **105**(2), 366–395 (2016)

Virtual Reality Enabled Robotic Surgery Training Through Telesurgery

Patel Akshet, Yiyu Cai, and Randhawa Princy

Abstract Telesurgery is a groundbreaking approach that brings together physicians and patients through the integration of wireless networking and robotics. By utilizing these technologies, telesurgery enables surgical procedures to be performed on remote patients, eliminating geographical constraints that hinder prompt and high-quality surgical interventions. This chapter explores the transformative potential of telesurgery, focusing on the advantages it offers in terms of accessibility, cost-effectiveness, and enhanced patient safety. This study's main objective is to provide evidence of the feasibility and effectiveness of telesurgery. To this end, a virtual environment is created utilizing the Niryo One Robotic Arm and the Oculus Quest Head Mounted Display (HMD) device. The Oculus Quest Controllers serve as the interface for operating the robotic arm within a Virtual Reality setting. The arm's movements are mirrored in real-time within a Gazebo Simulation, effectively representing a physical robotic arm. Furthermore, this chapter delves into the technical intricacies of establishing seamless communication between the HMD and the Robot Operating System (ROS) node. The crucial role of Wi-Fi connectivity in enabling reliable and efficient data transmission is emphasized. While this study focuses on the technical implementation of telesurgery using specific devices, it is essential to acknowledge the broader implications, challenges, and trends within this field. Issues such as connectivity stability, legal and regulatory considerations, patient privacy, and data security are identified as key challenges to be addressed for the widespread adoption of telesurgery. Additionally, emerging trends in telesurgery, including advancements in artificial intelligence, haptic feedback integration, remote collaboration and training, and augmented reality, are discussed as promising avenues for future development.

P. Akshet · R. Princy
Manipal University Jaipur, Jaipur, India

Y. Cai (✉)
Nanyang Technological University, Singapore, Singapore
e-mail: MYYCai@ntu.edu.sg

Y. Cai et al. (eds.), *Virtual and Augmented Reality Technology-Enhanced Learning*, Gaming Media and Social Effects, https://doi.org/10.1007/978-981-96-2332-7_6

1 Introduction

Telesurgery refers to surgical treatments that are performed remotely owing to advancements in robotic and computer technology and their applications in surgery [1]. The first trans-Atlantic telesurgery operation was reported in 2001 when surgeons in New York utilized remote-controlled robots [2] to take out a 68-year-old woman's gall bladder through laparoscopy in Strasbourg, France. Remote telesurgery overcomes all geographic barriers to specialist surgery. A person in need of surgery at the South Pole may get treatment through remote telesurgery (polar telesurgery). In a space station, an astronaut [3] may be operated on ("astrosurgery"). People in impoverished countries will also be able to benefit from specialist surgery without having to leave their homes. It has set a new standard in surgical technology and intervention, opening the possibility of operating on a patient at a distant location with more precision and accuracy. It comprises one or more arms operated by a surgeon, as well as a master controller in a distant location who has access to all information sent through telecommunications or over wireless networks. There are three main types of TeleSurgery Robots as follows [4]:

1. Endoscopic TeleSurgery Robots: Endoscopic telesurgery robots were created to operate faster than others. The first remote interventional robot-assisted surgery was a laparoscopic cholecystectomy using Zeus (Computer Motion, USA). Intuitive Surgical Company of the US created the Da Vinci surgical robot in 1999. It may be used for general, thoracic, cardiac, colorectal, gynecological, and urological operations. Cardiovascular Interventional surgery [27] or intracardiac interventional surgery [30] are also being intensively studied.
2. Neurological TeleSurgery Robots: The Socrates Robot Remote Cooperative System21 was the first neurosurgery telesurgery robot approved by the USFDA in the world. Its main job is to apply robot remote guidance, which aims to improve surgical care and training. Neuro-Arm, another neurosurgical telesurgery robotic device, can fuse patient MRI images with 3D force sensors and high-definition stereomicroscopy, but it is only used for short-distance surgery. Delays in long-distance surgery will remain a worry in the future. Beijing University of Aeronautics and Astronautics accomplished the world's first telerobotic-directed neurosurgery in 2003.
3. Orthopedic Telesurgery Robot: Robodoc from Curexo in the US, R10 from Mako-Surgical in the US, and Renaissance from Mazor in Israel are well-known orthopedic robotic systems, although there are limited reports on their early use in telerobot surgery. In 2006, Wang from China reported that the Beijing University of Aeronautics and Astronautics' orthopedic telesurgery robotic system was utilized to perform remote tibial fracture closure reduction and internal fixation with locking intramedullary nail surgeries.

In this research project, the Oculus Quest HMD was used to experience and actuate the robotic arm in VR and the same robotic arm can be visualized actuating in Gazebo, which is a simulator for ROS to verify the proof of concept.

1.1 Issues and Problems

- Connectivity and Reliability: Telesurgery heavily relies on wireless networking for communication between the surgeon and the remote robotic system. The challenge lies in ensuring stable and low-latency connections, as any interruptions or delays can have serious implications on patient safety and surgical outcomes.
- Regulatory and Legal Considerations: Telesurgery raises complex legal and regulatory issues, including licensure, liability, and jurisdictional challenges. Clear guidelines and standards need to be established to ensure compliance, patient safety, and accountability across different regions and healthcare systems.
- Patient Privacy and Data Security: Transmitting patient data and live surgical footage over networks introduces potential risks to patient privacy and data security. Robust encryption and privacy measures must be implemented to safeguard patient information from unauthorized access and potential breaches.
- Technical Challenges: Telesurgery requires sophisticated robotic systems and virtual reality interfaces that are accurate, reliable, and user-friendly. Overcoming technical limitations, such as latency, haptic feedback, and system integration complexities, is crucial to ensuring a seamless and efficient surgical experience for both the surgeon and the patient.

1.2 Motivation

Telesurgery allows patients who are unable to travel vast distances to get safe and precise surgical operations. This kind of surgery is becoming increasingly viable as robots and wireless communication technologies progress [5]. The advantages of Telesurgery are [6]:

- It can perform high-quality surgery in places that would otherwise be unable to afford it, including remote rural regions, war zones, and even outer space.
- Reduces or eliminates the necessity for long-distance travel, along with the financial load and potential risks associated with such travel.
- With today's 3D display technology, surgeons working in separate hospitals may share the same high-definition visual input.
- It enables surgeons from various medical institutes to collaborate in real-time.
- Accelerometer technology cancels out the operator's natural tremor in real-time, enhancing surgical precision and decreasing tissue damage in the vicinity.
- As a result, the patient's recovery time has sped up.

1.3 Objectives

The five objectives of this research project are as follows:

- To develop a VR application for the Oculus Quest by importing the Niryo One Robotic Arm into a VR environment.
- To map the inputs with the Quest Controllers to actuate the arm in VR.
- To run a Gazebo Simulation of the Niryo One Robotic Arm which represents the physical arm.
- To Connect the HMD with the ROS Server over TCP Protocol over the same Wi-Fi network.
- To Create a ROS Package for simulating the simulated arm in Gazebo using the messages sent by the HMD.

2 Literature Review

The authors of the paper [7] demonstrated the viability of doing surgery across long distances utilizing specialized asynchronous transfer mode (ATM) communications technology. The authors attempted remote robot-assisted laparoscopic cholecystectomy on a 68-year-old lady with a history of stomach discomfort and cholelithiasis to test the feasibility and safety in people. The surgeons were in New York, while the patient was in Strasbourg. A high-speed terrestrial network was used to connect the locations via ATM service. The procedure was completed satisfactorily in 54 min with no difficulties. Despite a round-trip distance of 14,000 km (about 8699.2 mi), transmission latency was 155 ms. Surgeons deemed the procedure safe and the system reliable. The patient resumed normal activities two weeks after surgery.

With the introduction of robotic surgery and the emergence of a commercial surgical system, the da Vinci™ robot by Intuitive Surgical®, the need for a reliable VR-assisted simulator became paramount. Such a simulator could potentially reduce the steep learning curve associated with these intricate procedures and lead to improved outcomes. At the time of this review, however, a valid VR-assisted robotic surgery simulator had not yet been developed. Nonetheless, many agencies and corporations were actively engaged in pursuing this transformative technology. The article explored the history and progress of VR simulation in surgical training, its promising applications in the realm of robotic-assisted surgery, and the remaining challenges that needed to be addressed for successful implementation.

The annual number of robotic surgical procedures is on the rise worldwide, necessitating specialized skills. To acquire these skills, virtual reality (VR) simulation has been increasingly utilized in the da Vinci Surgical System. This review examined the literature from 2007 to 2015, focusing on commercially available VR platforms. The study found that VR simulation offers several advantages for acquiring robotic surgical skills, with proficiency-based training being the most effective. VR training is comparable to dry laboratory simulation, and its future applications include credentialing assessment, advanced procedural-based training, and preoperative warm-up.

The article [8] examined the technologies and discoveries that led to the transformation of a commercial robotic surgical system, Computer Motion, Inc.'s Zeus, to completely and securely allowed minimally invasive human telesurgery done across extremely long distances. Because human lives were at risk, concerns of safety, error detection, and fail-safe operation were critical. As a result, it was critical that all the commercial product Zeus's safety features remained intact throughout this change. This article discusses the commercial robot's safety features and the real-time communications technology built for Operation Lindbergh, the first transatlantic human telesurgery. The authors of the article [9] proposed The Green Telepresence Surgery System that was made up of two parts: a surgical workstation and a remote worksite. The remote location has a 3-D camera system and sensory-input manipulators. Workstations have 3-D displays and force-feedback handles. Future medical instructors may 'fly' through and around organs using sophisticated computers and 3-D graphics. The surgical simulator is a stylized human abdomen containing critical organs. This apparatus simulates scalpels and clamps so students and clinicians can practice surgery. The authors of the paper [10] included a comprehensive overview of the field's key experiments, as well as a viable proposal for extending telemedicine beyond Earth orbit. The key obstacle to weightless surgery is also mentioned. Teleoperated surgical robots might transform health care in space and on Earth. Despite its advantages, teleoperation over long distances is problematic owing to latency. Advanced virtualization and augmented-reality technologies should help humans adapt. A three-layered architecture is suggested for exploratory missions to meet space safety rules and give the finest telepresence theoretically achievable. The suggested three-layered system uses only current technology to enable telesurgery and space travel. The authors of the paper [11] discussed the Telerobotic laparoscopic radical prostatectomy and analyzed and reviewed the da Vinci system. The authors of the paper [12] described a surgical telerobot that allows surgeons to do less invasive surgical operations with more dexterity. The reasoning for a complete seven-degree-of freedom master–slave system is discussed, as well as the underlying computational architecture and current clinical applications. The article [13] aimed to provide a better knowledge of the features of robotic tele-surgical systems, as well as the limiting variables, potential telesurgery services, and multi-modal sensory data transmission quality of service (QoS) needs. On this foundation, a network architecture based on the integrated edge and core cloud is described, as well as the associated research difficulties, outstanding topics, and enabling technologies in the 5G communication system. The authors of the paper [14] proposed AaYusH (Ethereum smart contract (ESC) and IPFS-based TS system) in this study. The ESC protocol can tackle the security and privacy difficulties in AaYusH, while the InterPlanetary File System (IPFS) protocol can alleviate the storage cost issues. The also demonstrated a real-time SC developed in Solidity and implemented in the Truffle suite. They used the open-source program MyThril to test AaYusH's security flaws and found none. They compadre the performance of AaYusH to that of a standard telesurgery system in terms of latency and data storage cost and found that it

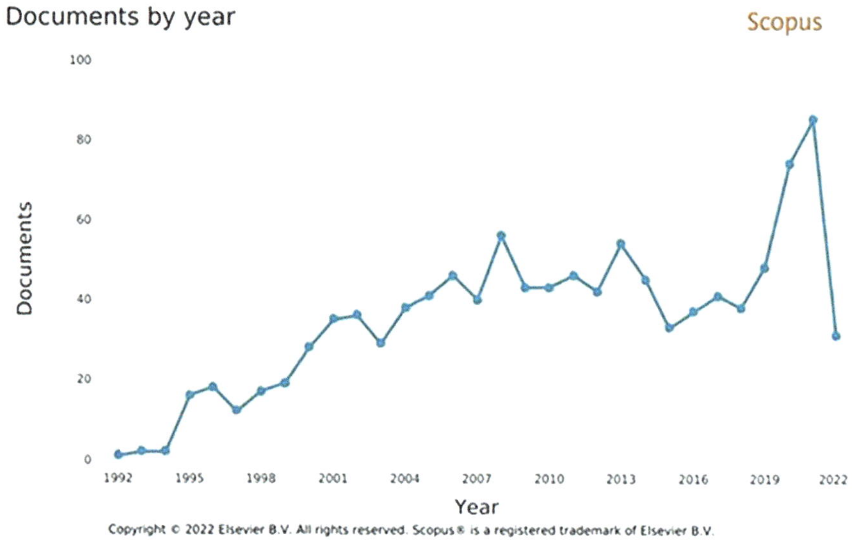

Fig. 1 Number of documents on Telesurgery since 1990

outperformed it. The authors of the paper [15] concluded that operators may undertake surgical activities with considerable transmission delays. The surgical platform investigated had shown itself to be successful for short latencies (300 ms or less). Slowing down motions for delays more than 500 ms reduced mistakes and simplified job planning. After performing the analysis on Scopus, it was observed that the first publication on TeleSurgery was published in 1992. Over the recent years, it is evident that TeleSurgery is being researched much more compared to the initial years. This is represented in a graphical format in Fig. 1.

It was observed that the researchers from the USA published and researched the most with more than 370 documents from the year 1990 with the keyword 'Telesurgery'. Followed by France and Germany with less than one hundred documents and research articles, respectively. This is represented in Fig. 2.

Vos Viewer was used to analyse and represent the keywords in pictorial form from all the research articles and publications since the year 1990. It has been observed that since 2014, the word TeleSurgery is being used more compared to the last decade and its pictorial representation is shown in Fig. 3.

The literature survey sheds light on the recent developments in the field of VR and TeleSurgery but also it highlights the drawbacks of this technology. The practice of tele-surgery, which is also known as tele-robotics, combines the benefits of robotic surgery, which include a magnified view, augmented reality, virtual reality, and improved ergonomics and dexterity, with the delivery of surgical care in remote areas and locations that are difficult to access, such as ships and spacecraft. There are still questions to be answered about price, availability, and the laws necessary to

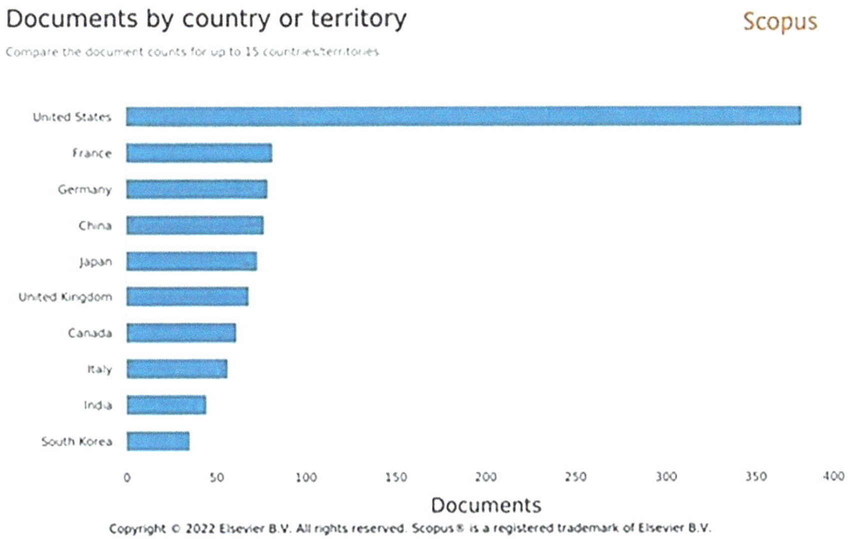

Fig. 2 Documents and articles by country or territory since 1990

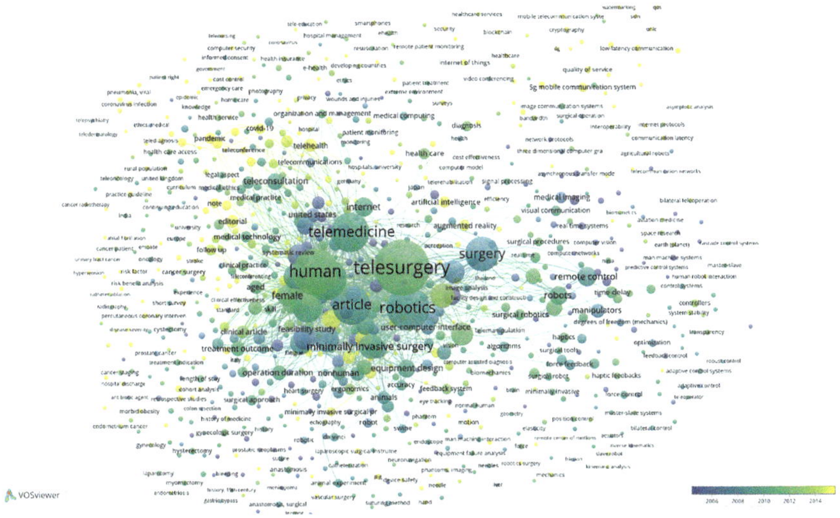

Fig. 3 Keywords used in the research publications since 1990

resolve legal and ethical concerns. The latency needs to be improved for real-time operation and its implementation over 5G and 6G network technologies.

3 Methodology

The research project was built using the following tools, open-source software packages, languages, and software applications:

(1) Unity 3D Game Engine—Software Application the Unity game engine was developed by Unity Technologies and may be used with games from a wide variety of different platforms. The Unity editor and games may be scripted in C# using the Mono framework, and the platform supports drag-and-drop functionality and 2D and 3D game development. Both Boo, which was phased out with Unity 5, and UnityScript, a Boo-based version of JavaScript, were deprecated in favour of C# in August 2017 [16].

(2) Robot Operating System (ROS)—Meta Operating System (OS) There are several precursors to ROS and many people have contributed to it. In 2007, a neighbouring robotics incubator named Willow Garage, Inc. provided substantial funding to further develop these concepts. Countless researchers contributed to ROS and its core software packages. The software, which was developed under the liberal BSD open-source licence, has found widespread use in robotics studies. The ROS ecosystem contains tens of thousands of users working on hobby projects and industrial automation systems [17].

(3) Oculus (Meta) Quest HMD—Hardware Tool for experiencing VR During October's Oculus Connect conference, Facebook debuted its newest VR platform, Quest (code named Santa Cruz). Quest is a standalone solution for full 6DoF virtual reality experiences that will be available in the spring of 2019 for $399. (Room-scale VR with 6DoF controllers). Oculus Go, the company's 3DoF standalone device, starts at US$199, while the high-end tethered Rift, which requires a PC, starts at USD 399. The CPU is based on Qualcomm's Snapdragon 835 platform (which is better than Go's Snapdragon 821 platform but worse than the current Snapdragon 845 platform). Because of differences in hardware, not all experiences will be transferred over from other platforms (tethered/PC) [18].

(4) Niryo One Robotic Arm (Unified Robotics Description Format- URDF File)—Open-Source Package Niryo One is a collaborative, open-source robot with six axes designed for use in French universities and research labs dedicated to vocational education and advancement. Its use is especially well suited to robotics and programming research in the context of Industry 4.0 [19]. The Niryo One Robotic Arm is shown in Fig. 4 [20]. The URDF file for this robotic arm is open source and available freely on the internet. The same can be imported to Gazebo and Unity 3D Game engine for further development and visualization.

(5) Unity Robotics Hub Package—Open-Source Package This is a common repository for robotic simulation tools, tutorials, resources, and documentation in Unity. Robots are anticipated to do more complicated tasks with total autonomy as Artificial Intelligence (AI) and Machine Learning (ML) advance. Robots are wanted to detect the environment, adapt to it, and execute actions inside it.

Fig. 4 Niryo one robotic arm

In industrial and commercial environments, robots are used for sophisticated assembly picking and packaging, last-mile delivery, and other tasks. Roboticists and developers require an experimental platform that precisely simulates the real-world environment as well as the robot's physical interactions with that environment to evaluate solutions before implementing them in a robot. They may assess robot performance in terms of localization, motion planning, and control using simulation [21].

(6) ROS Unity Integration—Open-Source Package A TCP endpoint that serves as a ROS node and enables message transfer between ROS and Unity. The messages being sent between ROS and Unity are anticipated to be serialized as ROS would serialize them on its own. To do this, ROS.msg files are used to build C# classes with serialization and deserialization methods using the MessageGeneration plugin (from the ROS TCP Connector source). The Unity scripts required to publish, subscribe to, or call a service are provided by the ROSConnection plugin (available from ROS TCP Connector) [22]. The flow diagram of the communication is shown in Fig. 5.

(7) The ROS Server, Subscriber and Publisher scripts communicate with the server endpoint over the TCP protocol over Wi-Fi network. The ROS nodes on the ROS network communicate with each other and subscribe and publish messages to ROS Server Endpoint.

(8) MoveIt—Open-Source Package Using MoveIt, an open-source robotic manipulation platform, applications can be created. Over 150 robots have been equipped with MoveIt, the most popular manipulation program. It is free for industrial, commercial, and academic usage since it is distributed under the provisions of the BSD license [23, 24]. MoveIt is an innovative software for mobile manipulation because it incorporates the most recent developments in

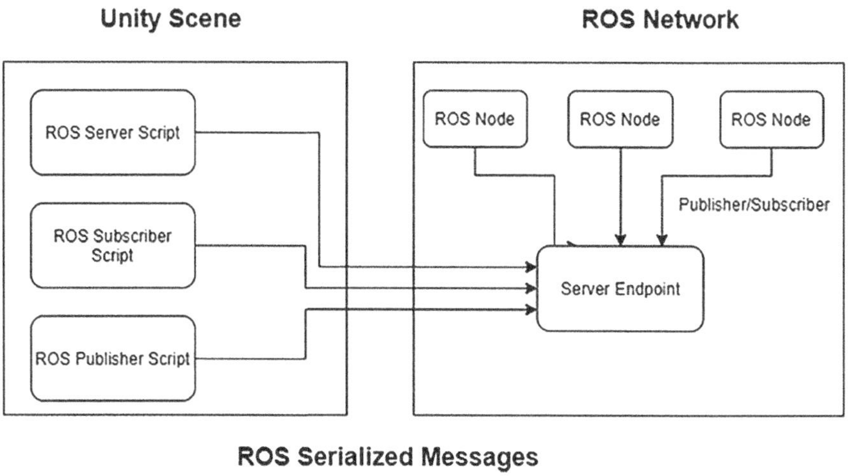

Fig. 5 Unity ROS communication flow diagram

motion planning, manipulation, 3D perception, kinematics, control, and navigation. Using mesh elements that have already been created in any CAD application, MoveIt enables the construction of the planning scenario environment and enables robot interaction. MoveIt comes with several tools that make it possible to execute motion planning to the desired posture with a great deal of freedom while optimizing the motion planning algorithm for the application. The flow diagram of MoveIt is shown in Fig. 6.

MoveIt interacts with motion planners using a plugin interface. This enables MoveIt to connect with and utilize various motion planners from several libraries, making MoveIt readily extendable. The interface to the motion planners is via a ROS Action or service (provided by the move group node).

(9) Gazebo [25]—Simulator for ROS (Open-Source) While ROS acts as the robot's interface, Gazebo is a 3D simulator. A potent robot simulator is produced by combining the two. A 3D scene with robots, obstacles, and many other items may be built on a computer using Gazebo. Gazebo also uses a physical engine for things like lighting, gravity, inertia, etc. The robot may be evaluated and tested in difficult or dangerous settings without jeopardising its safety. When compared to starting the whole scenario on the real robot, using a simulator often takes less time. Gazebo was first intended to assess robot algorithm evaluations. Testing the robot application is crucial for various applications, including those involving error handling, battery life, localization, navigation, and grasping. Gazebo was created and enhanced because of the need for a multi-robot simulator. The flow diagram [26] of the working of Gazebo as shown in Fig. 7. A launch file is used to create a Gazebo, and then a node associated with Gazebo is launched. Because Gazebo does not give much in the way of built-in support for closed-loop chains, we had to construct a control plugin to provide an interface for users to interact with the simulation.

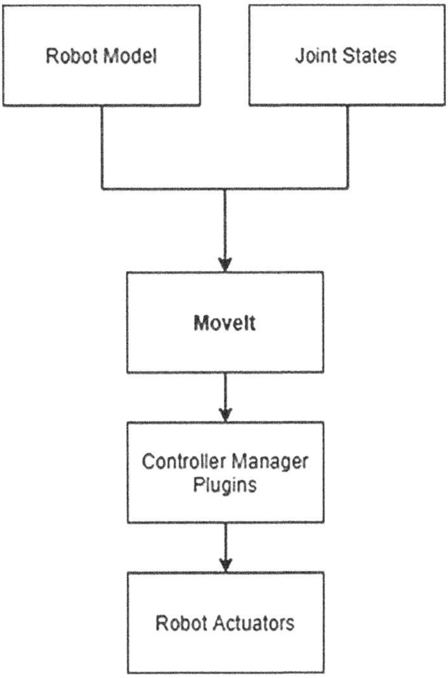

Fig. 6 Flow diagram of MoveIt

This was accomplished by making relevant rostopics. For example, the control plugin is responsible for reading the joint states of the simulation and for receiving topics that broadcast desired joint instructions to Gazebo. Additional parameters necessary for the plugin that are user dependant are uploaded and obtained from the ROS parameter server. Some examples of these additional parameters are PID gains and starting joint angles. Gazebo is used to simulate real-life scenarios for a ROS-based application.

(10) URDF Importer—Open-Source Package A robot described in URDF format may be imported into a Unity scene using the URDF Importer. The geometry, visual meshes, kinematics, and dynamic properties of a robot are defined by URDF. Using PhyX 4.0 articulation bodies, the importer parses a URDF file and imports it into Unity [28, 29].

The entire methodology followed for this research project is summarised in five steps and procedures. The flow chart represented in Fig. 8 with the following steps.

- Importing the Niryo One Robotic Arm in Unity3D
- Mapping the Inputs with the Quest Controllers
- Run Gazebo Simulation of Niryo One Robotic Arm
- Connect the HMD with the ROS Server over TCP Protocol
- Creating a ROS Package for Simulating the Arm in Gazebo.

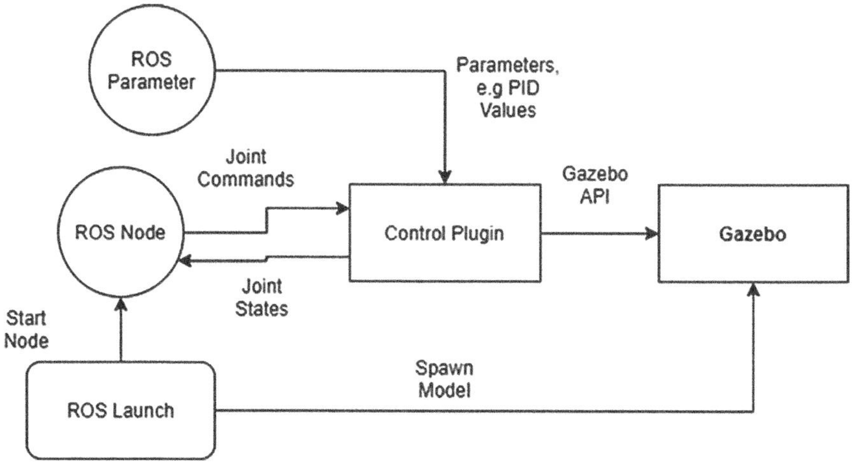

Fig. 7 Flow diagram of Gazebo

The Unity3D Game Engine is used to develop the VR application and the Oculus Quest as the HMD device. The Niryo One robotic arm is imported into Unity3D, and the controllers were mapped for the inputs of the arm to control it in VR. The Gazebo simulation was run to represent the physical robotic arm. The HMD was connected to the ROS server over Wi-Fi and a ROS package was created to simulate the arm in Gazebo based on the inputs received from the HDM via ROS Topic/niryo joints.

3.1 Importing the Niryo One Robotic Arm in Unity3D

Niryo One is a learning robot arm. With its three Dynamixel XL servomotors and NiryoSteppers, it improves robotics learning in an industrial context, its six axes allow it to mimic all advanced manoeuvres. It is an open-source robotic arm and the open-source package URDF Importer was used to import the arm in Unity3D. A robot described in URDF format may be imported into a Unity scene using the URDF Importer. A Robot's geometry, graphic meshes, kinematic, and dynamic properties are defined by URDF. Importer parses a URDF file and uses PhyX 4.0 articulation bodies to import it into Unity which is shown in Fig. 9.

3.2 Mapping the Inputs with the Quest Controllers

The niryo one controller.cs was edited to map the inputs from the keyboard to the Quest Controllers. Initially, the Niryo One arm was actuated using the up and down arrow key of the keyboard by selecting the joints using the left and right arrow keys.

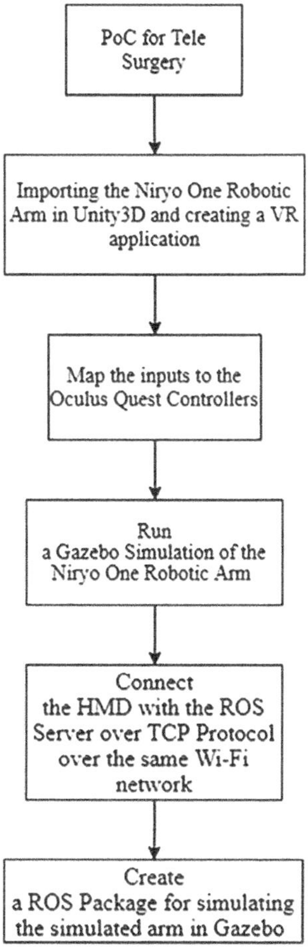

Fig. 8 Flowchart of the methodology

The same algorithm was mapped to the left and right trigger and grasp buttons of the controllers.

3.2.1 Run Gazebo Simulation of Niryo One Robotic Arm

The ROS Stack is built and tested on ROS Melodic, which needs Ubuntu 18.04 to function properly. A new workspace named 'catkin ws niryo ned' was created and the need repository was cloned into the src folder of this workspace. All the dependencies were updated, and the terminal was sourced. Using the following command, the gazebo simulation was run for the Niryo One Robotic Arm.

roslaunch nir yo_robot_gazebo gazebo.launch

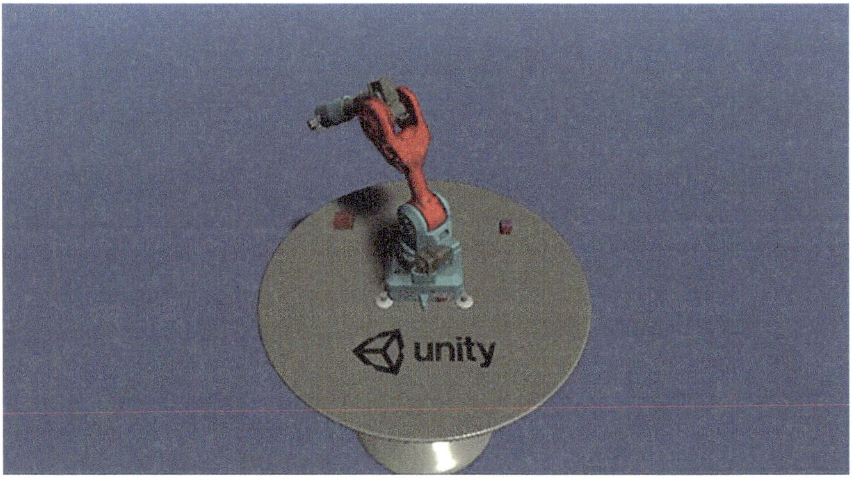

Fig. 9 Niryo one URDF model imported to unity3D

The niryo moveit should be initialized for the inverse kinematics calculations. This is done by using the following command:

roslaunch nir yo_moveit_demo.launch

The arm is simulated in Gazebo as shown in Fig. 10

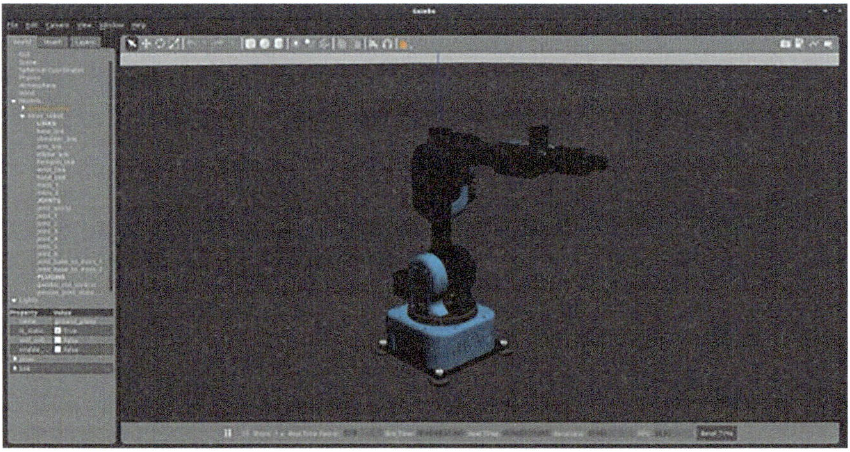

Fig. 10 Niryo one robotic arm simulated in Gazebo

Fig. 11 The joint angles messages received from the HMD

3.2.2 Connect the HMD with the ROS Server Over TCP Protocol

All message forwarding is handled by a TCP endpoint running as a ROS node, which allows Unity and ROS to communicate. You may publish, subscribe to, or contact a service using the TCP endpoint ROS node with the help of Unity's ROSConnection component. Unity and ROS should use the same serialization scheme for ROS communications as ROS uses internally. This is accomplished using the MessageGeneration plugin, which creates C# classes from ROS messages, including serialization and deserialization routines. A TCP endpoint that runs as a ROS node that allows messages to be sent between Unity and ROS. The messages transmitted between Unity and ROS are supposed to be serialized in the same way as ROS serializes them internally. The MessageGeneration plugin (from the ROS TCP Connector repo) can produce C# classes from ROS.msg files, including serialization and deserialization routines. The Unity scripts required to publish, subscribe, or call a service are provided by the ROSConnection plugin (available from ROS TCP Connector). This is shown in Fig. 11.

3.2.3 Creating a ROS Package for Simulating the Arm in Gazebo

A ROS Package was created for simulating the Arm in Gazebo by subscribing to the/ niryo joints topic and publishing the joint angles, velocity, acceleration, and effort on the topic/niryo one follow joint trajectory controller/command. The python script subscribes to the necessary topics and then publishes the values to the appropriate topics of the Gazebo simulation at a rate of 90 messages per second. The algorithm of the code is shown in Fig. 12. The necessary libraries and messages are imported, and the messages type are assigned to all the user-defined variables. The joint callback function is defined and the publish and subscribe functions were defined. To move the joints of the simulation running on gazebo, the joint's angles along with joint position, velocity, acceleration, and effort are published 90 times a second to the appropriate joints. This causes the simulation to synchronize with the VR actuation of the arm.

4 Implementation and Results

The application was built using the Unity 3D Game engine and the product was an.apk file. The file is installed on the Oculus Quest HMD. This section explains the walkthrough of the application and the results. The HMD and the laptop running the

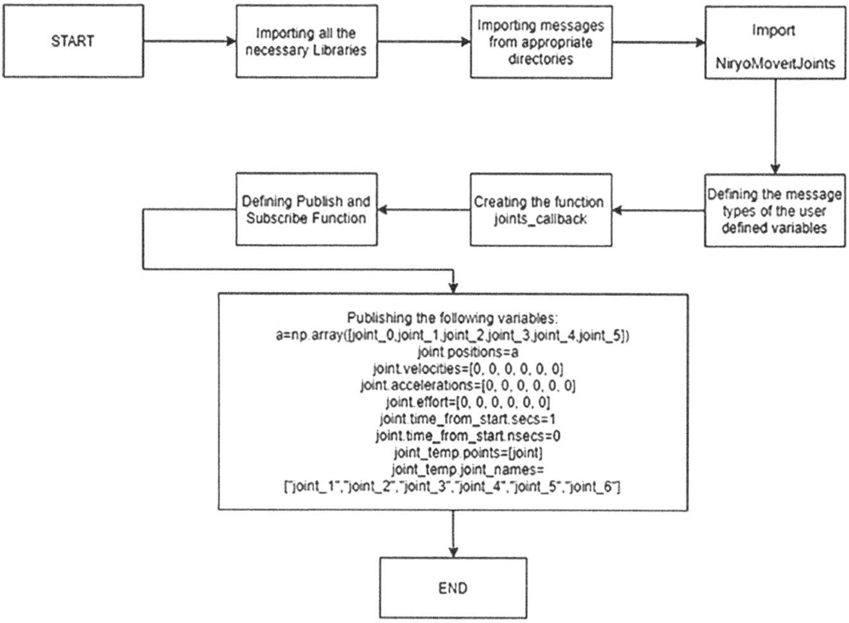

Fig. 12 Algorithm of the code flowchart

ROS Server and Gazebo were connected over the same Wi-Fi network over the TCP protocol.

4.1 *Walkthrough of the VR Application*

As the application is opened, a welcome screen appears with the message as shown in Fig. 13. The application then takes input from the user on whether the user is familiar with the HMD device controllers or not as shown in Fig. 14. If the user inputs that he/she is not familiar with the controllers, then the application throws out the following instructions for the left-hand controller as shown in Fig. 15. If the user inputs that he/she is not familiar with the controllers, then the application throws out the following instructions for the right-hand controller as shown in Fig. 16. If the user inputs that he/she is familiar with the controllers, then the application throws out the following instructions for the left-hand controller as shown in Fig. 17. If the user inputs that he/she is familiar with the controllers, then the application throws out the following instructions for the right-hand controller as shown in Fig. 18. Once the user is familiar with the user interface and controls, the user must input the IP Address of the ROS server to establish a wireless connection over the Wi-Fi network. This is shown in Fig. 19. After entering the IP Address, the robotic arm spawns and the controllers are used by the user to actuate the arm. This is shown in Fig. 20. The credit panel is shown in Fig. 21. The Niryo One Robotic Arm is imported into Gazebo for simulating the arm as a substitute to the hardware device and the arm is shown in Fig. 22. Once the HMD is paired with the laptop running the Gazebo simulation, the HMD sends the joint angles ninety times per second to the laptop running the ROS Server. The ROS package "unity ros" runs the script "joint controller.py." This script subscribes to the messages received from the HMD and publishes them to the joints of the Niryo One simulated arm running on Gazebo.

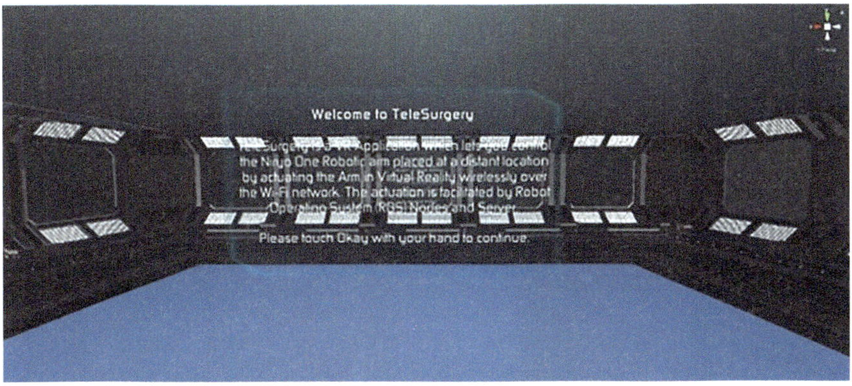

Fig. 13 TeleSurgery application welcome message

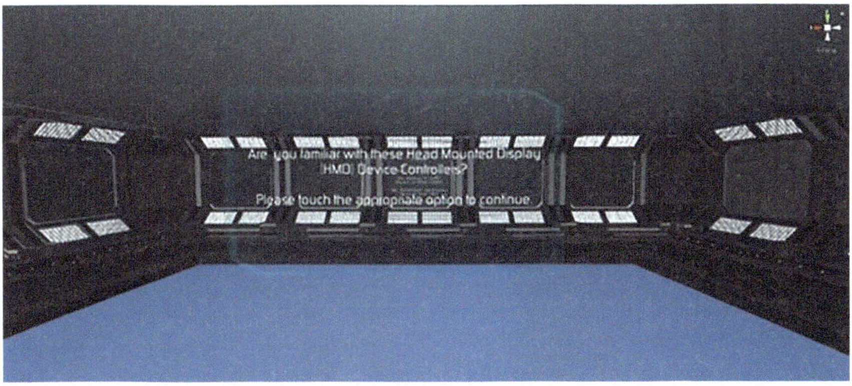

Fig. 14 Taking user inputs about the familiarity with the HMD device controllers

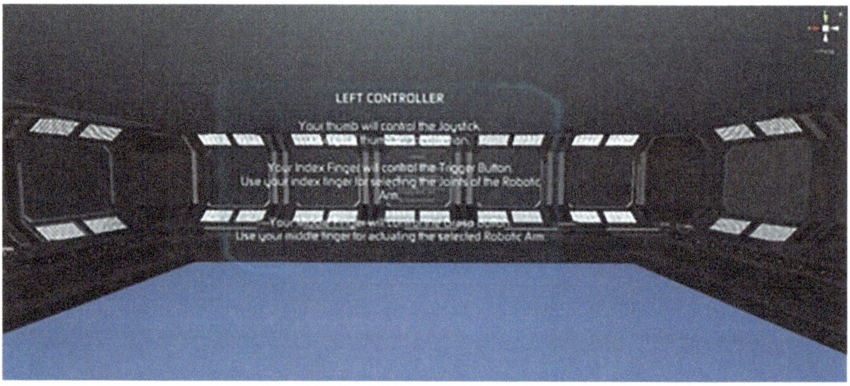

Fig. 15 Left hand controller instructions (for the choice No)

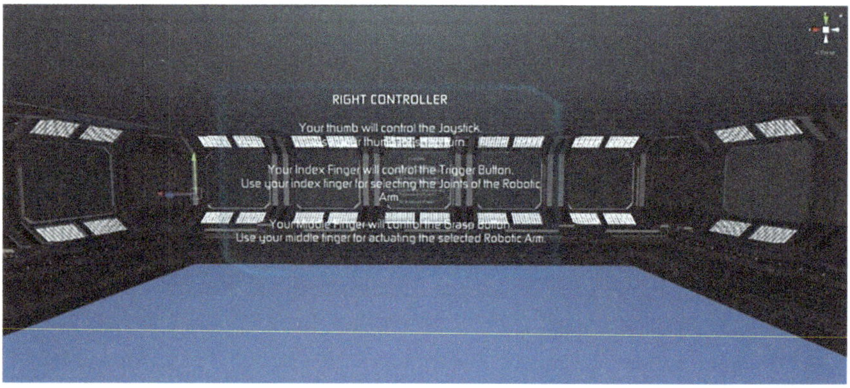

Fig. 16 Right hand controller instructions (for the choice No)

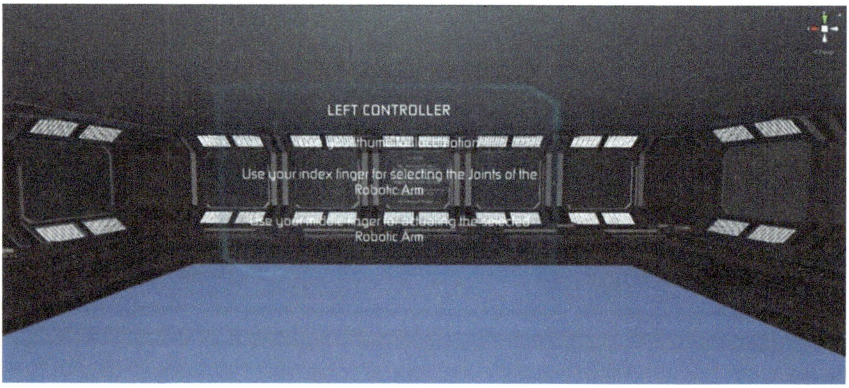

Fig. 17 Left hand controller instructions (for the choice Yes)

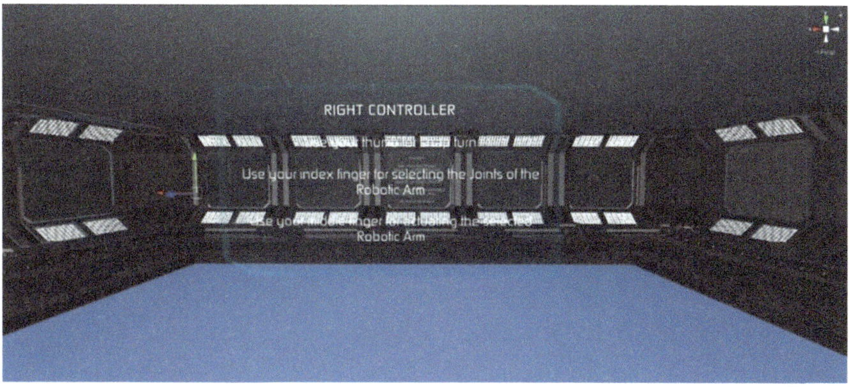

Fig. 18 Right hand controller instructions (for the choice Yes)

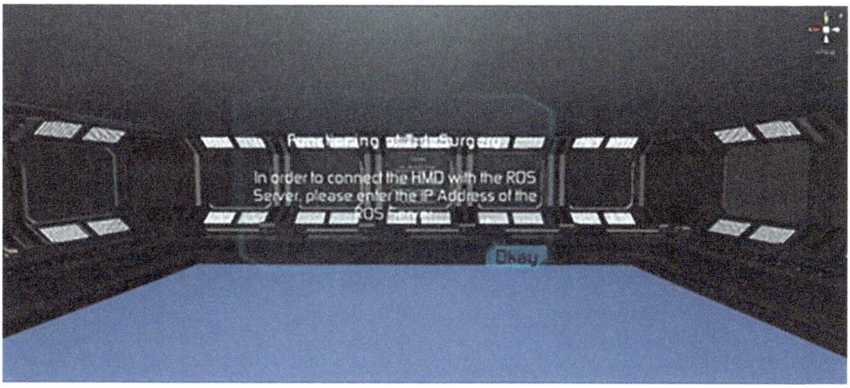

Fig. 19 Keypad instructions

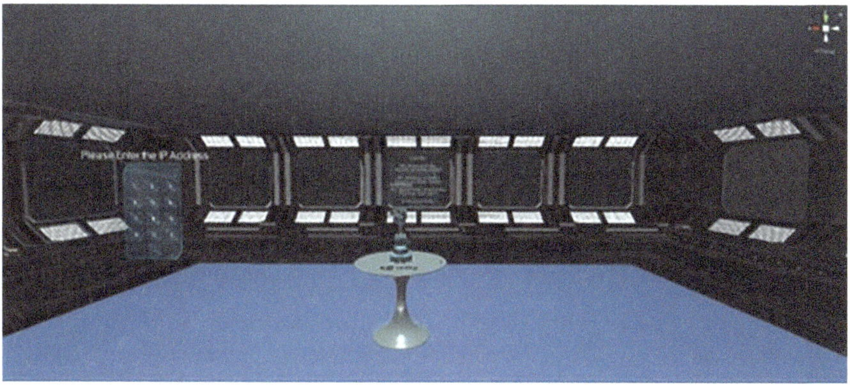

Fig. 20 Controlling the arm using the controllers

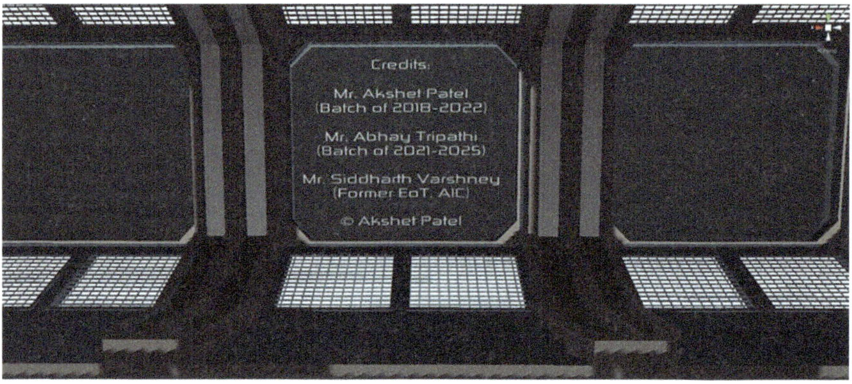

Fig. 21 Credit panel

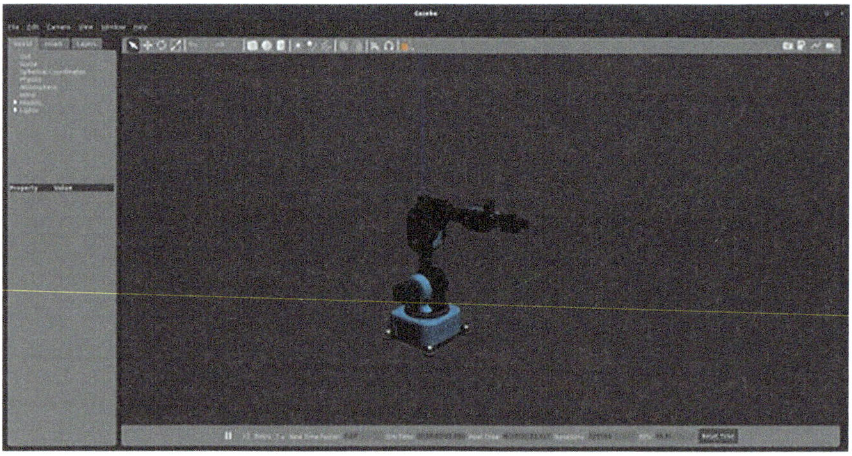

Fig. 22 Gazebo simulation of the Niryo one robotic arm

5 Conclusion and Future Scope

5.1 Contribution

This chapter presents the development and integration of a comprehensive system for visualizing and controlling the Niryo One Robotic Arm in virtual reality (VR) environment using the Oculus Quest controllers. The system allows for the manipulation of the arm within a VR simulation, closely mirroring its real-world movements. The process begins by importing the Niryo One Robotic Arm into the VR application developed specifically for the Oculus Quest. Mapping the inputs to the

Quest Controllers enables users to actuate the virtual arm within the VR environment. To create a realistic representation of the physical arm, a Gazebo Simulation is executed using the Unified Robot Description Format (URDF) file. This simulation accurately emulates the behavior and functionality of the Niryo One Robotic Arm. Establishing connectivity between the Oculus Quest Head Mounted Display (HMD) and the Robot Operating System (ROS) server is a crucial step in enabling real-time control of the simulated arm. The IP address of the ROS server is entered through a virtual keypad in the VR environment. The HMD and the laptop running the Gazebo simulation are connected over the same Wi-Fi network using the TCP protocol, facilitated by the ROS-TCP Connector. To facilitate communication between the HMD and the Gazebo simulation, an ROS package is developed. This package handles the transmission of messages from the HMD to control the simulated arm in real-time, ensuring a synchronized and interactive experience. In conclusion, this chapter highlights the successful development and integration of a system that combines the Niryo One Robotic Arm, VR technology, and the Gazebo simulation environment. The system allows for precise control and visualization of the arm in VR, offering potential applications in training, education, and remote robotic surgery.

5.2 Limitations

Like any other work in this domain, this research work has many limitations. The first and the most crucial drawback is the latency. For this architecture to work in real time, the latency should be as close to zero as possible. This will only be possible after 5G or 6G rolling out for the public or enterprises. Since the simulated robotic arm is running, the hardware device errors are neglected. For this technology to be used on a human, it must be approved by many international standards and clear many tests. The cost of implementing this technology is extremely high and it will take many years for it to thrive in the developing and underdeveloped nations. Bandwidth, which refers to the data transmitted every second, is also a major limitation as the current speeds are not enough to perform remote surgery overseas and on continents. Telesurgery has been used for astronauts on the moon and aquanauts at underwater research facilities. Although an agreement on acceptable latencies for operating has not yet been achieved, latency is now the main obstacle facing telesurgery. Despite this, hundreds of successful telesurgeries have already been carried out in various parts of the globe. Additionally, many clinics today cannot afford the prohibitively costly telerobotic surgical devices and networking infrastructure needs. Telesurgery is projected to grow more popular as technology advances, particularly the creation of faster computer networks with more bandwidth capacity, and prices decrease.

5.3 Future Scope

This is the Proof of Concept for telesurgery in a simulated environment. The same system can be scaled to operate on a human remotely by using high-precision robotic arms and 5G or 6G network communications. The future of telesurgery will be driven by technology, both in terms of network infrastructure and in terms of robotic-assisted surgery. The digital transition signifies a shift away from what has been the industry standard in robotic surgery and into a new age of Performance-Guided Surgery, which refers to next-generation technology that totally alters the concept of what is achievable in surgery. This VR app can be used when the world moves towards the metaverse, and it will facilitate the application of Digital Twins. It will provide a better viewing experience for remote operations.

Acknowledgements We are grateful to Manipal University Jaipur, India and Nanyang Technological University, Singapore for this opportunity to work on this project. We would also like to thank each and every person who was involved in the development of this project.

References

1. Choi, P.J., Oskouian, R.J., Tubbs, R.S.: Telesurgery: past, present, and future. **10**(5), 7–8 (2018). https://doi.org/10.7759/cureus.2716
2. Minkel, J.: First complete trans-Atlantic robotic surgery—Scientific American (2001). https://www.scientificamerican.com/article/first-complete-trans-atla/. Accessed 13 Jun 2022
3. Barrow, B.E.: ThinkIR: The University of Louisville 's institutional repository design and development of a multifunctional surgical device for ground and space-based surgical applications. By department of Bioengineering (2018)
4. Xia, S.B., Lu, Q.S.: Development status of telesurgery robotic system. Chinese J. Traumatol. English **24**(3), 144–147 (2021). https://doi.org/10.1016/j.cjtee.2021.03.001
5. Cahill, D.J.: Telesurgery: surgery in the digital age telesurgery: surgery in the digital age. Dartmouth Undergrad. J. Sci. **19**(3):12–16 (2017)
6. Robotic telesurgery—benefits, and challenges. Robotics Biz (2021). https://roboticsbiz.com/robotic-telesurgery-benefits-and-challenges/. Accessed 13 Jun 2022
7. Marescaux, J., Rubino, F.: Transcontinental robot-assisted remote telesurgery, feasibility and potential applications. Teleophthalmology **235**(4), 261–265 (2006). https://doi.org/10.1007/3-540-33714-831
8. Butner, S.E., Ghodoussi, M.: Transforming a surgical robot for human telesurgery. IEEE Trans. Robot. Autom. **19**(5):818–824 (2003). https://doi.org/10.1109/TRA.2003.817214
9. Satava, R.M.: Virtual reality, telesurgery, and the new world order of medicine. Comput. Aided Surg. **1**(1):12–16 (1995). 10.3109 /10929089509106821
10. Haidegger, T., Sa, ndor J., Benyo, Z.: Surgery in space: the future of robotic telesurgery. Surg. Endosc. **25**(3), 681–690 (2011). https://doi.org/10.1007/s00464010-1243-3
11. Rassweiler, J., Binder, J., Frede, T.: Robotic and telesurgery: will they change our future? Curr. Opin. Urol. 11(3):309–320 (2001). https://doi.org/10.1097/00042307200105000-00012
12. Guthart, G.S.: The IntuitiveTM telesurgery system: overview and application the intuitive TM telesurgery system: overview and application. Robot. Autom. 2000. Proceedings. ICRA '00. IEEE Int. Conf., vol. 1, no. February 2000, pp. 618–621 (2015)
13. Zhang, Q., Liu, J., Zhao, G.: Towards 5G enabled tactile robotic telesurgery. 1–7 (2001)

14. Gupta, R., Shukla, A., Tanwar, S.: AaYusH: a smart contract-based telesurgery system for healthcare 4.0. 2020 IEEE Int. Conf. Commun. Work. ICCWork. 2020 Proc. (2020). https://doi.org/10.1109/ICCWorkshops49005.2020.9145044
15. Rodr, E., Kypson, A.P., Moten, S.C., Nifong, L.W., Chitwood, W.R.: Robotic mitral surgery at East Carolina University. Int. J. 211–215 (2006). https://doi.org/10.1002/rcs
16. Unity (game engine)—Wikipedia. https://en.wikipedia.org/wiki/Unity_(game_engine). Accessed 21 Jun 2022
17. Morgan Quigley, W.D.S., Gerkey, B.: Programming Robots with ROS (2015)
18. Facebook Announces Stand-Alone 6DoF HMD, Oculus Quest. Accessed 22 Jun 2022
19. Niryo One. https://robots.ros.org/niryo-one/. Accessed 22 Jun 2022
20. Ned: robotics arm for education and research. https://niryo.com/product/nededucation-research-cobot/. Accessed 23 Jun 2022
21. Robotics Simulation—Unity. https://unity.com/solutions/automotive-transportationmanufacturing/robotics. Accessed 22 Jun 2022
22. ROS-Unity Communication. https://www.com/Unity-Technologies/UnityRobotics-Hub/blob/main/tutorials/ros/unity/integration/README.md. Accessed 23 Jun 2022
23. MoveIT For Robotic Manipulation Application—Robotnik®. https://robotnik.eu/moveit-manipulation-application/ Accessed 23 Jun 2022
24. MoveIt Motion Planning Framework. https://moveit.ros.org/. Accessed 23 Jun 2022
25. Robotic simulation scenarios with Gazebo and ROS. https://www.generationrobots.com/blog/en/robotic-simulation-scenarios-with-gazeboand-ros/. Accessed 23 Jun 2022
26. Gondokaryono, R.A., Agrawal, A., Munawar, A., Nycz, C.J., Fischer, G.S.: An approach to modeling closed-loop kinematic chain mechanisms, applied to simulations of the da Vinci surgical system. Acta Polytech. Hungarica **16**(8):29–48 (2019). https://doi.org/10.12700/APH.16.8.2019.8.3
27. Cai, Y., Chui, C.K., Ye, X., Fan, Z., Anderson, J.H.: Tactile VR for hand–eye coordination in simulated PTCA. Computers in Biology and Medicine 36(2): 167–180 (2006). https://doi.org/10.1016/j.compbiomed.2004.10.002
28. Unity-Technologies/URDF-Importer: URDF importer. https://www.com/UnityTechnologies/URDF-Importer. Accessed 23 Jun 2022
29. Learn How to Code in C# for Beginners—Unity Learn. https://unity.com/howto/learning-c-sharp-unity-beginners. Accessed 23 Jun 2022
30. Chiang, P., Zheng, J., Yu, Y., Mak, K.H., Chui, C.K., Cai, Y.: A VR Simulator for Intracardiac Intervention. IEEE Computer Graphics and Applications 33(1): 44–57 (2013). https://doi.org/10.1109/MCG.2012.47

Tree Pruning Simulation and Tree Growth Modelling

Zhen Xuan Tan, Chee Kiat Seow, Yiyu Cai, and Qi Cao

Abstract There currently exist many methods for 3D plant growth modelling. Both terrestrial laser scanning and UAV use sensors which are great ways to obtain the structural properties of plants. Structural properties such as volume and size of plants are crucial as they are valuable inputs that enable the research of plant growth, specifically on anomalies detection and the effects of before and after pruning. Computational modelling is widely used to simulate various phenomena for different applications. It is a fast and cost-effective way using computational modelling to better understand the cause and effect of pruning certain parts of the tree without having to go on site. Nevertheless, many computer pruning simulation programmes have a fixed plant growth model and do not have many options for user interactions and customization. Therefore, this book chapter proposes a plant pruning simulation utilizing L-Py as a solution that caters to various types of plants suiting users' needs. Different parameters can be set by the user to customize and conduct pruning on their own tree models. The proposed method can be employed for preliminary analysis and a learning platform for users to understand pruning management.

Keywords Tree pruning · 3D simulation · Growth modelling · L-Py

Z. X. Tan · C. K. Seow · Q. Cao (✉)
School of Computing Science, University of Glasgow, Glasgow, UK
e-mail: qi.cao@glasgow.ac.uk

C. K. Seow
e-mail: cheeKiat.Seow@glasgow.ac.uk

Y. Cai
School of Mechanical and Aerospace Engineering, Nanyang Technological University, Singapore, Singapore
e-mail: myycai@ntu.edu.sg

Y. Cai et al. (eds.), *Virtual and Augmented Reality Technology-Enhanced Learning*, Gaming Media and Social Effects, https://doi.org/10.1007/978-981-96-2332-7_7

1 Introduction

Over the years, while active efforts have been implemented by National Parks Board (NParks) of Singapore to strengthen the tree management regime, there are still hundreds of cases of tree failures which might lead to serious consequences each year. One specific example in the news was a fatal incident, where a woman was killed by a falling tree [1]. In 2020 alone, Singapore suffered from 339 cases of tree failure risks [2]. Tree failure risks referred to trees that cause harm to either people or properties could be reduced by proper tree pruning. Presence of tree pruning achieved proven success on safety and tree health. It is observed to enhance the aesthetics of trees [3].

Firstly, the act of tree pruning improves safety by removing branches that could fall and cause damage to their surroundings [4]. Secondly, tree pruning removes dead and dying tree branches, improving the overall tree health. Removal of dying branches will not only seal off the area, preventing insects and diseases from spreading [5], but also lessen competition for sunlight and enable other healthy tree branches to receive more [6]. By regularly pruning, trees are more likely to grow stronger and withstand potential bad weather conditions. Lastly, due to lesser competition for sunlight, branches are likely to produce high-quality flowers or fruit [7, 8], hence improving the aesthetics of the tree in line with the 'garden city' vision of the nation [9].

As such, it is particularly relevant to Singapore with the aim of achieving a greener and more sustainable city by 2030. There are currently about 7,000,000 trees in Singapore, and as part of the Green Plan, one of the goals is to plant 1,000,000 more trees between 2020 and 2030 [10]. To better maintain these trees, it would require many years of proper tree maintenance planning and fieldwork [11], alongside regular pruning which is an essential part of tree management regime [1].

Hence, it results in many studies and research that focus on developing different techniques and models specific to enhancing plant growth and maintaining plant health. The existing methods include terrestrial laser scanning [12–17], unmanned aerial vehicles (UAV) [12, 13, 18–21], etc. However, despite the precision of crop architecture measurement through terrestrial laser scanning and UAV, there exist shortcomings that are often restricted by cost and labour. Terrestrial laser scanning excels in speed and accuracy in data collection but requires the user to be physically on the field. UAV on the other hand does not require users to be on the field, but faces a flaw of limited viewing angle.

These disadvantages highlight the demand for an alternative method. It is necessary to leverage the latest technologies to increase the productivity of the tree pruning and monitoring. More recently, with the rapid advancement of artificial intelligence (AI) and automation, it is promising to explore machine algorithms approach [22] as a supportive tool to aid the monitoring of crop architecture.

In addition, the vision for Singapore to transform into a smart nation, as well as a 'garden city' which boasts abundant lush greenery, allows for data collection on trees planted locally. It will aid the exploration of machine algorithm approach of this study. NParks with support from the Government Technology Agency of Singapore have been collecting data on more than 500,000 trees [23]. With the newly added data collection, it paths a less labour-intensive, fast, and cost-effective opportunity to the method of computational modelling [24–28]. This method would allow for better understanding of the cause and effect of pruning specific parts of a tree.

While computational modelling has been studied and proven to be promising, a model that grants various customizations of trees has not been widely explored. Therefore, the tree pruning simulation and modelling are proposed in this paper. The proposed model focuses on utilizing computer simulations to allow customization of scenarios that aid users to determine the most effective way of tree pruning, for specific types of tree alongside simulating the effect of pruning on the tree growth. The organization of the remaining parts of the book chapter is as follows. Section 2 introduces the related works in the literature. Section 3 presents the proposed model for tree pruning and growth. It depicts the implementation of the tree pruning simulations. Section 4 discusses the experiment results. The chapter is concluded in Sect. 5.

2 Literature Review

2.1 Terrestrial Laser Scanning Approach

Light detection and ranging (LiDAR) active sensors have been adopted for scanning and estimating structural measurements, that are used to reproduce the 3D structure of an object [14]. LiDAR instruments can be operated on different platforms, such as terrestrial laser scanning as a land-based method, that uses the LiDAR technology to acquire three-dimensional axis data points: x, y, and z. It forms a dense point cloud by measuring the distance to the target through the laser scanning. Terrestrial laser scanning has been used in a variety of applications such as structural deformation in buildings and detecting change in plant growth [12].

The advantages of the terrestrial laser scanning method include high levels of accuracy with detailed data up to millimetre accuracy. It is not achievable by other methods, such as the technique of UAVs. It allows accurate acquisition of important plant parameters such as height. Due to its high level of accuracy and each terrestrial laser scanner having a short and limited working range, it results in the disadvantage of the terrestrial laser scanning method. The cost per area for data acquisition is very high. This method may also face obstruction by branches, twigs, or leaves, leading to inaccurate data [13].

Bogdanovich et al. [14] focused on capturing the temporal dynamics of a tree and analysing its structural features based on data collected from terrestrial laser scanning. Tree crown voxelization was used to evaluate and compare point clouds of crowns of all trees for both the Years 2015 and 2018. It was shown that the addition of the point cloud method to estimate the dimensions of the tree, allows detection of pruning and fertilization aftereffects. However, this method introduced over-heads for slow-growing trees, requiring more time to collect data.

Another method was conducted on terrestrial laser scanning to provide precision tree models in a relatively short span of time that segmented the point cloud and reconstructed 3D tree models from the terrestrial laser scanning (TLS) data [15]. In order to achieve that, multiple terrestrial scanners were utilized to scan a single tree from multiple angles. A buddle of small local subsets of the point cloud was added step by step as connected surface patches to build the global model. As the advantages, the study obtained a precise and comprehensive point cloud model which was then reconstructed to form a flexible cylinder that resembles a real tree. The volume and size distributions can then be estimated. However, the cost, time, and hardware requirements are some drawbacks to this method.

Friedli et al. [16] presented how scanned data collected from crops, mainly Maize (Zea mays), soybean (Glycine max), and wheat (Triticum aestivum) using terrestrial laser scanning is ideal for measurement of plant canopy height. The scanner was mounted on a tripod with about 3.5 m height, for distances up to 10 m between scanner and canopy. The scanning for the same point was conducted from different positions with regular time intervals per scan. Height analysis was carried out after the point cloud transformation to the fixed coordinate system. Whilst the results were accurate, this method is limited only to the plant growth aspect of canopy height, as the disadvantages of the method.

Moreover, a study utilizing terrestrial laser scanning approach to estimate leaf area, leaf area density, and vertical leaf area profiles of mango, avocado, and macadamia tree crowns [17]. A terrestrial laser scanner was used to collect data pre-pruning and post-pruning. To calculate leaf area density, data were collected using a laser scanner with a height of approximately 1.5 m and laser wavelength of 1550 nm in the spectrum, sending laser pulses up to a distance of 350 m. Thereafter, the data collected were classified into point clouds. Detailed classification parameters were then retrieved and analysed. From the 3D geometrical properties, the return of laser pulses was classified into branches or leaves. The point clouds for both leaves and branches were estimated, that was followed by the segmentation algorithm to classify each point using multi-scale measures of the point cloud dimensionalities. Good accuracy is achieved for this method, as the leaf area changes retrieved were in line with the expected changes. However, its limitations include the cost, labour, and subject knowledge.

2.2 UAV Approach

UAVs, also known as drones, are aircrafts that are flown autonomously, without the need for pilots, and can be equipped with LiDAR to accurately gather data [12]. Due to being lightweight and capable of travelling at high speed, UAVs can cover a large area for data collection, which also allows users to spot anomalies much quicker.

This is especially beneficial for spacious and large environments such as forests and farms [13].

Due to over-saturation in the current market, the cost of drones had been observed to keep decreasing, and it became more cost effective. It has also improved in term of physical attributes such as weight, due to improvements in technology over the years. This enables drones to be cheaper and faster alternatives as compared to crewed aircraft. With their agile flying ability, it is ideal for managing forest-like environments to survey and measure the canopy height of the trees using cameras or other types of sensing. However, there exist potential inexorable challenges such as weather vulnerability or lighting conditions which might cause damage to the drones, further incurring costs [18].

The use of data collection by a UAV was investigated to detect change applied to a Eucalyptus plantation [19]. UAVs were flown before and after each stage of pruning for data collection. Two independent point clouds were captured by two flights before the tree pruning and after each of five stages of tree pruning. Each flight was at about height 40 m above ground level with 2.8 m/s speed. The collected point cloud was pre-processed to produce individual tree segments. The extraction of tree-level features was then conducted. The differences of the data acquired at various stages could be observed. This allowed for a more cost-effective way of determining the optimal pruning regimes in modern forest management. However, this method is not 100% accurate as there still exist a number of false positives and false negatives based on the data collection.

Jiminez et al. researched on quantifying the pruning impacts on olive tree architecture using UAV-based 3D modelling for different pruning techniques as an alternative to intensive fieldwork [20]. The method consisted of combining UAV technology with advanced object-based image analysis (OBIA) methodology to monitor hundreds of olive trees that were pruned differently. The method automatically identified all trees and calculated their growth information, such as projected areas, tree heights and crown volumes, study dates. It was observed that the predicted values by the method were generally in line with the results of tree pruning and the tree development occurred over one year. As the advantages, this procedure resulted in high accuracy in most cases of his experiments. However, there were some exceptions whereby some of 3D point clouds were incorrectly generated during the automatic photo-reconstruction stage that required further labour-intensive visual analysis to correct it.

The utilization of Multi-Spectral UAV Imagery was to obtain the structural properties and pruning effects on plants [21]. Tree height, crown width, crown perimeter, and plant-protective cover (PPC) measurements were taken from multiple lychee trees after the pre-processing of the images taken from the UAV. The changes in tree structure were compared before and after a tree pruning using multi-spectral images taken by UAV. It helped predict and estimate the tree structure models. Additionally, geographic object-based image analysis (GEOBIA) was used to automatically delineate the individual tree crowns. This GEOBIA approach proved to be useful for lychee trees in delineating the measurements of different trees. However, different

GEOBIA approaches might not work for other crops and must be designed specifically for each crop which might be tedious and time-consuming. The flying heights of UAV may also cause some impact to the tree structural measurements.

2.3 Computational Modelling

Computational modelling was made possible by advancements of technology in the last few decades that used a computer to simulate the behaviour or result of a real-life system. Computational modelling using 3D simulation, virtual reality, augmented reality, and serious games are well investigated recently [29–33].

The creation of an interactive pruning simulation for apple trees was explored by Yang et al. [24]. The simulation was built using Qt framework and OpenGL graphics library. The morphological tree structure is described by the hierarchical data first. It is followed by building the 3D tree model with OpenGL library. It next develops an interactive interface of tree pruning simulation using the QT framework. By setting the type of pruning, the before and after pruning 3D model were displayed. Users were allowed to do defined point pruning or conditional pruning with the results visualized in milliseconds. However, this simulation was designed specifically for apple trees. It would be good if more different types of trees could be supported in future work.

Another interactive plant pruning system was designed based on GreenLab Model, simulating plant growth at organ level [25]. Every node was an organ during each growth cycle. Biomass within the plant was computed; new organs were created; and their properties will change. Users can remove organs by clicking on them and pressing delete. After removing the organ, the available biomass would be distributed which caused other tree organs to grow differently, mimicking a real-life pruning scenario. However, this simulation does not allow much user interaction and customization.

Auzmendi and Hanan explored using L-Studio to modify plant architectures and visualize field data [26]. Lpfg, a simulator of L-Studio, was used to read and write Multi-scale Tree Graphs (MTG) to translate them into L-String. The architecture in the field was coded as an MTG and was read by Lpfg to form the 3D architecture of the plant. By inserting a cut module, Lpfg was able to obtain the coordinates of the internodes to cut the branches at that point. The node was removed from the plant simulating pruning. However, similarly to [19, 20], this simulation does not allow much user interaction and customization. It could be addressed in their future works.

The uses of pruning simulation of virtual trees were compared to the pruning skills of a human expert [27]. An automated method for simulating pruning of trees was introduced to maximize the light exposure within the tree crown. Results showed that this method was comparable to the human expert with similar light exposure in the end. This shows the potential of computer simulation for tree pruning. However, this method does not include the long-term growth of the plant. As a part of the future works, the method could be tested considering the long-term plant health and development.

A software application named EduAPPLE was developed for tree training simulation and tree pruning teaching. To prove that the plant pruning simulation could contribute to getting real-life tree pruning skills, an experiment was conducted on two groups of students: one group was tasked to use the programme; the other group acted as the control and tested their knowledge of pruning [28]. The results show that the plant pruning simulation can be an asset in the tree training and pruning learning process. The outcome of the experiment was that students who used the simulation learned about pruning were much quicker than the control group. Adverse weather conditions were also negligible as they were using a simulation. However, the sample size is small; hence, might not represent the population.

3 Proposed Methodology

In this research, a 3D plant pruning simulation model is proposed with friendly user interactions and flexibility of customization. It is able to support various types of plant pruning, including sympodial trees, monopodial trees, etc. For monopodial trees, the terminal bud continues growing even if the lateral bud branches out. However, for sympodial trees, the growth of lateral buds is dominant. The apical bud stops growing while the lateral buds start to grow. The simulation model should support the customization of tree properties, models, factors, settings, and inputs, such as branches or tree types. It is also able to visualize the plant pruning at different locations on the trees and visualize the tree growth after pruning. In this section, the relevant software tools to be used in this research are introduced first, followed by the design of the proposed plant pruning simulations.

3.1 Software Tools Used in This Research

Lindenmayer system (L-system) is a recursive framework in which an input string of characters is matched against rules which are evaluated repeatedly [34]. Results can be used to visualize organic growth in computer graphics. The common uses of L-system are to model the growth of plants and complex fractals. The main components of a L-system include:

- Variables: Characters or elements that can be changed.
- Axiom: The starting L-string.
- Production Rules: Set of rules applied to the axiom during each iteration.
- Iterations: Number of iterations the programme runs.

For example, an L-system model with two variables, X and Y; the starting axiom is X; two production rules: (1) $X \rightarrow XY$ and (2) $Y \rightarrow X$; and three iterations. In this case, X is in Iteration 0. During the next three iterations, the production rules will modify the string to find the next strings recursively, through actions of adding string, removing string, replacing string, etc. It produces the string of XY in Iteration 1, according to the production rule $X \rightarrow XY$. In Iteration 2, X is replaced by XY, while Y is replaced by X, that derives the string of XYX, according to both production rules. In Iteration 3, X is replaced by XY; Y is replaced by X; another X is replaced by XY as well, that derives the string of $XYXXY$, according to both production rules.

L-String is a string of alphabets and symbols used in the L-system. For plant modelling, each alphabet can represent different parts of a plant, such as stem, leaves, flowers. During each iteration, each part of the plant can be affected by the production rule set in the L-system, e.g., stem growing leaves, stem growing longer, and so on.

Turtle graphics are vector graphics that allow users to 'control' a turtle by giving it a set of instructions [35]. By giving the turtle instructions via our L-string, the turtle will then be able to create shapes and pictures that we desire. Figure 1 shows an example of how the turtles can be used together with L-strings to form a 3D model that resembles a tree branch. In turtle graphics, $F(x)$ tells the turtle to draw a cylinder of x length at the current location. There are also symbols, including f, $+$, $-$, etc. to represent different directions to move the turtle. By using these symbols and drawings, we can use them to model our trees.

OpenAlea is an open-source project aimed at the plant research community [36] that allows users to create models with a visual programming interface to conduct plant modelling. It provides different modules to analyse and model plant growth. One such module is L-Py, an integrated visual development environment created by the OpenAlea community. It is a programme that allows Python programming language to construct L-systems models. The interface for L-Py enables users to create parameters and input the plant model, including axiom, production rules, etc.

PlantGL 3D Viewer, another module included in the OpenAlea, provides a graphic toolkit for the user to create the 3D models of plants [37]. It allows users to control and manipulate geometric models of various plant parts, such as tissues, organs. PlantGL will be used to complement L-Py to provide the user the pruning simulation. The

Fig. 1 Using L-string and turtles to form a branch

F(x) → Draw a cylinder of x length

f → Move forward without drawing

\+ → Turn left

\- → Turn right

[] → Create Branch

Axiom = F(5)[+F(2)[+F(1)]F(1)]F(3)[-F(1)]F(1) →

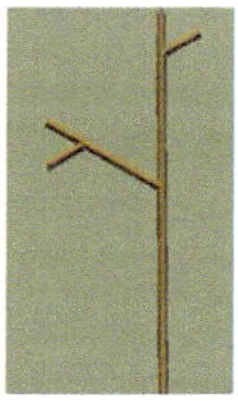

interface for PlantGL enables users to interact with the model generated by the L-system such as different camera angles and lighting options.

3.2 *Design Process Flow*

L-Py is the selected software tool for modelling of the tree in this study. L-Py is a simulation software provided by OpenAlea to create plant models based on L-Systems. By using PlantGL 3D Viewer, another module provided by OpenAlea, we can visualize and model the code we wrote in L-Py. Through introducing production rules every iteration in L-Py, we can model the tree recursively.

For the proposed plant pruning simulations in this research, the complete design process is shown in Fig. 2. Before running the programme, there are several production rules to be set, including tree apex growth, tree branch growth, leaves growth and adding lateral buds, etc. The process begins when the user enters all the parameters, including the type of trees, number of branches, etc. into our tree models created in L-Py. When the simulation is run in each iteration, every production rule affects the L-String. This allows the tree to grow recursively in the simulations. During each iteration, the 3D model of the tree formed by the current L-String will also be shown in PlantGL. When the user decides to prune the tree, they can select any desired branch in the model. After the user prunes the branch, the branch will be removed from the 3D model and the L-String. An updated model of the tree will also be shown to the user after the pruning. Once the user runs the next iteration, all the production rules are applied again to the current L-String to make a new L-String. The cycle would repeat till the maximum iteration number is reached.

3.3 *Modelling of Trees*

3.3.1 Production Rules in Each Iteration

Figure 3 shows the overview of the whole tree model production rules that would run during every iteration to continuously grow the tree. Before running the software programme, there are several production rules set: tree trunk growth, tree branch growth, leaves growth, and adding lateral buds, etc. During every iteration, the production rules check the current iteration's L-string; modify it and grow the tree.

Shown in Fig. 4, we can see one of the production rules for leaves during each iteration, the software programme checks for the age of every leave in the L-string. If the leaves are recently added, the colour given to them will be green. If the leaves are old, the leaves will be coloured yellow by the production rules. Lastly, if the leaves have reached the end of their lifespan, they will be removed from the tree model.

Fig. 2 Proposed design flow

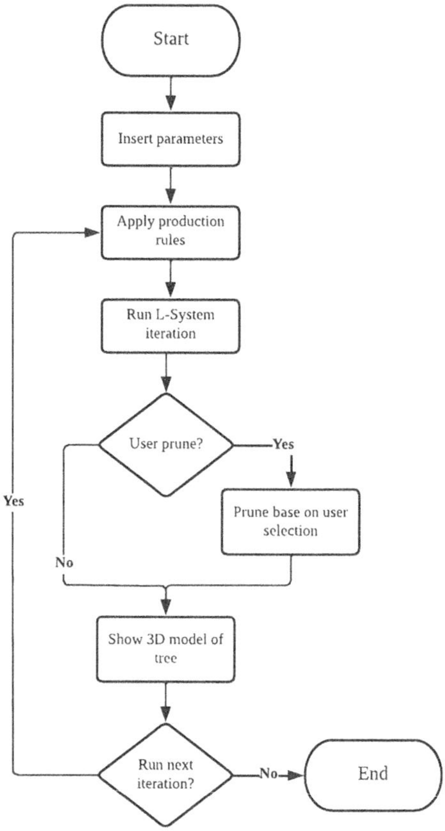

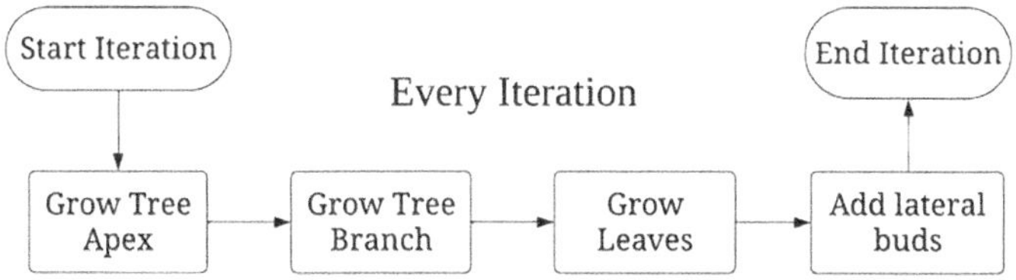

Fig. 3 Production rules overview

Shown in Fig. 5, the production rules for branches check every branch and its age in the L-String. If the branch has reached the age to branch out, new lateral branches and leaves will grow from the current branch. The current branch would also grow in diameter and length regardless of whether new branches have grown.

The production rules for the tree trunk can be seen in Fig. 6. The tree trunk, depending on the type of trees chosen, will be grown differently. However, just like the production rules for branches, the tree trunk would grow and await the iteration

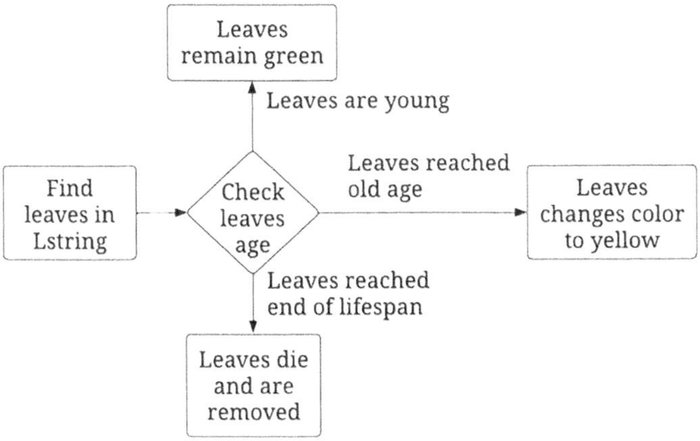

Fig. 4 Production rules for leaves

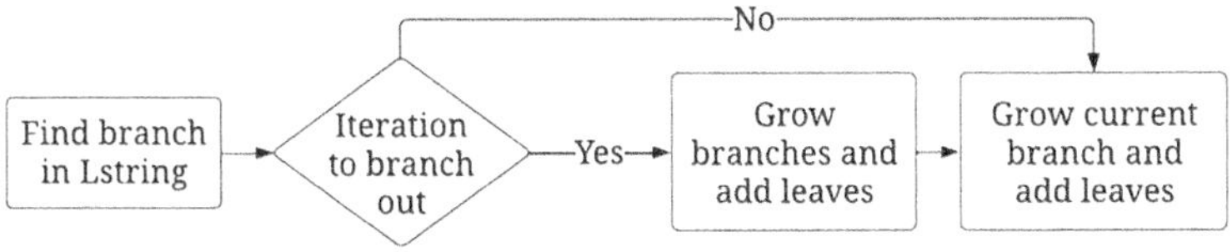

Fig. 5 Production rules for branches

to branch out to create new lateral branches. After checking if it is the iteration to branch out, the tree trunk would grow in diameter and length, similar to a real tree.

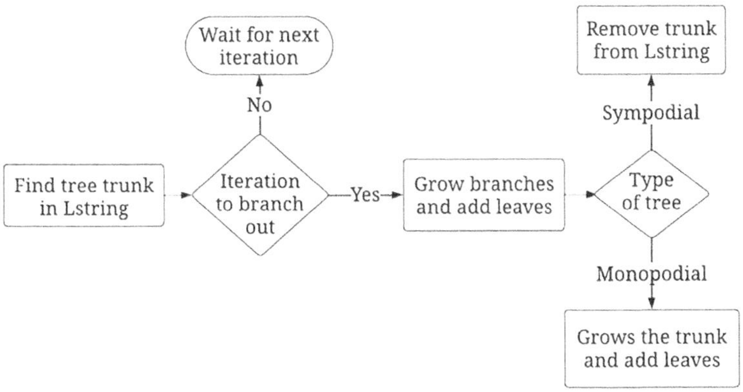

Fig. 6 Production rules for tree trunk

Fig. 7 A monopodial (left) tree and a sympodial (right) tree

3.3.2 Branching Structure

Users are allowed to select various parameters that will affect the tree growth model, such as the branching structure of the tree. The options for branching structure are sympodial or monopodial.

The visual difference between these two different tree types can be seen in Fig. 7. The sympodial tree trunk stops growing and splits into lateral branches, while for the monopodial tree, the tree trunk continued to grow instead.

3.3.3 Number of Lateral Branches

Another parameter for the users to select is the number of lateral branches that the tree should grow at each branching point. Shown in Fig. 8, we can see the tree with three lateral branches selected on the left, compared to the tree with five lateral branches selected on the right. Not only the number of lateral branches from the trunk is affected, but also the lateral branches of individual branches. This parameter, combined with the type of tree structure, allows the modelling of many different species of trees.

3.4 Tree Branch Transformation

3.4.1 Apical Dominance

Apical buds are the buds on the plant where growth and elongation occur compared to lateral buds, which are buds that are not growing but have the potential to form

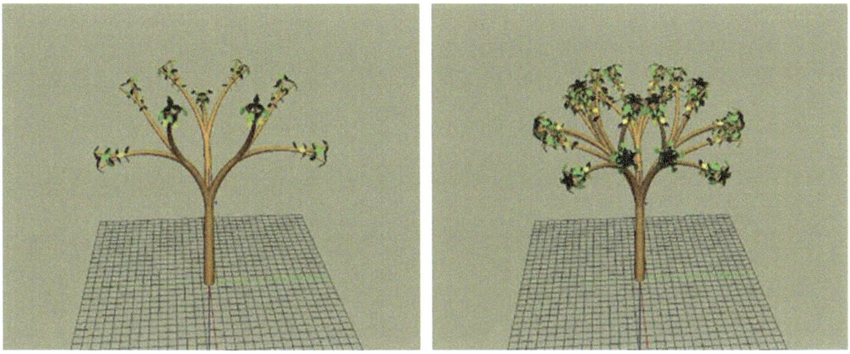

Fig. 8 Tree with three lateral branches (left); tree with five lateral branches (right)

branches. Apical dominance levels represent how dominant the apical bud of a plant is, which affects the growth of lateral buds and how the plant grows. For lateral buds to grow into branches, the apical bud is pruned, and the apical dominance of the tree is low.

Apical dominance is implemented into L-Py by a production rule adding leaves and lateral buds whenever a branch node is added. An apical dominance parameter allows the user to select different levels of apical dominance and would, hence, affect how many lateral buds sprouting into branches when the apical bud is pruned. The lateral buds will take in information of where the user has pruned and what level of apical dominance is set.

3.4.2 Pruning Recommendations

By calling L-Py's built-in function 'GetPos' on every node in the L-string, we can get the x, y, and z coordinates of every part of the tree. This is especially useful to give pruning recommendations for branches that fall out of a specified boundary. The users are able to set the parameters for the chosen pruning method. The branches that are recommended to prune will be highlighted to the user.

Crown Thinning

Crown thinning is one of the common methods of pruning that allows for more sunlight and nutrients to reach the other branches by selective removal of the branches. This is done to prevent overcrowding of branches which could lead to nutrient deficiency of the tree. To implement it, users are able to select the fraction of tree branches to remove. It can be seen in Fig. 9 where users select crown thinning and branches that are recommended for pruning highlighted in blue for the user.

Fig. 9 Recommendation pruning using crown thinning method

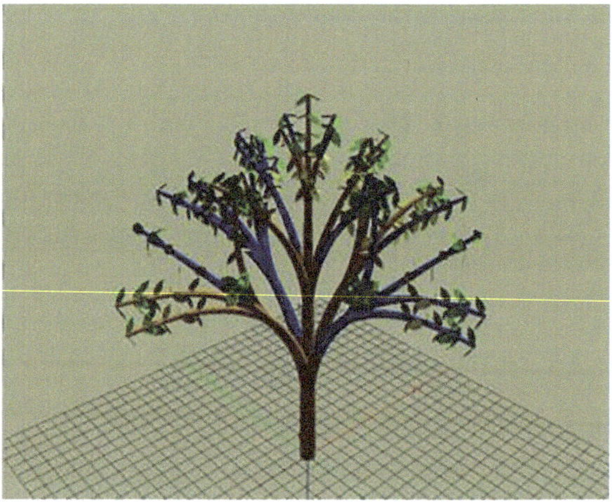

Fig. 10 Recommendation pruning using crown reduction method

Crown Reduction

Crown reduction removes the end of branches that are too long in order to form a new crown for the tree. It reduces the height and spread of the tree crown by removing the ends of the branches. To implement the crown reduction, the user selects the x, y, and z boundary for the tree and removes the branches that exceed the parameters. Shown in Fig. 10, the branches that are recommended for pruning are highlighted in blue for the user when the user selects crown reduction.

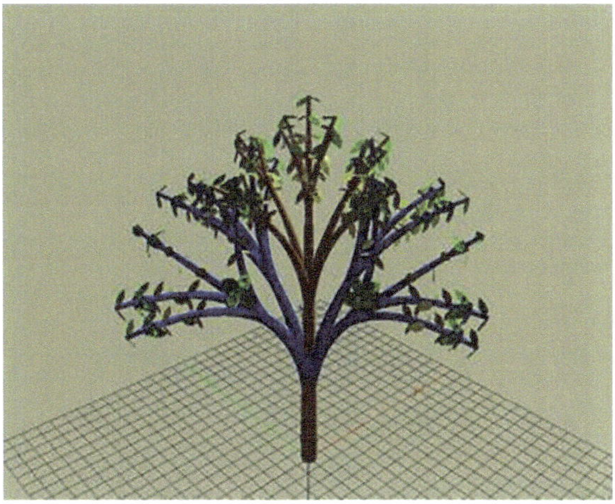

Fig. 11 Recommended pruning using crown raising method

Crown Raising

Crown raising is another common pruning method that increases the clearance space between the ground level and the lower branch of the tree. This method of pruning is commonly done to remove branches that can block the way of pedestrians as well as block the vision of drivers. To implement crown raising, the user can set the minimum height of the tree and remove the branches below the minimum height. Figure 11 shows the branches that are recommended for pruning will be highlighted in blue for the user.

3.5 User Interactions

3.5.1 Point and Click Pruning

Using the 'MouseIns' module in L-Py, an '*X*' will be inserted into the L-String where the user has selected. A production rule written in L-Py will then convert the '*X*' into a Cut module, replacing the '*X*' with '%'. Using this cut module, it allows the user to delete the branch that was selected removing part of the L-String. Point and click pruning can be used together with pruning recommendations to provide user remove branches highlighted to the user.

3.5.2 Settings of Parameters

A set of the parameters are available for users to adjust the production rules of the model in the simulations, shown in Table 1. These parameters should be set before

Table 1 Setting of parameters for the 3D model simulations

Parameters of the model	Descriptions
Branching structure	Select either sympodial tree or monopodial tree
Tree branching	Set how many lateral buds grow after each branching
Apical dominance	Set how many sets of lateral buds to grow after pruning
Crown reduction	Select crown reduction and input the *x*, *y*, and *z* coordinates for the pruning recommendation. Uncheck crown reduction if not applicable
Crown raising	Select crown raising; input the minimum height for pruning recommendation. Uncheck crown raising if not applicable
Crown thinning	Select crown thinning and input the ratio for the pruning recommendation. Uncheck crown thinning if not applicable
Maximum iterations	How many iterations the L-system should run
Skip iteration	Select skip iteration and input the iteration to skip until. Select Animate and click stop when animation stopped

running the simulation programme. Users can set the corresponding parameters such as branch structures, lateral buds to grow, set crow reduction, raising or thinning. The change of the parameters setting can generate different simulation outputs. The last two parameters that can be set by the user are the max iterations that the tree can grow to and the skip to iteration number from the scalar parameter tab. By ticking the checkbox and selecting which iteration to skip to, user can directly skip to the desired iteration without pressing the next iteration button repeatedly.

3.5.3 Pruning the Tree

Users are able to select a branch node to prune in the tree model. The selected branch node should turn red in the software simulation, indicating it to be pruned. Users can click and confirm the pruning on this branch node. Following such steps, other branches of the tree can be pruned one by one. The overall procedure of the tree pruning simulation is shown in Fig. 12a–e.

4 Experiment Result Analysis

Experiments have been performed to evaluate the performance of the proposed tree pruning simulation models. The experiments are conducted on Windows 10 operating system and the computer contained an Intel 4 Cores i7-8550U CPU @ 1.80 GHz,

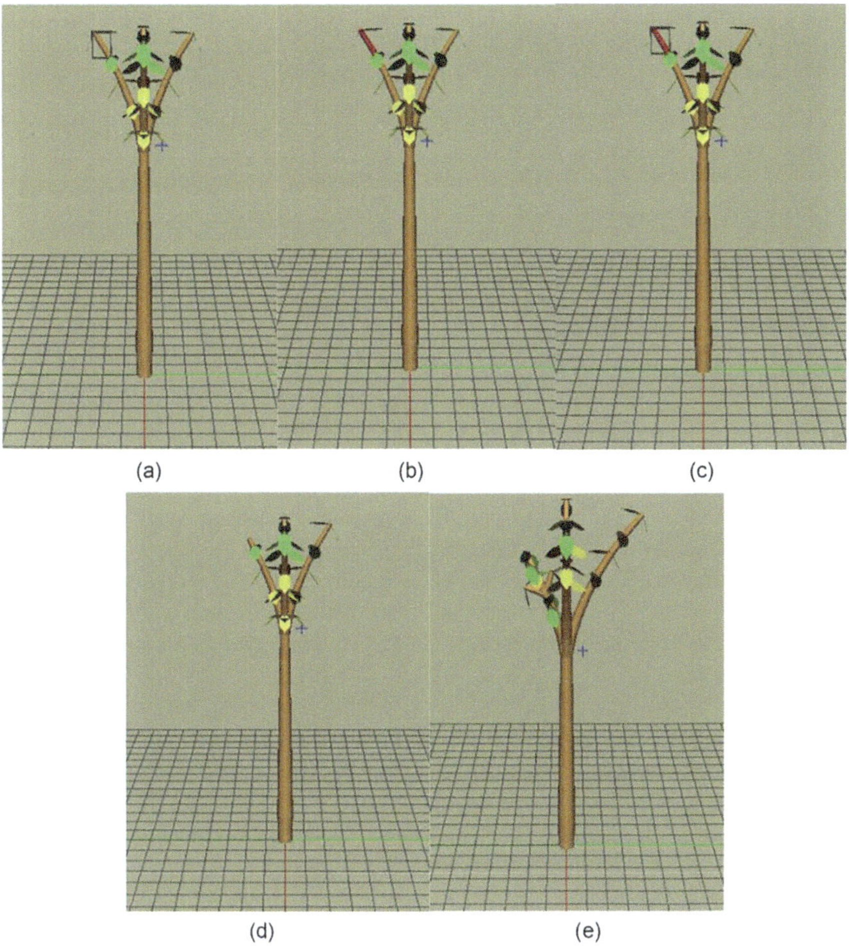

Fig. 12 Steps for simulating the tree pruning process: **a** selection on tree branch; **b** red visual cue indicating tree branch; **c** confirmation to select the red tree branch; **d** removing the selected tree branch; **e** lateral buds growing the pruned area

and a NVIDIA GeForce GTX 1050 GPU. The simulation software program used is Openalea L-Py 3.9.1.

The average total time taken for L-Py to run the software codes and PlantGL 3D Viewer to display the 3D model can be computed in the unit of seconds. The time taken is measured using the 'profile' feature of L-Py which records how much time is being spent in each part of the software programme. Lateral branches are also set to 4 for this measurement.

Tables 2 and 3 show the average time with different numbers of iterations in the simulation for the sympodial trees and monopodial trees, respectively. It is observed that the average time of the sympodial trees is short than that of the monopodial trees

Table 2 Average runtime for a sympodial tree in different max iterations

Iterations	Type of tree	Time taken (s)
10	Sympodial	0.035
20	Sympodial	0.062
30	Sympodial	0.21
40	Sympodial	0.91

Table 3 Average runtime for a monopodial tree in different max iterations

Iterations	Type of tree	Time taken (s)
10	Monopodial	0.06
20	Monopodial	0.18
30	Monopodial	1.22
40	Monopodial	9.5

under the same number of iterations. It illustrates that the growth patterns of different types of trees will affect the average growth time.

As the iterations continue, the time taken to generate the 3D model will exponentially increase. This is because more and more plant parts, such as branch, leaves, trunk, will have to go through production rules as the iterations go on. This can be seen from the line chart in Fig. 13, where the *x*-axis is the number of iterations, and *y*-axis is the average time taken.

We observe that monopodial trees require longer time to generate the 3D model as its trunk grows continuously unlike the sympodial tree. This in turn leads to the creation of more branches and leaves, making the L-String longer. Due to the long L-String, the monopodial tree will take longer to generate in PlantGL.

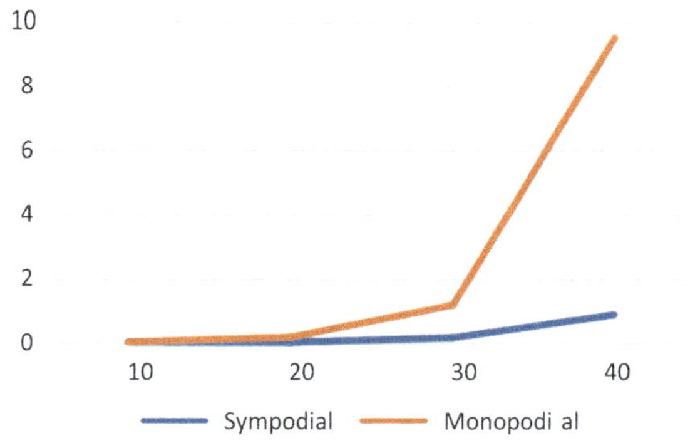

Fig. 13 Time taken for different iterations in L-Py simulation

Table 4 Comparison of tree pruning simulation runtime

Pruning type	L-PY and PlantGL 3D Viewer in this paper (ms)	QT framework and OpenGL graphics library [24] (ms)
Time taken for point and click pruning of branch PA = 2	1	1
Time taken for point and click pruning of branch PA = 3	1	2
Time of conditional pruning for all branches PA = 3	2	3
Time of conditional pruning for branches PA = 3 and 1/3 of the branches of PA = 2	3	3

In [24], Yang et al. created an interactive pruning simulation for apple trees. The simulation was built using Qt framework and OpenGL graphics library. Users can select between point and click or conditional pruning. Physiology age represents which order the branch is, e.g., trunk is PA = 1; branch from trunk is PA = 2; etc.

Table 4 shows the comparison between our proposed method and the use of Yang's method [24] for pruning in terms of pruning type (PA). It is observed that the time taken of both pruning simulations is quite similar. L-py and PlantGL 3D Viewer compared with Qt framework and OpenGL graphics library are tied for point and click pruning of branches.

It is also observed that the L-Py and PlantGL 3D Viewer in our proposed method is faster than the Qt framework and OpenGL graphics library in [24], in terms of time taken for point and click pruning PA = 3. However, OpenGL graphics library is faster in conditional pruning for all branches PA = 3 and 1/3 of the branches of PA = 2. This could be due to the extra number of nodes in the tree simulation created by L-Py which increased the time taken to remove these nodes.

It is visualized from Fig. 14 that the time taken by both pruning simulations is quite similar. The approach of L-py and PlantGL 3D Viewer compared with the approach of Qt framework and OpenGL graphics library is tied for point and click pruning of branch and the time taken for conditional pruning of all branches.

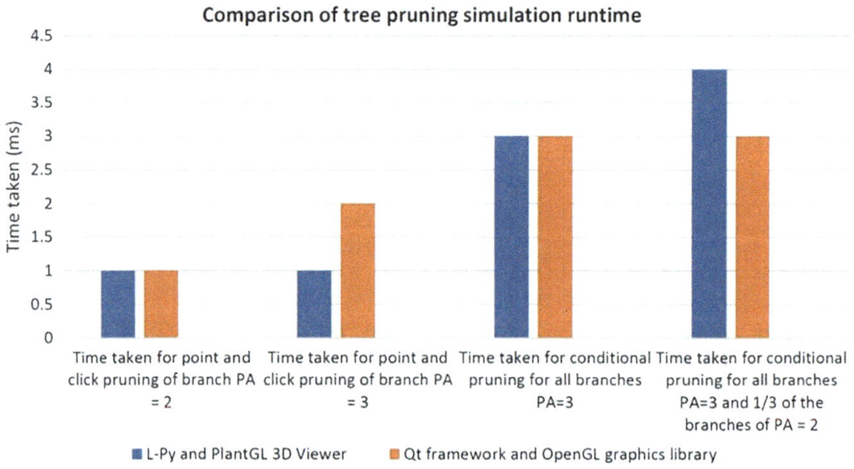

Fig. 14 Bar chart comparisons of simulation runtime

5 Conclusion

To address the challenges in the existing methods for plant growth modelling, a tree pruning simulation modelling method is proposed in this paper. Several software tools are utilized to model various production rules of tree growth, including leaves, branches, and tree trunk. The interactive tree pruning model built in this paper provides a user-friendly interface and flexibility for customization of the modelling. Two types of trees are able to be simulated with the average time being compared. The proposed method is also outperformed with another method reported in the literature. It provides a useful software simulation tool to the government authority or relevant organizations to model and visualize the plant pruning and their growth.

There are some limitations currently in the proposed method. In our study, the software programme runtime performance is conducted based on a single tree. Hence, it may also be good to consider the modelling of multiple trees to investigate relationship of multiple trees in the same area to test out its possible fields of applications, which may be beneficial for forest-like environments.

Besides achieving the effects on growth after tree pruning, the prospects of modelling trees in certain environmental factors, such as sunlight, gravity, should be looked into as well. Future experiments can be conducted to add these as factors to model the tree growth. This will allow the study of tree growth and pruning effects in different climates.

It is also important to delve into the aspect of allowing the user to import real-world data, such as climate and soil data to the simulation and experiment with different real-life tree models. It can allow the proposed model to simulate more accurate and realistic growth patterns. Factors such as importing configurations, tree models, and pruning techniques and be explored in future works.

Currently, the types of trees supported by the model are limited. More tree types could be added into the model in future. As another part of the future work, the simulated tree pruning and growth models would be compared to the measured results from the real trees that provide the performance validation of the model. We also plan to conduct the usability testing of the software simulator by inviting participants from the potential user groups, to obtain their feedback and opinions to further improve the 3D modelling and simulations.

Acknowledgements The first author would like to express his deepest appreciation to his internship supervisor, Dr. Like Gobeawan at the Department of Fluid Dynamics, Institute of High Performance Computing (IHPC), A*STAR, for providing great feedback for this study. Her constant guidance and encouragement of independent learning have allowed him to think critically and effectively mitigate any issues encountered during the research.

References

1. Iau, J., Wong, S.: Woman killed by falling tree in Marsiling Park; last inspection by NParks in April 2020. The Straits Times (2021)
2. Staff Writer: Tree failure incidents fell to 340 cases in 2020, from 420 in 2019: NParks. Yahoo! News (2021)
3. Bedker, P.J.: How to Prune Trees. Forest Service (2016)
4. Nowak, D.J.: Street tree pruning and removal needs. J. Arboric. **16**(12), 309–315 (1990)
5. Johnson, D.L.: Pruning. In: Kuser, J.E. (eds) Urban and Community Forestry in the Northeast. Springer (2007). https://doi.org/10.1007/978-1-4020-4289-8_14
6. Dhillon, W.S., Thakur, A.: Canopy management and effects of pruning on flowering tendencies in fruit trees. In: Souvenir of National Seminar-cum-Workshop on Physiology of Flowering in Perennial Fruit Crops, pp. 182–201. (2014)
7. Mika, A.: Physiological responses of fruit trees to pruning. Hortic. Rev. **8**, 337–378 (1986)
8. Strnad, D., Kohek, Š., Benes, B., Kolmanič, S., Žalik, B.: A framework for multi-objective optimization of virtual tree pruning based on growth simulation. Expert Syst. Appl. **162** (2020)
9. Garden city vision is introduced. Web link: https://eresources.nlb.gov.sg/his-tory/events/a7fac49f-9c96-4030-8709-ce160c58d15c
10. Home. Green Plan. Web link: https://www.greenplan.gov.sg/
11. Fini, A., Frangi, P., Faoro, M., Piatti, R., Amoroso, G., Ferrini, F.: Effects of different pruning methods on an urban tree species: a four-year-experiment scaling down from the whole tree to the chloroplasts. Urban Forestry Urban Greening **14**(3), 664–674 (2015)
12. Vosselman, G., Maas, H.G.: Airborne and Terrestrial Laser Scanning. CRC Press (2010)
13. Van Leeuwen, M., Nieuwenhuis, M.: Retrieval of forest structural parameters using LiDAR remote sensing. Eur. J. Forest Res. **129**(4), 749–770 (2010)
14. Bogdanovich, E., Perez-Priego, O., El-Madany, T.S., et al.: Using terrestrial laser scanning for characterizing tree structural parameters and their changes under different management in a Mediterranean open woodland. For. Ecol. Manage. (2021)
15. Raumonen, P., Kaasalainen, M., Åkerblom, M., et al.: Fast automatic precision tree models from terrestrial laser scanner data. Remote Sensing **5**(2), 491–520 (2013)
16. Friedli, M., Kirchgessner, N., Grieder, C., et al.: Terrestrial 3D laser scanning to track the increase in canopy height of both monocot and dicot crop species under field conditions. Plant Methods **12**(1), 1–15 (2016)
17. Wu, D., Phinn, S., Johansen, K., Robson, A., Muir, J., Searle, C.: Estimating changes in leaf area, leaf area density, and vertical leaf area profile for mango, avocado, and macadamia tree crowns using terrestrial laser scanning. Remote Sensing **10**(11) (2018)

18. Tang, L., Shao, G.: Drone remote sensing for forestry research and practices. J. For. Res. **26**(4), 791–797 (2015)
19. Wallace, L., Watson, C., Lucieer, A.: Detecting pruning of individual stems using airborne laser scanning data captured from an unmanned aerial vehicle. Int. J. Appl. Earth Obs. Geoinf. **30**, 76–85 (2014)
20. Jiménez-Brenes, F.M., López-Granados, F., De Castro, A.I., et al.: Quantifying pruning impacts on olive tree architecture and annual canopy growth by using UAV-based 3D modelling. Plant Methods **13**(1), 1–15 (2017)
21. Johansen, K., Raharjo, T., McCabe, M.F.: Using multi-spectral UAV imagery to extract tree crop structural properties and assess pruning effects. Remote Sensing **10**(6) (2018)
22. Ministry of Trade and Industry (MTI): Nparks to transform Singapore Landscape Sec-tor with new 10-Year plan. Microsite News Article (2019)
23. Government Technology Agency: The inside story of how NParks mapped 500,000 trees in Singapore on trees.sg. (2018)
24. Yang, L., Chen, J., Hua, J., Kang, M., Dong, Q.: Interactive pruning simulation of apple tree. In: International Conference on Computer and Computing Technologies in Agriculture, pp. 604–611 (2015)
25. Hua, J., Kang, M., de Reffye, P.: An interactive plant pruning system based on GreenLab model: Implementation and case study. In: IEEE International Conference on Computer Science and Automation Engineering, pp. 185–188 (2011)
26. Auzmendi, I., Hanan, J.: Using L-studio to visualize data and modify plant architecture for agronomic purposes: visualization and modification of plant architecture with L-studio. In: 6th International Symposium on Plant Growth Modeling, Simulation, Visualization and Applications, pp. 43–49 (2018)
27. Kolmanič, S., Strnad, D., Kohek, Š., Benes, B., Hirst, P., Žalik, B.: An algorithm for automatic dormant tree pruning. Appl. Soft Comput. **99** (2021)
28. Kolmanič, S., Tojnko, S., Unuk, T., Kohek, Š: The computer-aided teaching of apple tree pruning and training. Comput. Appl. Eng. Educ. **25**(4), 568–577 (2017)
29. Cai, Y., Goei, S.L.: Simulations, Serious Games and Their Applications. Springer, Singapore (2013)
30. Cai, Y., Goei, S.L., Trooster, W.: Simulation and Serious Games for Education. Springer, Singapore (2016)
31. Xu, D., Li, Z., Cao, Q.: Object-based illumination transferring and rendering for applications of mixed reality. Vis. Comput. **38**(12), 4251–4265 (2022). https://doi.org/10.1007/s00371-021-02292-2
32. Cai, Y., Van Joolingen, W., Walker, Z.: VR, Simulations and Serious Games for Education. Springer (2019)
33. Cai, Y., van Joolingen, W., Veermans, K.: Virtual and Augmented Reality, Simulation and Serious Games for Education. Springer, Singapore (2021)
34. Lindenmayer systems. Web link: https://cgjennings.ca/articles/l-systems/
35. turtle—Turtle graphics. Web link: https://docs.python.org/3/library/turtle.html
36. Christophe, P., Samuel, D., Frédéric, B., Christian, F., Christophe, G.: OpenAlea: a visual programming and component-based software platform for plant modeling. Funct. Plant Biol. **35**(10), 751–760 (2008)
37. Christophe, P., Frédéric, B., Christophe, N., Jérôme, C., Christophe, G.: PlantGL: a Python-based geometric library for 3D plant modelling at different scales. Graph. Models **71**(1), 1–21 (2009)

Educational Metaverse for Learning of CPU Scheduling Algorithms in Operating Systems

Eng Choon Ang, Yiyu Cai, Min Wang, and Qi Cao

Abstract Computer science (CS) students in higher education often face challenges in learning theory of central processing unit (CPU) scheduling algorithms in the operating system (OS) course. There are various common CPU scheduling algorithms, including first-come first-served (FCFS), shortest-job-first (SJF), round-robin (RR), and priority scheduling algorithms. Metaverse and three-dimensional (3D) interactive digital media technologies including virtual reality (VR) are ideal to resolve the limitations of the conventional teaching methodologies. Moreover, the recent COVID-19 pandemic has created new norms in terms of learning modality. This study uses the educational metaverse in visualizing and interacting with the operations of four types of CPU scheduling algorithms to enhance student learning effectiveness. In this research, the associations between CPU scheduling algorithm and a car workshop servicing center are utilized, as there are some similarities on the scheduling and functioning between these two. The educational metaverse is developed using Unity3D in teaching CPU scheduling algorithms theory, which is an interactive content creation platform for the game engine. Besides the keyboard control for gameplay on the computers, mobile touchscreen control is incorporated to allow the playing of the educational metaverse game using mobile platforms, such as tablets or mobile phones. It provides user interactions of the game on various platforms conveniently and increases the frequency of game participations. Twenty students participate in the experiment evaluation of the serious game. Ten students form the control group that uses conventional learning methods to learn about the concept

E. C. Ang
Singapore University of Social Sciences, Singapore, Singapore
e-mail: ecang001@suss.edu.sg

Y. Cai
School of MAE, Nanyang Technological University, Singapore, Singapore
e-mail: myycai@ntu.edu.sg

M. Wang
Guangdong University of Foreign Studies, Guangzhou, China
e-mail: wangmin@gdufs.edu.cn

Q. Cao (✉)
School of Computing Science, University of Glasgow, Glasgow, UK
e-mail: qi.cao@glasgow.ac.uk

Y. Cai et al. (eds.), *Virtual and Augmented Reality Technology-Enhanced Learning*, Gaming Media and Social Effects, https://doi.org/10.1007/978-981-96-2332-7_8

of CPU scheduling, while the remaining ten students form the experimental group that learns through the serious game. The experiment results show the experimental group achieves better performance. The educational metaverse provides an alternative approach to the conventional classroom learning that enables better engagements to students and enhancement of learning efficiency.

Keywords Computer science education · Scheduling algorithms · Educational metaverse · Serious games

1 Introduction

Operating system (OS) is a core and essential subject for computer science (CS) in most universities in the world [1]. OS is widely used in computers and embedded systems, in the form of various types, such as Windows OS, Linux OS, and Mac OS. CS students usually have difficulty in learning the theory in the OS subject [2, 3]. There are various scheduling algorithms for selecting the corresponding processes in the short-term queue of the OS for being executed by the central processing unit (CPU). Many processes are managed and selected by the OS by means of the scheduling algorithms, according to certain criteria or goals. These are related to the concepts and different ways to handle multiple processes concurrently by the CPU for real-time applications with context switching. Usually, CS undergraduates learn the four types of CPU scheduling algorithms that are commonly used in the OS subject, namely first-come first-served (FCFS), shortest-job-first (SJF), round-robin (RR), and priority scheduling [4]. Operating systems shall decide which process will be executed first. Criteria used to make these decisions are waiting time, response time, turnaround time, and throughput. Processes have varied arrival time and burst time. OS has to make decisions to achieve good balances of various criteria. Relevant mathematical calculations are required to compare or evaluate how each scheduling algorithm performs. In order to perform these mathematical calculations for each scheduling algorithm, the Gantt charts need to be drawn first by students as well as for better visualization purposes. By presenting these concepts in lecture with the conventional teaching, it does not provide intuitive grasp of the materials and students tend to be struggling to focus for the long period. There is a mismatch between the algorithms' theory and their functionalities in the real world [5].

Educational games represent a wide variety of digital learning games. It becomes accessible for more learners to promote learning. Nowadays, various video games are popular to students in universities. Therefore, students can adapt to educational games easily through the integration of learning and play. Educational games allow students to learn in a familiar and playful environment that promotes interactive learning [6]. Meta-analysis is performed to compare outcomes related to serious educational games, serious games, and educational simulations [7]. It is reported on higher cognitive gains in learning with educational games. Educational games provide situations of exploration, discovery, observation, construction of knowledge, and most importantly, providing practical experience of abstract concepts.

In a well-designed educational game, the player can spend up to 80% of their time failing. However, it is the act of failing that motivates the players to keep going and try again [8]. University students in CS show tenacity in a gamified mode by completing the gamified activities, surpassing expectations even after reaching the targeted learning goal [9]. A form of metaverse educational games is created to make players have a sense reality in the present environment [10]. Traditional 2D technology does not provide depth to the game objects. Three-dimensional (3D) and metaverse technology offer unique perception of spatial depth experiences on the screen.

As the operating system subject is considered a tough subject in the universities [2], the technology-enhanced learning can be utilized to help students in learning the tough subjects. The educational games empowered by the metaverse technologies can provide more realistic and immersive learning experiences to students. It could provide better visualization and user interactions in the learning journey.

The objectives of this research are to design and build an educational metaverse game to interact with the operations of the four types of CPU scheduling algorithms in the OS subject. The proposed methodology aims to enhance the learning and performance of CS students on CPU scheduling algorithms. There will be two modes in the educational metaverse game: tutorial mode and practice mode. The tutorial mode is designed to improve players knowledge on these scheduling algorithms. The practice mode is designed for players to apply the knowledge that they have learnt in the metaverse game.

The organization of the rest parts of the book chapter is as follows. Section 2 presents the related work in the literature. Section 3 introduces the proposed methodology of the educational metaverse game. Section 4 describes the experiment results. This chapter is concluded in Sect. 5.

2 Related Works

Reports suggest the digital gaming industry could grow by 13% from 2018 to be worth US$300 billion by 2025 [11]. The video game sector has obtained tremendous technological advancement in recent years. Traditional 2D technology does not provide depth to the game objects. On the other hand, 3D and metaverse technologies provide unique perceptions of spatial depth experiences on the screen. Building on the traditional two-dimensional (2D) modality, many video games can be played in a stereoscopic 3D or even in a head-mounted-display (HMD)-based virtual reality environment. The 3D environment is much more realistic and attractive than the 2D version [12]. In particular on the 3D technology, it can be achieved by using stereoscopic displays to generate spatial depth off the screen with 3D pop-up visualizations. Thus, players may think that some game objects seem to be nearer and touchable. At times, additional hardware, like the usage of dedicated 3D-capable glasses and

HMD, is required to achieve these realistic and immersive experiences. Games using 3D format appear better to improve on the learning experience, although there are potential downsides in prolong use of 3D technologies for certain players, such as discomfort, eye fatigue, dizziness, headache, disorientation, or motion sickness [13, 14]. Some users can gradually get used to full 3D immersive scenes after multiple tryouts or practices, while other users may have difficulties overcoming the 3D sickness in immersive 3D. One possible mitigation way is to provide alternative options to those users with semi-immersive scenes or 2D display for the same games.

Related work on learning games has been reported in the literature, such as educational metaverse games, game genres, serious games, and game development. 3D games provide the greatest effect and serious educational games are effective. Serious games were reported in education and learning of various areas [14–16]. A study with 387 participants was carried out to determine if a serious game could enhance sustainability marketing outcomes [17], where findings showed that the serious game significantly increased knowledge, value-in-behavior, and sustainable behavioral intention after one week. The research analyzed the effect of reward based and meaningful game mechanics on the educational point of view. According to Muratet et al. [6], the characteristics of serious games can be classified by game genre, game mode, and game complexity. The accurate determinations of these types of characteristics are important in the success of serious games in achieving learning objectives [6]. Some factors for successful learning games include generating realistic solutions, focusing on current topics, challenging with multiple levels, interesting, creative, and innovative [18].

COVID-19 pandemic has developed a new normality of learning approaches [19]. The need for distance learning creates a brand-new set of behavioral norms and standards for socializing [20]. Online interactions are encouraged over face-to-face meetings, and public gatherings are replaced by private video chat sessions. Home-based working and learning are also recommended. Professional meetings are frequently conducted online. Some of the tertiary learnings have been implemented online courses and webinars for a certain of time [21].

Educational metaverse games consist of different underlining technologies, including virtual reality (VR), augmented reality (AR), or mixed reality [22, 23]. It is one of the effective ways to teach complex concepts by allowing learners to interact with virtual objects in-game scenes. It enhances students' ability to analyze and solve problems. VR creates a virtual and imaginary space to provide players with a sense reality in the present environment [16, 24]. The immersive and interactive 3D metaverse games provide students with the flexibility to self-study or group study at their own pace. The educational metaverse games are used to provide situations of exploration, discovery, observation, construction of knowledge, and practical experience of abstract concepts. Unity3D tool is one of the commonly used development engines for educational metaverse games that can be visualized through the use of different types of devices, such as HMD, mobile devices, tablets, personal computers, or Cave Automatic Virtual Environment (CAVE) equipment. Unity3D tool is convenient in use and for designing educational games in both 2D and 3D rendering.

OS is the essential system software running in the computer systems. OS performs memory management, process management, CPU time resources, and storage spaces, etc. As a tough subject, the teaching of the OS subject has been discussed in some prior works. In a survey paper [2], teaching approaches of OS subjects have been discussed covering three types of delivery modes: face-to-face, online, or blended, as well as various learning approaches, such as real OS, graphical simulators, learning environments, collaborative learning, games, etc. The OS subject was taught in a traditional face-to-face setting versus an online manner where the learning impact on student success was evaluated and compared [25]. Project assignment-based method was presented to teach the OS subject to reinforce the students' learning on the programming language [26].

Many visualization tools were developed in the last two decades to ease the learning of CS subjects. Reported in [5], web games using JavaScript and HTML5 were used as visualization aids in teaching scheduling algorithms. It is to encourage students' interaction, maintain their attentions, and understand the theory of scheduling algorithms better. It recommends a few suggested the future works on better 3D graphical animations, scoring systems, and clearer sounds [5]. An educational simulator, named as simulator to operating system—process scheduling education was introduced [27], that allows students to choose the types of computer system and scheduling algorithms. It is followed by understanding the functionalities of the various scheduling algorithms. The web games and kinesthetics learning activities were both developed to enhance the understanding of CPU scheduling algorithms [28], where the web games were developed using object-oriented programming techniques. Suggested future works would be integrating with better graphical animations to attract students and introducing a scoring system to promote competitive spirits.

One of the important skills required in CPU scheduling algorithms is decision making [29]. According to Suter et al. [30], action, simulation, or strategy games genre are one of the best ways to learn decision making skill. Games provide players with special zones where divergent rules and laws are made culturally possible. These temporary spaces created by the games can position players into a 'magic circle' or as a kind of utopia of society [31].

3 Proposed Methodology

In the planning phase, we first analysis four common types of CPU scheduling algorithms including first-come first-served (FCFS), shortest-job-first (SJF), round-robin (RR), and priority scheduling. It is to promote learning for the CS freshman undergraduates by developing an interactive learning platform to visualize and interact with the operations of these four types of CPU scheduling algorithms in the OS subject. This learning platform will take place anywhere virtually to significantly close the learning gaps. Next, the learning of these four scheduling algorithms has been correlated to a real-world application as the car workshop servicing center. Various cars act like the processes in the operating systems. The car workshop acts like the CPU to handle the car requests according to the decision derived from the scheduling algorithms. The educational metaverse game is designed for teaching the four types of CPU scheduling algorithms. It is useful in helping students understand scheduling algorithms. There are three key enhancers in using animation for visualizing the various algorithms. Firstly, the cost of development is reduced. Second, there is the avenue to allow students to pause the metaverse game to review any game scene. Thirdly, visualization is interactive to better help understand concepts quickly.

3.1 Scheduling Algorithms

The CPU scheduling algorithms make decisions which process in the ready queue of the operation systems is selected and allocated to the CPU [31]. In the operating systems, processes which have been obtained all necessary resources except for the CPU time will arrive at the ready queue, as shown in Fig. 1. The CPU is always busy executing tasks of different processes. For the process executed by the CPU, there are three possible destinations:

(i) If all tasks of this process complete the executions by the CPU successfully, this process will be moved to the exit state, where it will be terminated next.
(ii) If this process needs to wait for the input/output events or other events, the CPU cannot continue the execution for this process. This process will be moved into the wait queue.
(iii) If the execution of this process times out, or the CPU receives interrupts or system calls from other module, the CPU has to pause the execution for this process. This process will be moved back the ready queue.

For the processes in the ready queue, the short-term scheduler will utilize the scheduling algorithm to select one of them whenever the CPU is freed up. The selected process by the scheduling algorithm will be dispatched to the CPU for the execution of their tasks.

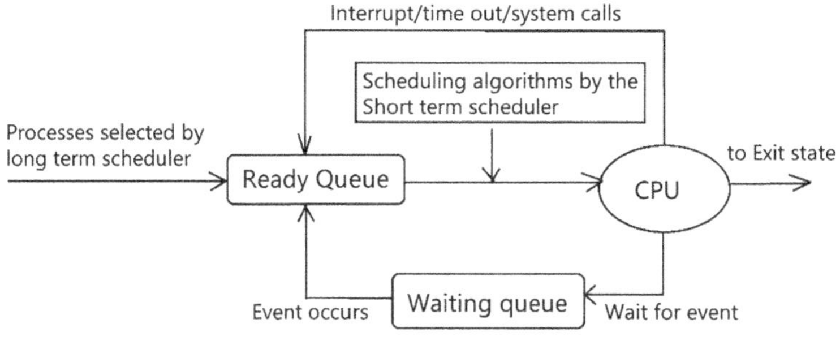

Fig. 1 Processes in the ready queue to be selected by scheduling algorithm

There are various scheduling algorithms to help the operating systems select a process to dispatch to the CPU for executions. Different universities may teach different types of scheduling algorithms in the operation systems subject, four such example types of scheduling algorithms including FCFS, SJF, RR, and priority scheduling.

The purpose of scheduling algorithms is better utilizing the CPU resources, to keep the CPU as busy as possible. The performance metrics to evaluate the efficiency of each scheduling algorithm consist of the average turnaround time, average waiting time, and average response time, etc. As such, it is able to obtain good throughput with a greater number of processes completing their executions in fixed time durations.

For each process arriving at the ready queue, the information of arrival time and burst time will be recorded by the operation systems. The burst time is referred to as the serving time needed to complete all tasks of a process by the CPU. With each process in the ready queue being selected by the scheduling algorithms and executed by the CPU, its finish time will be recorded by the operating system. The average turnaround time for the scheduling algorithms denotes how much time is needed to get all the processes in a queue finishing the tasks [31]. The finish time and arrival time of all processes that are in the ready queue waiting to be selected by the OS are utilized to derive the average turnaround time of the scheduling algorithms. With the timing information, the average turnaround time can be computed as shown in Eq. (1).

$$\text{Average turnaround time} = \frac{\sum_{k=1}^{n}(\text{Finish time}_k - \text{Arrival time}_k)}{n}, \quad (1)$$

where n denotes the number of processes in the ready queue; finish time_k denotes the finish time of the kth process in the ready queue. Arrival time_k denotes the arrival time of the kth process into the ready queue. The shorter the average turnaround time, the more efficient the scheduling algorithm to manage multiple processes to be executed by the CPU. It is due to the executions of the processes can be finished in the shorter time when they arrive at the ready queue.

The average waiting time of the scheduling algorithm is referred to as the average time of all processes waiting in the Ready queue [31]. It can be computed based on the turnaround time and burst time of each individual process in the ready queue, as well as the total number of the processes. The average waiting time can be calculated as shown in Eq. (2).

$$\text{Average waiting time} = \frac{\sum_{k=1}^{n}(\text{Turnaround time}_k - \text{Burst time}_k)}{n}, \quad (2)$$

where n denotes as the number of processes in the ready queue; turnaround time_k denotes the turnaround time of the kth process in the ready queue. Burst time_k denotes the burst time of the kth process into the ready queue. Similarly, the shorter the average waiting time, the more efficient the scheduling algorithm. It reflects processes spend less time in the waiting in the ready queue.

The average response time denotes the average time of all processes in the ready queue being selected by OS to receive the first service [31]. The average response time of the scheduling algorithm can be computed as shown in Eq. (3).

$$\text{Average response time} = \frac{\sum_{k=1}^{n} (\text{First response time}_k - \text{Arrival time}_k)}{n}, \quad (3)$$

where n denotes as the number of processes in the ready queue; first response time$_k$ denotes the first response produced by the CPU for the kth process in the ready queue. It represents the CPU can respond to the processes in the ready queue quickly.

When all processes have been dispatched to the CPU, the average turnaround time, average waiting time, and average response time will be calculated and evaluated. The four scheduling algorithms will be briefly introduced in the next sub-sections.

3.1.1 FCFS Scheduling Algorithms

The FCFS scheduling is the simplest scheduling algorithm. It is a non-preemptive scheduling algorithm. For the FCFS scheduling, the process that arrives at the ready queue earliest will be selected by the short-term scheduler and dispatched to the execution of the CPU. Once the selected process is dispatched to the CPU, the CPU will continue executing the tasks of this process until the completion, if there is no interrupt, system call, or event waiting case. The next processes will be selected one by one according to their arrival time in the ready queue, using the same FCFS scheduling.

The FCFS scheduling is easy for implementation. But there is room for improving the performance. If a process with a very long burst time arrives first in the ready queue, and some other processes with shorter burst time arrive next, the process with the very long burst time will be selected by the FCFS scheduling. In this case, those processes with shorter burst time have to wait in the ready queue for a long time. It is most likely to increase the average turnaround time, average waiting time, and average response time, which affects the performance of the scheduling. Comparatively, the performance of the FCFS may be worse than the other three scheduling algorithms.

3.1.2 SJF Scheduling Algorithms

In order to address the disadvantages of the FCFS scheduling, the SJF scheduling enables the operating system to select the process with shortest burst time first, even if it is not the first arriving process in the ready queue. The selection of the SJF is based on the length of the burst time of each process in the ready queue. The process with the longest burst time will be the last one to be selected to dispatch to the CPU. In this case, the scheduling performance can be improved, with relatively shorter average turnaround time, average waiting time, and average response time.

3.1.3 RR Scheduling Algorithms

The round-robin scheduling tries to treat each process more fairly, by providing equal opportunity and equal time slice to each process in the ready queue. Each process takes turns to be selected and dispatched to the execution of the CPU. When the time slice expires, the CPU will pause the execution of this process and be shifted to the end of the queue. The RR scheduling algorithm will select the next process in the queue to dispatch to the CPU. The duration of the time slice can be set to different values. The recommended setting of the duration of the time slice is longer than the burst time of 80% of processes in the ready queue [31]. An advantage of RR scheduling is that a shorter average response time can be achieved. A potential disadvantage of the RR scheduling is that more CPU context switches are required due to processes taking turn to be dispatched to the CPU.

3.1.4 Priority Scheduling Algorithms

Under certain circumstances, some processes need to be executed first by the CPU. The priority scheduling algorithm can cater for such requirements to have the processes with highest priority to be selected first, while those processes with lowest priority will be selected last. Each process is assigned a priority value by the operating system. For those processes with the same priority, there are several ways to handle. The selection will be either based on their arrival time in the ready queue or based on RR scheduling. Priority scheduling has advantages to deal with the real-world scenarios where processes come with different priorities. The potential problem of priority scheduling is starvation, where those processes with lowest priority may not be selected. The solution for starvation is aging, where the priority of processes will be increased as the time progresses.

In the learning of the scheduling algorithms, if the number of processes increases, the evaluation and comparisons of the performance of each scheduling will become complicated. Usually, students are asked to draw the scheduling diagram according to the number of processes, their arrival time, and their burst time. With the diagram drawn, the calculations of the average turnaround time, average waiting time, and average response time can be computed correctly. The learning of the scheduling algorithm involves mathematics, logical thinking, and theory of operating system. It will be a good use case for the educational metaverse game to better help students in learning.

3.2 Game Design

In the service center of a car workshop, there will be many cars arriving. Due to the space or manpower constraints of the car workshop, most of cars need to wait in the queue which is ready for the servicing by technicians. Usually, the car workshop will use the FCFS scheduling. It may not always be the optimal scheduling. If different scheduling algorithms are tried and compared in the real world, it may involve certain costs and time. But it is possible to implement and compare various scheduling algorithms in an educational metaverse game. The computation of the performance of the scheduling algorithms can also be easily implemented in the metaverse game. It provides a unique feature compared to the real-world application of the scheduling algorithms.

The development of the educational metaverse game requires a powerful software development tool. Unity3D is chosen as the main development tool, which is a leading tool to perform 3D simulations and interactive digital game engine development. The developed metaverse games can be deployed into different platforms across HMD, computers, tablets, mobile phones, or web portals.

Constructivism suggests knowledge is constructed by learners in personal ways based on personal experiences [32]. Based on constructivist perspective, a suitable learning environment can promote the discovery, reflection, negotiation of meaning among learners, and communication of knowledge between learners [33]. One way to promote retention of a concept is to introduce an association to which students can relate. This can act as the foundation for subsequent topics to be built on. We can make use of the associations between a CPU and car workshop servicing center in teaching the CPU scheduling algorithms, which are required to handle different processes that are represented by 'cars.' These 'cars' will arrive at different timings (i.e., different arrival time), with different 'repair time of each car' (i.e., different burst time). There is always queuing waiting to be processed for both cases. Due to this familiarity, students are presented with framework through association to understand and remember [34].

Based on current trends, players often use mobile devices to play the game, as compared to using computers and tablet devices. Therefore, beside keyboard control, this game has incorporated mobile joystick control as shown in Fig. 2 to allow mobile touchscreen platform to play the game. With this mobile joystick, users can utilize the educational metaverse game conveniently and increase the frequency of game participation. The choice of devices depends on individual preferences and the availability of devices. This game also allows students to have freedom to decide when to play and on what devices to play [13].

The overall flowchart of the game design is shown in Fig. 3. There will be two modes in this metaverse game: Tutorial mode and practice mode. The tutorial mode will be designed to improve players' knowledge of these algorithms. The practice mode will be designed for users to apply the knowledge that they have learnt into scheduling the virtual car workshop in the game. This metaverse game follows the cycles of the experiential learning process by starting off with a game menu for the users to select which mode. Experiencing phase is carried out when the learners are using the tutorial mode to learn the various theory. Through reflection and conceptualization of the concepts, they will use the practice mode to engage in active experimentation to enforce their learning. The garage is used as a game menu background.

Fig. 2 Game menu with the mobile joystick

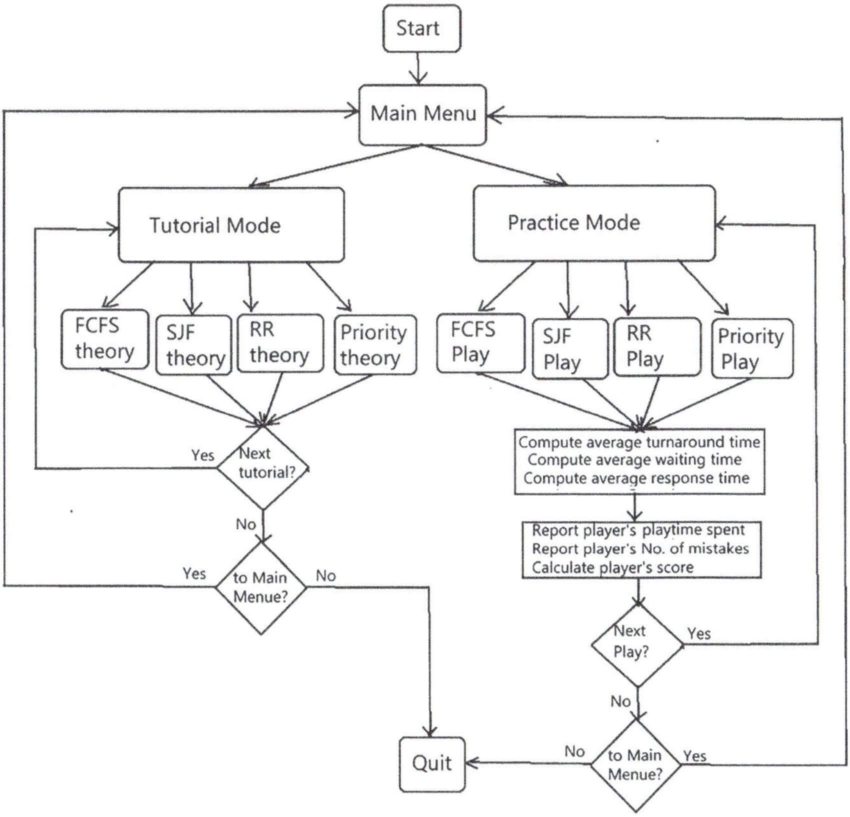

Fig. 3 Flowchart of educational metaverse game

The content in the tutorial mode, together with voice-over text audio, is displayed to the players to let them grasp the foundation on CPU scheduling algorithms. There are five options in the tutorial mode, namely introduction, FCFS, SJF, RR, and priority scheduling. Players can learn the scheduling algorithm one by one. For the practice mode, players apply their knowledge on the four types of CPU scheduling algorithms in the games. Players can select which scheduling algorithm is used in the gameplay. In the developed metaverse game, cars arriving at different timing with different repair or execution time that represents the different burst time. Players need to conduct the scheduling according to the algorithm selected. If players performed the wrong selection, the metaverse game will prompt players to see that the wrong car is selected, as shown in Fig. 4. The play time of each player is counted, and mistakes are also counted, that are utilized to derive the final score of the player in the current session. Upon the completion of the gameplay, the educational metaverse game will compute the total and average of the turnaround time, waiting time, and response time according to the car sequences chosen by the player.

The creation of the UI controls the flow of the games and displays key variables to the users prominently. For the RR scheduling, players are allowed to input the desired values of the time slice, i.e., the quantum time in the game, as shown in Fig. 5. For clicking of the car that the user wants to service, the execution time of the car will be reduced by the amount of the time quantum. The completion of the car servicing will occur when the execution time has reduced to zero.

For the priority scheduling, players are allowed to set different values of the priority for the various car servicing tasks shown in Fig. 6, which are represented as the priority of the programs to be executed by the CPU. They can visually understand the outcomes from these changes in values to link the difficult concepts into practical

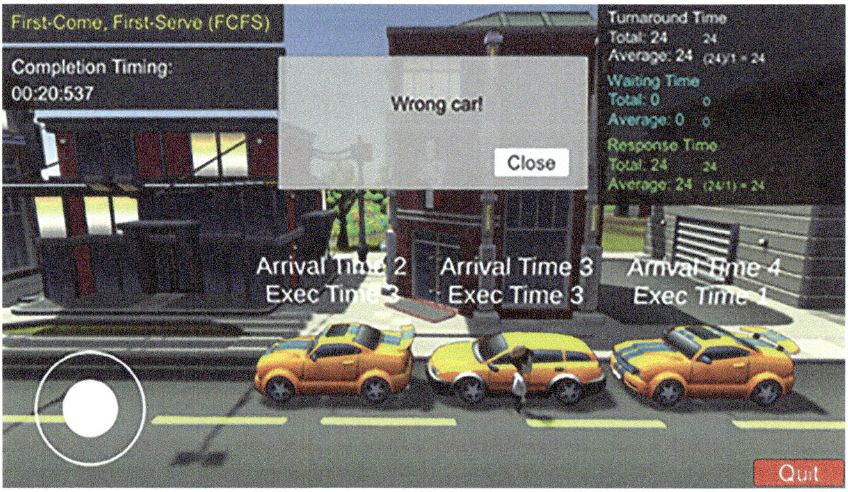

Fig. 4 Game scene of the practice mode in FCFS scheduling

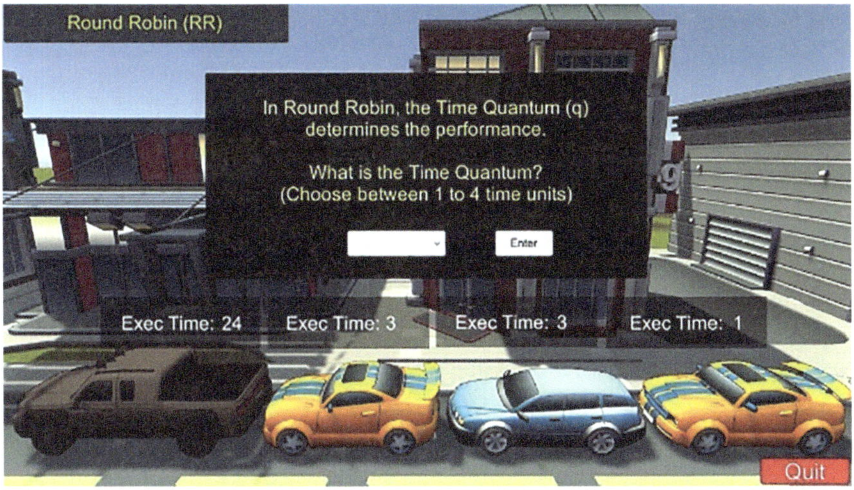

Fig. 5 Developed metaverse game for RR

sense. Similar to the rest of the game, users will be prompted when they have made a wrong decision on the selection of the car to service.

The developed educational metaverse game is hosted at the online portal. It is to ease the players in experiments and in the schools, where they do not need to install any application or software onto their mobile phones or computers. Some players have the concerns that installing applications in the mobile phone or computers may bring in potential security risks. With the game hosting online, players can launch

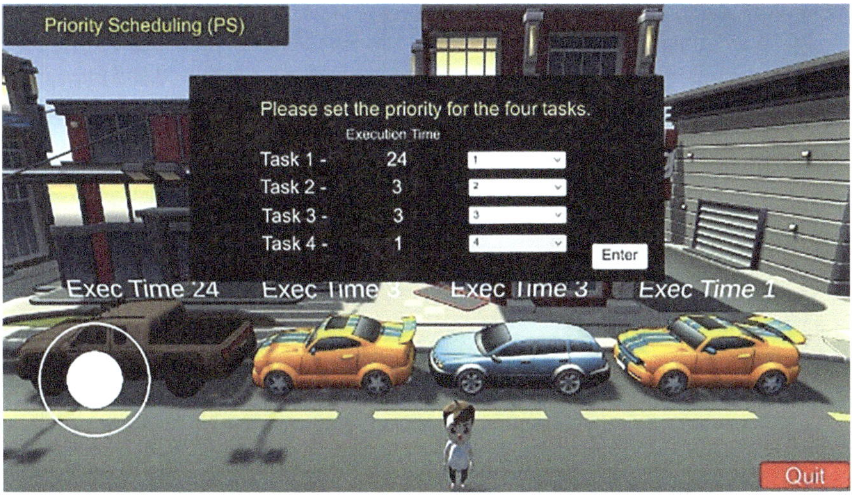

Fig. 6 Developed metaverse game for PS

and play the game online. This ease of use will motivate users to play this game and learn at the same time.

4 Experiment and Results

Experiments of the serious game are carried out to evaluate the effectiveness of the project. Twenty students from engineering disciplines are invited to participate in the evaluation of the project. One group of ten students is selected to use conventional learning methods to learn about the concept of CPU scheduling, where a lecture is conducted in a classroom setting using the electronic lecture notes. There are four calculation examples given in the lecture notes, each of which is for one scheduling algorithm. The processes in these four examples have different arrival time, burst time, or priorities. They represent the 'control' group.

While another group of ten students will utilize the VR serious game of this project and act as 'experimental' group. In order not to have the players to install any software or games on their own computers or devices, the serious game is hosted by WebGL. The URL of the WebGL is given to the players in the 'control' group.

Quiz of ten multiple-choice questions is conducted to assess their understanding of the knowledge, as shown in Table 1. User evaluation is conducted by extracting primary data via the use of the quiz. Those we have completed the learning via project game are also given feedback forms to provide them with an avenue to voice out their valuable comments.

As shown in Fig. 7, the overall quiz result from the 'Control' group is generally low, with mean value at 3.5 out of 10 and the median is also 3.5 with result range at 3. Only one question is answered correctly by 70% of the group, and the other 3 questions have 50% correct response.

On the other hand, the overall quiz result from the 'experimental' group is significantly higher, as shown in Fig. 8, with an average mark of 7 out of 10 and the median of the result is 7 with result range at 2. Six questions are answered correctly by 70% or more in this group.

The comparison of the quiz results from both groups is shown in Fig. 9. It is observed that the experimental group performs much better than the control group. From these results, we can deduce that the project game on CPU scheduling algorithm helps learners to understand the concept better as the 'experimental' group has a higher average score and narrower distribution score range (lower variability). Moreover, the results from the 'experimental' group are clustered around the mean, which can be interpreted as more accurate and reliable learning methodology to use over the conventional method.

Survey questionnaire with both quantitative and qualitative questions was carried out with the 'experimental' group. Survey result is as shown in Table 2. These questions are the quantitative questions that indicate the user's level of agreement pertaining to the project game. Open-ended qualitative questions like 'other comments' added breadth and depth to the information we would like to collect.

Table 1 Ten multiple-choice questions in the quiz

S/N	Survey questions
1	The amount of time to execute a particular process is known as ______ (a) Throughput (b) Waiting time (c) Turnaround time (d) Response time
2	Which type of CPU scheduling handles processes based on the arrival order? (a) First-come, first-served (b) Shortest-job-first (c) Round-robin (d) Priority
3	Which of the following will happen when optimum shortest-job-first (SJF) scheduling is achieved for a given set of processes? (a) Minimum throughput (b) Maximum total waiting time (c) Maximum turnaround time (d) Minimum average waiting time
4	Which of the following CPU scheduling is same as priority scheduling where priority is the inverse of predicted next CPU burst time? (a) First-come, first-served (b) Shortest-job-first (c) Round-robin (d) Priority
5	Time quantum, *q*, should be ___ compared to context switch time (a) The same (b) Smaller (c) Larger (d) With random value
6	Which of the following statement is not one of the scheduling algorithm optimization criteria? (a) Minimize average response time (b) Minimize average turnaround time (c) Minimize average throughput (d) Minimize average waiting time
7	Which is referred to total amount of time a process has been waiting in the ready queue? (a) Throughput (b) Turnaround time (c) Response time (d) Waiting time
8	Which is referred as the amount of time to execute a particular process? (a) Waiting time (b) Turnaround time (c) Response time (d) CPU utilization

(continued)

Table 1 (continued)

S/N	Survey questions
9	Time at which the process arrives in the ready queue is referred as ___ (a) Turnaround time (b) Arrival time (c) Completion time (d) Waiting time
10	Which of the following is a criterion to evaluate a scheduling algorithm? (a) CPU utilization: keep CPU utilization as high as possible (b) Throughput: number of processes completed per unit time (c) Waiting time: amount of time spent ready to run but not running (d) All of the above

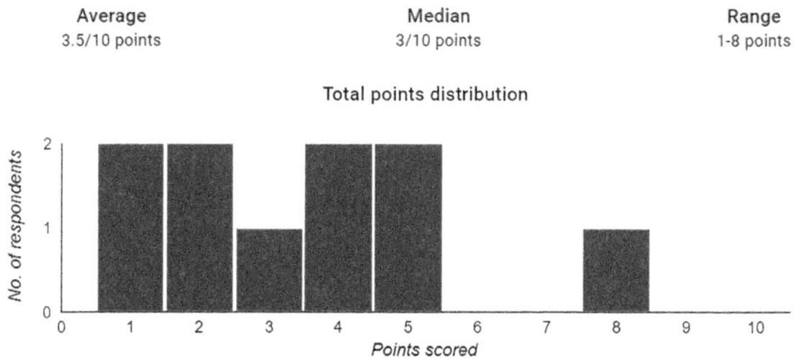

Fig. 7 Overall quiz results from 'control' group

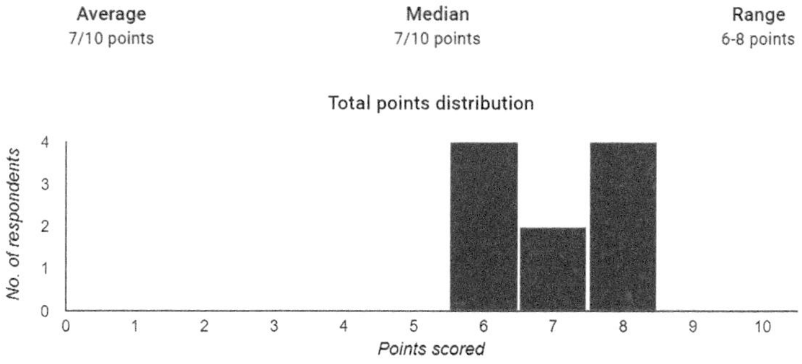

Fig. 8 Overall quiz results from 'experimental' group

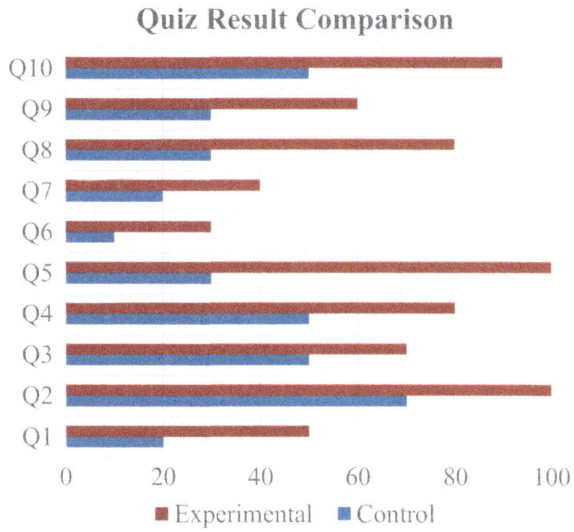

Fig. 9 Comparison of quiz results of 'control' and 'experimental' groups

Table 2 Percentage of students' responses on agree or strongly agree in the survey questionnaire

S/N	Survey questions	Agree or strongly agree (%)
1	I like to learn new skill by playing serious games	100
2	I understand the CPU Scheduling concept by playing the serious game	90
3	I am able to differentiate between the four basic types of CPU Scheduling methods	100
4	I am more confident in applying the CPU Scheduling concept	100
5	I will recommend this serious game to others who would like to learn CPU Scheduling	70

5 Conclusions

This educational metaverse game provides a greater level of autonomy for the students in their learning. They can take control of their learning by selecting the time, location, and the learning topic, which leads to higher intrinsic motivation. This serious game is also a useful avenue to studying the concept of CPU Scheduling by adopting the 'trial and error' strategy. Users have unknowingly retained the knowledge of this difficult concept by making mistakes and learning to overcome them. Instead of forcefully 'downloading' of knowledge by users, they can memorize the knowledge in a fun and relaxing way. Students from 'experimental' group feedback with comments, liked 'This game makes learning in a fun way,' 'The game is related to our study,' and 'Better understanding to the topic.' Users will feel a sense of competence when they are able to complete the four different games on the types of CPU scheduling. The games are set to be relatively easy to achieve this target. Unfortunately, the project game is perceived to be less fun by few students in the 'experimental' group. Comments like 'The game is too simple' or 'Not challenging' indicate that the game is fairly monotonous. More tasks of different levels of difficulty should be made available to make the game more challenging but yet attainable. If all games are set to be easy, the users will feel bored before they complete the games. If these games are too difficult, users may give up and be unable to meet the learning objectives. We must strike a balance between these two extremes. Two more difficult levels should be included in the future development of this project.

Hence, a metaverse game with 3D animation can create an innovative learning environment for students in their learning of CPU scheduling algorithms. Currently, there is not much research on educational metaverse games or serious games as the teaching methodology for CPU scheduling algorithms. By developing the educational metaverse game in simulation game genre, it provides an active learning environment to students that can be a complementary approach to the conventional classroom learning that enables better engagements to students and enhancement of learning efficiency.

References

1. Otero, R.R., Aravind, A.A.: MiniOS: an instructional platform for teaching operating systems project. In: 46th ACM Technical Symposium on Computer Science Education, pp. 430–435 (2015)
2. Pamplona, S., Medinilla, N., Flores, P.: A systematic map for improving teaching and learning in undergraduate operating systems courses. IEEE Access **6** (2018)
3. Jong, B.S., Lai, C.H., et al.: Using game-based cooperative learning to improve learning motivation: a study of online game use in an operating systems course. IEEE Trans. Educ. **56**(2), 183–190 (2013)
4. Omotehinwa, T.O.: Examining the developments in scheduling algorithms research: a bibliometric approach. Heliyon **8**(5) (2022). https://doi.org/10.1016/j.heliyon.2022.e09510
5. Alkhazl, F.S.: Applying Visualization for Teaching Scheduling Algorithms in Computer Science. Bluffton College (2017)
6. Muratet, M., Torguet, P., et al.: Towards a serious game to help students learn computer programming. Int. J. Comput. Games Technol. (2009)
7. Lamb, R., Annetta, L., et al.: A meta-analysis with examination of moderators of student cognition, affect, and learning outcomes while using serious educational games, serious games, and simulations. ScienceDirect J. (2018)
8. McGonigal, J.: Reality is Broken: Why Games Make Us Better and How They Can Change the World. Penguin Books, New York (2011)
9. Ibanez, M.B., Di-Serio, A., Delgado-Kloos, C.: Gamification for engaging computer science students in learning activities: a case study. IEEE Trans. Learn. Technol. **7**(3), 291–301 (2014)
10. Onu, P., Pradhan, A., Mbohwa, C.: Potential to use metaverse for future teaching and learning. Educ. Inf. Technol. (2023). https://doi.org/10.1007/s10639-023-12167-9
11. GlobalData.: Video games—thematic research (2019). Retrieved from https://store.globaldata.com/report/gdtmt-tr-s212-video-games-thematic-research/
12. Takatalo, J., Kawai, T., et al.: User experience in 3D stereoscopic games. Media Psychol. **14**(4), 387–414 (2011). https://doi.org/10.1080/15213269.2011.620538
13. Cheung, S.Y., Ng, K.Y.: Application of the educational game to enhance student learning. Front. Educ. **6** (2021).https://doi.org/10.3389/feduc.2021.623793

14. Tan, Q.P., Huang, L., et al.: Serious game for VR road crossing in special needs education. Electronics **11**(16) (2022). https://doi.org/10.3390/electronics11162568
15. Cao, Q., Png, B.T., et al.: Interactive virtual reality game for online learning of science subject in primary schools. In: IEEE International Conference on Teaching, Assessment, and Learning for Engineering, pp. 383–389 (2021). https://doi.org/10.1109/TALE52509.2021.9678916
16. Cai, Y., van Joolingen, W., Walker, Z.: VR, Simulations and Serious Games for Education. Springer, Singapore (2019). https://doi.org/10.1007/978-981-13-2844-2
17. Whittaker, L., Russell-Bennett, R., Mulcahy, R.: Reward-based or meaningful gaming? A field study on game mechanics and serious games for sustainability. Psychol. Mark. **38**(6), 981–1000 (2021)
18. Stevenson, D.E., Wagner, P.J.: Developing real-world programming assignments for CS1. In: 11th Annual SIGCSE Conference on Innovation and Technology in Computer Science Education, pp. 158–162 (2006)
19. Tesar, M.: Towards a post-Covid-19 'New Normality?': physical and social distancing, the move to online and higher education. Policy Future Educ. **18**(5), 556–559 (2020)
20. Abel, T., McQueen, D.: The COVID-19 pandemic calls for spatial distancing and social closeness: not for social distancing!. Int. J. Public Health **65**(3) (2020)
21. Cao, Q., Lim, L.H.I., et al.: Experiences in Python programming laboratory for Civil Engineering students with online collaborative programming platform. In: 14th Annual International Conference of Education, Research and Innovation (2021)
22. Shin, D.: The actualization of meta affordances: conceptualizing affordance actualization in the metaverse games. Comput. Hum. Behav. **133** (2022).https://doi.org/10.1016/j.chb.2022.107292
23. Chua, H.W., Yu, Z.: A systematic literature review of the acceptability of the use of Metaverse in education over 16 years. J. Comput. Educ. (2023)
24. Cai, Y., Cao, Q.: When VR Serious Games Meet Special Needs Education: Research, Development and Their Applications. Springer, Singapore (2021). https://doi.org/10.1007/978-981-33-6942-9
25. Buckley, I.A.: The impact of teaching operating systems using two different teaching modalities. Int. J. Adv. Comput. Sci. Appl. **11**(12) (2020)
26. Freedman, R.: Using an operating systems class to strengthen students' knowledge of C++. In: 51st ACM Technical Symposium on Computer Science Education, pp. 947–953 (2020)
27. Paschoal, L., Biazotto, J., et al.: Towards an educational simulator to support the CPU scheduling algorithm education. In: International Symposium on Computers in Education, pp. 1–6 (2019)
28. Fatimah, A., Norah, A., Young, G.: Two alternative modules for teaching scheduling algorithms in computer science using web games and kinesthetic learning activities. In: International Conference on Frontiers in Education, pp. 116–221 (2017)
29. Tahar, B., Wassila, D.: Using serious game to simplify algorithm learning. In: International Conference on Education and e-Learning Innovations, pp. 1–5 (2012)
30. Suter, B., Kocher, M., Bauer, R.: Games and Rules. Publisher: Transcript Publishing (2018). https://doi.org/10.25969/mediarep/11746
31. Silberschatz, A., Galvin, P.B., Gagne, G.: Operating System Concepts, 10th edn. Publisher: Wiley (2018)
32. Paris, S., Byrnes, J.: The constructivist approach to self-regulation and learning in the classroom. In: Zimmerman, Schunk (eds.), Self-regulated Learning and Academic Theory, Research and Practice, pp. 169–199. Springer-Verlag, New York (1989)
33. Bednar, A.K., Cunningham, D., et al.: Theory into practice: how do we link? In: Anglin, G. (ed.) Instructional Technology: Past, Present and Future, pp. 88–101. Libraries Unlimited Inc., Englewood, CO (1990)
34. Estell, J.K.: Introducing CPU Scheduling Algorithms: The Photocopier Scenario. Bluffton College (1997)

Designing Immersive Experiences Grounded in Culturally Relevant Teaching and Learning Strategies

Tutaleni Iita Asino, Oluwafikayo Adewumi, Ayodeji Ibukun, Tilanka Chandrasekera, Nicole Colston, Cynthia Orona, Allison Howard, Piper Bott, Katie Allen, and Brandon Scott

Abstract Emerging technologies can engage and draw students to STEM subjects however, what cannot be overlooked are the inequalities embedded in these technologies. This chapter explores the concept of digital decolonization as a way to readdress digital divides. We write through the context of an ongoing research project that aims to develop an immersive afterschool experience for Native American students, grounded in culturally relevant teaching and learning strategies.

Keywords Digital Decolonization · Designing Immersive Experiences · Culturally Relevant Teaching · Virtual Reality

1 Introduction

Emerging technologies can be used to engage and draw students to STEM subjects. This is because emerging technologies such as virtual reality and augmented reality provide a realistic understanding of abstract concepts and generate interests which 'creates the desire to learn content and skills within the STEM context' [1]. Despite their benefits, what cannot be overlooked are the inequalities embedded in these technologies, hence necessitating a need for digital decolonization and a readdressing of the digital divide. The term digital decolonization has numerous definitions; however, they all reference efforts to bridge the gap in digital practices, access, and representation as a means of limiting or mitigating inequalities that often have historical roots. This can mean creating a space where everyone irrespective of race, ethnicity, or culture can easily and safely practice digitalization by including themselves in the digital space (i.e., a space where digital designers are inclusive). Digital divide in the context of this chapter refers to the gap between the access to modern information and communications technology, as well as representation. To examine the

T. I. Asino (✉) · O. Adewumi · A. Ibukun · T. Chandrasekera · N. Colston · C. Orona · A. Howard · P. Bott · K. Allen · B. Scott
Oklahoma State University, Stillwater, OK, USA
e-mail: tutaleni.asino@okstate.edu

Y. Cai et al. (eds.), *Virtual and Augmented Reality Technology-Enhanced Learning*, Gaming Media and Social Effects, https://doi.org/10.1007/978-981-96-2332-7_9

divide, we need to look through the lens of equity to examine the gaps, differences in access to resources especially technologies in schools and home environment, and most important to note the influence of culture in education, and its influence on the underrepresentation particularly in STEM fields [2].

This chapter explores the aforementioned concepts through the context of an ongoing research project that aims to develop an immersive afterschool experience for Native American students, grounded in culturally relevant teaching and learning strategies. A report by NCWIT [3] suggests that existing barriers can be diminished by increasing Native American students' exposure to STEM fields, particularly in grades K-12. The literature further reports on some of the underutilized systemic strategies that have been documented to work with Native American students and their learning environments, such as collaborative learning, more inclusive and diverse learning environments, and using culturally relevant pedagogy, research, and evaluation processes [4]. Therefore, when working in VR, AR, or other emerging technologies, it is important to develop content that is culturally relevant and embodies the worldview that students embrace, through concepts that are used in daily lives and in a hands-on-situated learning background.

The Native American Middle-school Students Afterschool STEM (NAMSAS) Program which is the context of this chapter focuses on the design, development, and evaluation of structured afterschool STEM programs that incorporate virtual reality (VR), augmented reality (AR), and 3D printing through spatial design activities that are relatable to both girls and boys. The target audiences for this project are middle-school students (10–15 y/o) and youth educators in afterschool programs in three Native American communities in the United States of America's state of Oklahoma: Pawnee Nation, Citizen Potawatomi Nation, and the Chickasaw Nation. The project objective is to increase STEM career interests and motivations of Native American middle-school students through participation in seven education modules that teach students how to utilize VR, AR, and 3D printing technologies to solve spatial design problems. While VR can be interpreted as immersive three-dimensional computer-generated environments [5], AR can be conceptualized as overlaying virtual objects over the physical environment [6].

A version of the NAMSAS curriculum was piloted in 2018 with a small number of female students from the Pawnee Nation during a two-week summer program, with the generous financial support of the National Center for Women & Information Technology (NCWIT). The pilot project provided enough evidence of student positive learning outcomes which justified the expansion of the NAMSAS program. Seven design studio modules will be disseminated on a bi-weekly basis in the fall and spring semesters over four years and are implemented through a problem-based learning (PBL) approach that culminates in mini hackathons at the end of each semester and a larger hackathon at the end of the project. The project deliverables include: (a) professional development opportunities for youth educators from Native American communities, (b) afterschool programs for middle-school students, (c) three technology centers for community hackathons, including equipment kits (three laptops/headsets, three iPads, 3D printer), and (d) online curricular resources for the annual continuation and broader dissemination of the NAMSAS curriculum. The project

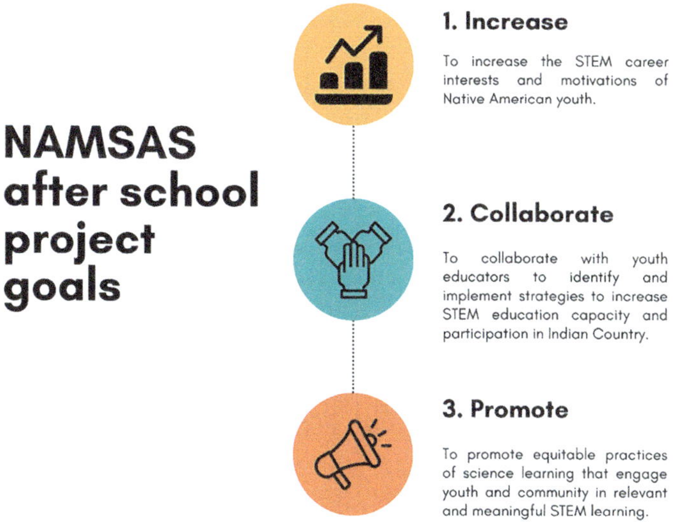

Fig. 1 NAMSA project goals

develops strategic partnerships with leaders and youth educators from afterschool programs in three Native American communities, as well as the Center for Sovereign Nations at Oklahoma State University, to promote problem-based learning (PBL) in a technology-rich afterschool environment. The project goals are shown in Fig. 1.

2 Background

In the USA state of Oklahoma, Native Americans are underrepresented in higher education and STEM degree programs in Oklahoma and across the nation [7, 8]. Afterschool experiences offered by Native American Tribal Nations can support the early-age STEM learning necessary to promote students' STEM interest and motivation. Increasing the STEM education capacity of afterschool programs will require both technology-rich experiences and inclusive learning practices. Developing skills and experiences using immersive technologies like VR and AR will prepare students for the contemporary workforce. Furthermore, subject areas such as architecture and spatial design have been known to attract both boys and girls, therefore encouraging a more diverse STEM-oriented workforce in the future.

3 Literature Review

A review of literature for this chapter led to an emergence of a few themes regarding the incorporation of virtual reality (VR), augmented reality (AR), and 3D printing through spatial design activities in developing student's interest in STEM subjects. Specifically, from the articles reviewed, the findings indicate that after-school programs can increase Native American participation in early-age STEM education, spatial design curriculum using VR/AR technologies to develop critical spatial abilities, and merging indigenous pedagogies and epistemologies with rich STEM experiences can foster a sense of inclusion, and participation.

3.1 Afterschool Programs and STEM Education

Afterschool programs increase Native American participation in early-age STEM education. There is a need to develop the STEM education capacity of Native American afterschool programs in Oklahoma. The state has the second largest Native American population of any state, with 39 federally recognized Tribal Nations headquartered. Statewide there are over 130,000 Native American students attending Oklahoma public schools and 403 Title VI Education programs in Oklahoma public schools. Current research indicates that traditional classroom learning alone is not sufficient to teach students about science or the opportunities for STEM careers [9]. Afterschool STEM programs have also been shown to increase the motivation and engagement in students, especially in rural communities [10], and promote early-age STEM interests as well as career thinking [11, 12]. Afterschool experiences provide for the convergence of indigenous knowledge and contemporary science not often reflected in school science practices [13]. According to the National Center for Women & Information Technology (NCWIT), existing barriers can be diminished by increasing Native American students' exposure to STEM fields, particularly in grades K-12. STEM education needs to begin early in students' educational careers and endure throughout. Students tend to make decisions on future career pathways in STEM-related disciplines long before they enter college [14, 15]. Upper elementary and middle-school students particularly are at a critical stage in forming their views on future STEM careers [16, 17]. Stevenson and Wagner [18] state that middle school is a critical time for students, especially girls when making decisions on their future. Yet, many students lose interest in science-related areas as early as elementary school, with girls being more impacted by this than boys [19–21]. Finally, socio-cognitive research suggests that when students participate in STEM activities at the K-12 level, it affects self-efficacy and motivation that influence success in future careers [9, 22]. Therefore, it is important to provide early-age opportunities to develop positive attitudes toward STEM learning and STEM-related careers.

3.2 Utilization of Spatial Design and VR/AR Technology Skills

Spatial design and VR/AR technology skills are vital to the ICT workforce. Spatial design involves the use of thinking and visualizing space and is used as a general umbrella term commonly associated with the domains of architecture, landscape architecture, landscape design, and interior design. Spatial thinking is a crucial skill for student success in many STEM degrees [23] and it is defined as a combination of three elements: concepts of space, tools of representation, and processes of reasoning [24]. Even though not traditionally considered as STEM fields, domains such as architecture and interior design often work in collaboration with other traditional STEM fields such as engineering disciplines. A bill of Congress [25] has recognized architecture as a STEM subject, emphasizing the increasing role of spatial design in the synthesis of art and science. Furthermore, spatial skills are utilized in a variety of STEM disciplines that could range from a biologist focusing on the double helix of a DNA structure, to an engineer designing and manufacturing 3D structures, to an environmental scientist modeling geographic systems. Research suggests that spatial skills can be enhanced through spatial design education, specifically architectural education [26, 27]. A number of studies have employed VR and AR environments for spatial ability improving exercises [28–31]. Some studies suggest that the training exercises to improve certain abilities are capable of transferring to improve other skills [32] and can last even months afterward [33]. Even limited amounts of training in spatial abilities can be very effective to the extent that even 12 min of instruction is capable of improving spatial abilities which would be equivalent to three years of untutored development [34]. These methods of improving spatial intelligence are supported by theories in neuroplasticity and offer guidance for developing spatial design curriculum using VR/AR technologies to develop critical spatial abilities.

3.3 Technologically Enhanced Learning Environments

Technology-rich and inclusive learning environments support community STEM capacity. 'Ethnocomputing,' or culturally responsive computing, is a pedagogical approach that values student knowledge, recognizes different knowledge systems, and prioritizes community relationships through a sense of 'design agency' that explicitly connects computational thinking to cultural dimensions. Open-ended design tools in out-of-school settings are needed to develop 'supportive space for youth to explore and work out their own identities and stances on technology' [35, pg. 245]. Design studio-based education is where a professional situation is reproduced in an academic context [36] and offers a prime example of a collaborative, multi-sensory, learner-centered, and experiential problem-based learning/teaching environment [37]; one capable of supporting best practices for teaching math and science [38]. Supported by a technology-rich environment, design studio education is capable of supporting several underutilized systemic strategies that have been documented to work with Native American students and their learning environments, such as collaborative learning, more and diverse learning environments, and using culturally relevant pedagogy, research, and evaluation processes [4]. Culturally relevant theoretical frameworks, such as Funds of Knowledge [39] and First Peoples Principles of Learning [13], inform appropriate strategies for connecting culture and STEM learning in a culturally responsive science curriculum [40]. Merging indigenous pedagogies and epistemologies with rich STEM experiences can foster a sense of inclusion, greater appreciation for science, and more informed decisions about careers [41].

4 Method

This study employs a design-based research (DBR) approach to explore ways to engage and increase interest in STEM by Native American students in communities in the midwest of the US. A design-based research is an 'educational practice through iterative analysis, design, development, and implementation, based on collaboration among researchers and practitioners in real-world settings, and leading to contextually sensitive design principles and theories' [42, p. 6]. DBR is not merely about iterations and evaluations. Research using the DBR methodology is also based on true tested prior research and theory, while at the same time building or contributing to 'theories of teaching and learning, and produce instructional tools that survive the challenges of everyday practice' [43].

The afterschool program provides and engages Native American students in technology-rich experiences, while recognizing and acknowledging cultural components of science learning. The project research plan provides evidence for the expansion of the program and advanced knowledge about Native American youths' perspectives, experiences, and early career thinking related to STEM degrees. The research plan includes quantitative data collection related to NAMSAS students' spatial thinking and computing abilities, as well as their attitudes about STEM careers and motivations. The guiding research questions focus on three core areas of inquiry: early-age career thinking, STEM content and skills, and afterschool STEM capacity.

RQ1: In what ways does participation in the NAMSAS Afterschool program influence students' career interests and aspirations related to college and STEM careers?

RQ2: What impact do VR/AR learning experiences have on students' spatial thinking and computing confidence?

RQ3: In what ways does the NAMSAS program influence youth educators' attitudes, confidence, and ability to teach STEM-related topics in their communities?

The method is characterized by iterative approaches of implementing and engaging with innovations in a real-world environment. The data collection for this study involved different interactions that will take a span of four years. At present, the research team has completed three phases.

5 Data Collection

5.1 Pre-Post Surveys

Students at the afterschool program who participated in the study completed a pre-post survey. The pre-survey includes the Native Self-Actualization Placement Assessment (NSA-PA) [44] and/or equivalent measures of cultural worldview. Similar to Aschbacher et al. [45] Is science me? questionnaire, the pre-survey generates a STEM identity profile [46] that provides an empirical and multi-dimensional baseline assessment across factors commonly known to influence individual STEM career thinking (i.e., career awareness, academic preparation and achievement, identification with STEM careers, self-efficacy, external environmental factors, attitudes about science, interest and enjoyment levels, social support and sense of belonging, and motivation for and identification with science). Pre-post items then assess changes in spatial thinking ability using instruments like the Mental Rotations Test (MRT-A) [47]. Student performance tasks and work products provided additional evidence of spatial thinking outcomes. Changes in students' attitudes about STEM learning and careers were assessed using items from the Career Interest Questionnaire (CIQ) [48], the Children's Attitudes Toward Technology Scale (CATS) [49], the Changes in Attitudes about the Relevance of Science-Version A (CARS) [50], and Draw-a-Scientist Test (DAST) [51].

5.2 Afterschool Students Focus Groups

Focus groups with afterschool student participants provided qualitative evidence of changes in STEM career thinking and motivation characterizing the dialectical experience of STEM identity development. Students completed focus groups at the end of the fall and spring terms during the first year of implementation at each afterschool site. A semi-structured interview schedule will account for culturally relevant considerations (e.g., communal community structure or sense of connectedness) in career thinking [52]. As a component of the afterschool program, individual students completed a career-life assessment activity, which required them to get the perspectives of one family member and one community across 8 dimensions identified in the Career-Life Planning Model for 1st Nations People [52]. These career-life plans and focus groups provide a unique opportunity to inspire students to think about STEM careers and for researchers to better understand from a youth perspective the influence of family and community on early-age career thinking.

5.3 Pre-Post-Post Surveys

Youth educators participating in the project completed a pre-post-post survey. The survey will be administered before and after the educator workshop and subsequently at the end of each project implementation year. The design will capture changes in teaching self-efficacy after training and multiple implementations using the Science Teaching Efficacy Belief (STEBI-B) [53]. The context beliefs about teaching science [54] will help identify the personal agency belief patterns of the teachers and their perceptions about the afterschool program strengths and weaknesses. Additional items will be included related to the use of VR/AR technologies, personal identification with science, and attitudes about higher education and STEM careers.

5.4 Youth Educators Focus Groups

Focus groups were conducted with youth educators during the summer workshop to facilitate better understanding of the STEM capacity and experiences of each afterschool program. Semi-structured interviews will be developed and analyzed using Bronfenbrenner's ecological model to identify the factors and systems contributing to the K-12 STEM learning ecosystem [55]. Additional focus groups will be conducted annually with repeat participants during NAISEF to capture changes over time and identify program sustainability needs.

5.5 Narrative Programming Reports

Youth educators will provide narrative programming reports each semester using FlipGrid, an online video discussion platform. The narrative programming reports will be archived and available to all youth educators, creating a space for exchanging (and researching) implement approaches, program experiences, and community reactions. Analysis will focus on the advantages, strategies, and challenges of PBL on students learning. Narrative programming reports offer the opportunity to compare teacher and student reports for varying perspectives on inclusion and engagement aspects of the PBL experiences.

5.6 STEM Engagement Observations

Project researchers will observe the hackathons using the STEM Engagement Observation Protocol (STELAR instrument). Graduate students will be trained as observers and score individuals' engagement in technology and collaborative learning during the hackathons. Observations about sequential segments of activity will provide evidence of what is being done, with what resources, and how students are able to navigate, use, and benefit from the technology-rich experiences offered in the afterschool program. The protocol captures active and passive involvement by NAMSAS students, as well as student affect and whom students interact (e.g., community members, expert visitors, or family members) during the hackathon.

6 Findings

6.1 First Iteration

This stage of the project was a three-day workshop for afterschool educators. During the workshop, afterschool educators were introduced to the history and artifacts from the Spirit Lodge in the Craig Mound, a ritual and ceremonial center. In addition, as an introduction to spatial design in VR/AR, afterschool educators replicated the virtual representation of the Spiro mound. Survey was conducted prior to the start of the workshop, and daily reflections were collected from participants at the end of each workshop.

6.2 Second Iteration

During this phase, we designed seven different modules on immersive technology in connection with the 'Prophecy of the Seven Fires of the Anishinaabe' as a cultural relevance. Each module was explored bi-weekly. By the seventh module, which is the last module and titled 'the Hackathon project,' students collaborate with elders in redesigning their cultural center using SketchUp.

6.3 Third Iteration

This phase was the latest iteration of the program which adopted all the previously designed modules from the second iteration. The immersive technology involved in this iteration was connected to the seven surviving clans of the Otoe-Missouria tribe. These are the bear, beaver, elk, eagle, buffalo, pigeon, and owl tribes. The clans' tribal creation story was described through seven educational modules as one of many versions that described the origin of these clans. Each module was explored bi-weekly and in the seventh module, which is the last module the students applied what they have learned to improve their classroom space. The students collaborated with one another to redesign their classroom space successfully using SketchUp.

7 Discussion

This discussion is based on the findings we reported on the three iterations of the afterschool program.

7.1 *First Iteration*

During the first iteration, the goal was to better understand the educators' prior STEM teaching experiences and instructional comfort with STEM education, as well as teaching beliefs about technology. 86% felt comfortable or somewhat comfortable implementing STEM and technology activities. These findings overall demonstrate how the amount of experience of the teachers, commitment to STEM education, and available resources varied across each afterschool program.

7.2 *Second Iteration*

The goal of the second interaction was to engage the afterschool students with immersive technology in connection with the 'Prophecy of the Seven Fires of the Anishinaabe' as a cultural relevance. Findings from the second iteration were based on observation of student's activities, focus group discussions, and their performances from the first module to the last module. Therefore, the themes from this iteration focused more on students' interaction with the technology devices adopted for the study.

Curiosity: A curious environment prepares students for learning and prepares them to ask questions. The children were curious about the activities they were doing. A kid once asked the question, 'are we doing that next,' when introduced to a new activity using the augmented reality app on the tab, he asked another question, 'I wonder what would happen if I do it all at once' (referring to his rocket coloring activity). The more the students got curious, the more they wanted to engage and explore the devices.

Creativity: Providing the students the opportunity to work independently increases their creativity. Increased curiosity leads to increased creativity; every time the students worked on a new module, we noticed increased creativity; they brought up new ideas and even discussed with peers how to successfully execute the new ideas. For instance, when we presented students with coloring activities at the beginning of the modules, there was evidence of disengagement; however, when they had the opportunity to bring their work to life through an AR application (Quiver), students colored differently. Another evident example of increased creativity was how varied students' actions were when asked to choose an action; they chose different activities, built differently on the Sketup app, and owned their design by adding different components. In addition, when working on the SketchUp app, they designed using different features; their work was always different.

Motivation: Self-motivation is an essential aspect of learning. Providing students with the opportunity to own their learning influenced their motivation level. On one of our visits to the afterschool center, we employed a coloring activity as a starter for the module. From observation, students showed disengagement and a lack of interest in the coloring activity because students are increasingly drawn to digital and interactive forms of entertainment and learning [56]. Coloring with traditional materials might hold a different appeal than digital or multimedia activities; however, when the instructor introduced the AR application (Quiver), they colored some more to view their coloring using the AR application. Witnessing technology integration, the students rekindled their enthusiasm for the coloring task, engaging more fervently, driven by the prospect of experiencing their creations in augmented reality. This vivid example underscores the transformative potential of technology in enhancing a student's motivation to participate actively in the learning process. Technology plays a significant role in students' motivation to engage significantly in activities.

Computing Confidence: At the beginning of the implementation stage, the graduate assistants had to help with computing activities like logging on to the computer and showing students some keys on the keyboard. However, by the fourth module, students shared their ideas and started working independently on the computer; students exhibited a significant boost in their computing confidence, marking a stark contrast to their initial engagement. Field notes captured expressions of newfound self-assuredness and enthusiasm. Some statements documented during this phase included phrases like 'wow' and 'It wasn't challenging.' Notably, in the fourth module, one student's exuberance culminated in an enthusiastic exclamation, 'this is amazing,' followed by another's declaration, 'This is cool.' These verbal expressions are compelling evidence of the burgeoning computing confidence among the students."

7.3 Third Iteration

In the third iteration, outcome measures were aggregated from the observation of the students all through the seven modules related to the creation story of the seven clans. The outcome measures were based on the three aforementioned research questions (RQ):

RQ1: In what ways does participation in the NAMSAS afterschool program influence students' career interests and aspirations related to college and STEM careers?

On average about fifteen Native American students participated in each of the seven modules. The students were very active and excited throughout the class sessions from Introduction to VR/AR visualization to the concepts of human-centric designs. The students gained tangible skills and introductory knowledge linked to three main technological tools: Virtual reality (VR), augmented reality (AR), and 3D printing, along with their basic applications. The students were excited to learn how these tools can be used for supporting indigenous history about 'creation story of the seven clans' and tribal artifacts, including bears, beavers, and owls, among others. The program allowed the students to expand their career choices in STEM-related careers such as architecture, civil engineering, interior design and mechanical engineering, and gaming. The students' very high level of excitement was visible in the 'wow' and 'how cool' remarks made by majority of them when they experienced the use of the Oculus VR goggles, the five bears printed in orange using the 3D printer, and the 3D images coming into life from colored picture sheets using Quiver mobile AR application. This 'wow' factor enriched the students' minds, enhanced their creativity, and increased their aspirations toward making informed career choices in college. It was interesting to observe how the students reacted positively to learning about these emerging technologies which exposed them to relevant and useful real-life career and professional jobs they can choose in the future.

RQ2: What impact do VR/AR learning experiences have on students' spatial thinking and computing confidence?

The impact VR/AR learning experience had on the students' spatial thinking and computing confidence was tremendous. This was visible right from the moment when they saw and passed around the 3D bears; to illustrate this, their eyes lit up and they were eager to learn more on how to print similar artifacts using the 3D printer. In addition, wearing the VR goggles gave them the perception of the depth and how to move around in a VR space. The colored pages exercise allowed them to witness how 3D objects came to life and rotate when the mobile device that was used to scan the colored paper. The students learned about the isometric view and the three types of perspective view. The students had the opportunity to use an aging simulation gerontology suit (G.E.R.T Suit) to help develop solutions through the empathic design process. The empathic design process has been known to spark innovation. All this knowledge was very useful and beneficial to spatial thinking ability. It was evident they were all having a good time playing and having fun with their suits. The ease with which they related with the models and drawings showed that they were very confident and comfortable with 3D spatial drawings as they progressed in their VR learning experience.

RQ3: In what ways does the program influence afterschool educators' attitudes, confidence, and ability to teach STEM-related topics in their communities?

The afterschool educators were very open, supportive, and receptive to the method of STEM teaching introduced by the program facilitators. This method involved using technological tools as teaching aids. The educators loved the way their students were interacting with the VR goggles and GERT suits as they showed interest in the questions asked by their students and the answers provided by the facilitators. They were happy and excited about the different components from the program which could be adopted for their teaching and improving their overall students experience in the classroom.

8 Conclusions

The program's influence on participants' interest in STEM became increasingly apparent as the research progressed. Over the course of multiple sessions, during which students created a dorm room designs, explored advanced design applications beyond the computer's primary use, and brought their designs to life through 3D printing, their enthusiasm for experiencing various technologies intensified noticeably. Notably, students' sustained interest was evident through inquiries about upcoming sessions, reflecting a continued curiosity and engagement with the program's activities.

The heightened engagement extended beyond the participants themselves. Instructors also demonstrated a strong commitment to the program, actively engaging in students' activities and wholeheartedly fostering student involvement. In some sessions, instructors even volunteered to participate directly in student activities, exemplifying their keen interest and dedication to the program's success.

References

1. Hsu, Y.S., Lin, Y.H., Yang, B.: Impact of augmented reality lessons on students' STEM interest. Res. Pract. Technol. Enhanc. Learn. **12**(1), 1–14 (2017)
2. Williams, D.H., Shipley, G.P.: Cultural taboos as a factor in the participation rate of Native Americans in STEM. Int. J. STEM Educ. **5**(1), 1–8 (2018). https://doi.org/10.1186/s40594-018-0114-7
3. Intel & National Center for Women & Information Technology (NCWIT): Collaborating to Grow the Pathway of Native Americans in STEM (2016). Retrieved 23 Feb 2023, from https://ncwit.org/resource/nativeamericanwhitepaper/
4. Bowman, N.R.: Cultural differences of teaching and learning: a Native American perspective of participating in educational systems and organizations. Am. Indian Quart. **27**(1), 91–102 (2003)
5. Bryson, S.: Approaches to the successful design and implementation of VR applications. Virtual Reality Appl. 3–15 (1995)

6. Fischer, J., Cunninham, D., Bartz, D., Wallraven, C., Bulthoff, H., Stasser, W.: Measuring the discernability of virtual objects in conventional and stylized augmented reality. In: Proceedings of Eurographics Symposium on Virtual Environments, pp. 53–61.
7. National Science Board (NSB): Science and Engineering Indicators 2014. National Science Foundation (NSB 14-01), Arlington VA (2014)
8. National Science Foundation (NSF): Women, Minorities, and Persons with Disabilities in Science and Engineering (NSF 17-310) (2017). Retrieved 23 Feb 2023, from https://www.nsf.gov/statistics/2017/nsf17310/static/downloads/nsf17310-digest.pdf
9. Dierking, L.D.: A comprehensive approach to fostering the next generation of science, technology, engineering, and mathematics (STEM) education leaders. New Educ **6**(3–4), 297–309 (2010)
10. Chittum, J.R., Jones, B.D., Akalin, S., Schram, Á.B.: The effects of an afterschool STEM program on students' motivation and engagement. Int. J. STEM Educ. **4**(1), 1–16 (2017)
11. Zoldosova, K., Prokop, P.: Analysis of motivational orientations in science education. Int. J. Sci. Math. Educ. **4**(4), 669–688 (2006)
12. Bell, P., Lewenstein, B., Shouse, A.W., Feder, M.A.: Learning Science in Informal Environments: People, Places, and Pursuits. National Academies Press, Washington (2009)
13. First Nations Education Steering Committee: Science First Peoples Teacher Resource Guide: Grades 5 to 9. First Nations Education Steering Committee & First Nations Schools Association, West Vancouver, BC (2016)
14. George, P.S.: The Middle School--and Beyond. Association for Supervision and Curriculum Development, VA (1992)
15. Sadler, P.M., Sonnert, G., Hazari, Z., Tai, R.: Stability and volatility of STEM career interest in high school: a gender study. Sci. Educ. **96**(3), 411–427 (2012)
16. Maltese, A.V., Tai, R.H.: Pipeline persistence: examining the association of educational experiences with earned degrees in STEM among US students. Sci. Educ. **95**(5), 877–907 (2011)
17. Christensen, R., Knezek, G.: Contrasts in student perceptions of STEM content and careers. In: Society for Information Technology & Teacher Education International Conference, pp. 2048–2053. Association for the Advancement of Computing in Education (AACE) (2013, March)
18. Reisberg, R., Ziemer, K.S., Knight, M., Wong, P., Swan, A., Erkut, S., Camesano, T.: A Model STEM Team Collaboration: Four Schools for WIE. Women in Engineering ProActive Network (2006)
19. Abell, S., Lederman, N.G. (eds.): Handbook of Research in Science Education. Lawrence Erlbaum, Mahwah, New Jersey (2007)
20. Davison, D.M., Miller, K.W., Metheny, D.L.: What does integration of science and mathematics really mean? Sch. Sci. Math. **95**(5), 226–230 (1995)
21. Lederman, N.G., Niess, M.L.: The nature of science: naturally? Sch. Sci. Math. **97**(1), 1 (1997)
22. Bandura, A.: Social cognitive theory: an agentic perspective. Annu. Rev. Psychol. **52**(1), 1–26 (2001)
23. Newcombe, N.S.: Picture this: increasing math and science learning by improving spatial thinking. Am. Educ. **34**(2), 29 (2010)
24. DeSouza, A., Downs, R.: The Nature of Spatial Thinking. Learning to Think Spatially: GIS as a Support System in the K-12 Curriculum (2006)
25. ED Gov: CARL D. Perkins Career and Technical Education Act of 2006 (2006). Available at https://www.govinfo.gov/content/pkg/COMPS-3096/pdf/COMPS-3096.pdf
26. Stringer, P.: Drawing training and spatial ability. Ergonomics **18**(1), 101–108 (1975)
27. Sarı, R.M.: The effect of architectural education on the spatial intelligence progress American. J. Sci. Res. **17**, 94–102 (2010)
28. Chandrasekera, T., Yoon, S.-Y.: Adopting augmented reality in design communication: focusing on improving spatial abilities. Int. J. Architectonic Spat. Environ. Des. **9**(1), 1–14 (2015)
29. Rafi, A., Samsudin, K.A., Ismail, A.: On improving spatial ability through computer-mediated engineering drawing instruction. J. Educ. Technol. Soc. **9**(3), 149 (2006)

30. Durlach, N., Allen, G., Darken, R., Garnett, R.L., Loomis, J., Templeman, J., Wiegand, T.E.V.: Virtual environments and the enhancement of spatial behavior: towards a comprehensive research agenda. Presence: Teleoperators Virtual Environ. **9**(6), 593–615 (2000)
31. Sorby, S.A.: Developing 3-D spatial visualization skills. Eng. Des. Graph. J. **63**(2) (2009)
32. Wright, R., Thompson, W.L., Ganis, G., Newcombe, N.S., Kosslyn, S.M.: Training generalized spatial skills. Psychon. Bull. Rev. **15**(4), 763–771 (2008)
33. Terlecki, M.S., Newcombe, N.S., Little, M.: Durable and generalized effects of spatial experience on mental rotation: gender differences in growth patterns. Appl. Cogn. Psychol. **22**(7), 996–1013 (2008)
34. Rovet, J.: Cognitive and neuropsychological test performance of persons with abnormalities of adolescent development: a test of Waber's hypothesis. Child Develop. 941–950 (1983)
35. Kafai, Y., Searle, K., Martinez, C., Brayboy, B.: Ethnocomputing with electronic textiles: culturally responsive open design to broaden participation in computing in American Indian youth and communities. In Proceedings of the 45th ACM Technical Symposium on Computer Science Education, pp. 241–246 (2014, March)
36. Masdéu, M., Fuses, J.: Reconceptualizing the design studio in architectural education: distance learning and blended learning as transformation factors. ArchNet-IJAR **11**(2), 6 (2017)
37. Kurt, S.: An analytic study on the traditional studio environments and the use of the constructivist studio in the architectural design education. Procedia Soc. Behav. Sci. **1**(1), 401–408 (2009)
38. Zemelman, S., Daniels, H., Hyde, A.: Best practice: today's standards for teaching and learning in America's schools. Educ. Rev. (2005)
39. Stevens, S., Andrade, R., Page, M.: Motivating young Native American students to pursue STEM learning through a culturally relevant science program. J. Sci. Educ. Technol. **25**, 947–960 (2016)
40. Stephens, S.: Handbook for Culturally Responsive Science Curriculum (2001)
41. McMahon, T.R., Griese, E.R., Kenyon, D.B.: Cultivating Native American scientists: an application of an Indigenous model to an undergraduate research experience. Cult. Sci. Edu. **14**, 77–110 (2019)
42. Wang, F., Hannafin, M.J.: Design-based research and technology-enhanced learning environments. Educ. Tech. Res. Dev. **53**(4), 5–23 (2005). https://doi.org/10.1007/BF02504682
43. Shavelson, R.J., Phillips, D.C., Towne, L., Feuer, M.J.: On the science of education design studies. Educ. Res. **32**(1), 25–28 (2003)
44. Brown, S.A.S.: Cultural Worldview Assessment Placement: Establishing the Validity of the Native Self Actualization Placement Assessment (NSA-PA). ProQuest (2007)
45. Aschbacher, P.R., Li, E., Roth, E.J.: Is science me? High school students' identities, participation and aspirations in science, engineering, and medicine. J. Res. Sci. Teach. **47**(5), 564–582 (2010)
46. Angle, J., Colston, N., French, D.: STEM Identity Profile of National Lab Day Students. School Science and Mathematics Association. Little Rock, AK (2018, October)
47. Peters, M., Laeng, B., Latham, K., Jackson, M., Zaiyouna, R., Richardson, C.: A redrawn Vandenberg and Kuse mental rotations test-different versions and factors that affect performance. Brain Cogn. **28**(1), 39–58 (1995)
48. Tyler-Wood, T., Knezek, G., Christensen, R.: Instruments for assessing interest in STEM content and careers. J. Technol. Teach. Educ. **18**(2), 345–368 (2010)
49. Frantom, C.G., Green, K.E., Hoffman, E.R.: Measure development: The children's attitudes toward technology scale (CATS). J. Educ. Comput. Res. **26**(3), 249–263 (2002)
50. Siegel, M.A., Ranney, M.A.: Developing the changes in attitude about the relevance of science (CARS) questionnaire and assessing two high school science classes. J. Res. Sci. Teach. **40**(8), 757–775 (2003)
51. Laubach, T.A., Crofford, G.D., Marek, E.A.: Exploring native American students' perceptions of scientists. Int. J. Sci. Educ. **34**(11), 1769–1794 (2012)
52. McCormick, R.M., Amundson, N.E.: A career—life planning model for first nations people. J. Employ. Couns. **34**(4), 171–179 (1997)

53. Enochs, L.G., Riggs, I.M.: Further development of an elementary science teaching efficacy belief instrument: a preservice elementary scale (1990)
54. Lumpe, A.T., Haney, J.J., Czerniak, C.M.: Assessing teachers' beliefs about their science teaching context. J. Res. Sci. Teach. **37**(3), 275–292 (2000)
55. Basham, J.D., Israel, M., Maynard, K.: An ecological model of STEM education: operationalizing STEM for all. J. Spec. Educ. Technol. **25**(3), 9–19 (2010)
56. Bogiannidis, N., Southcott, J., Gindidis, M.: An exploration of the possible educational opportunities and the challenges at the intersection of the physical and digital worlds occupied by 10–14 year-old students. Smart Learn. Environ. **10**(1), 1–21 (2023)

Evaluating Learning and Engagement in Immersive Spaces

Tutaleni Iita Asino, Nicole Colston, and Clement Abai

Abstract This chapter discusses the application of an evaluative framework for understanding the use of AR/VR as a supplement to a library program promoting youth and family participation in outdoor learning called the Backyard Explorers (BYE) program. A design-based research (DBR) approach was utilized in designing an immersive learning space called Virtual Citizen Science Expo (VCSE) that contains BYE rooms.

Keywords Augmented Reality · Virtual Reality · Design Based Research · Citizen Science · Evaluating Learning · Emerging Learning Spaces

1 Introduction

Immersive learning environments which often use virtual reality (VR) are becoming common across the education landscape. This is in large part because AR and VR have been widely demonstrated to have positive effects on students' motivation, presence, and engagement [1, 2]. Given their prevalence in education, there is a need for us to develop new methods for evaluating engagement in these hybrid spaces and to produce research that informs educators about how to design and plan to use immersive technologies [3]. This chapter discusses the application of an evaluative framework for understanding the use of AR/VR as a supplement to a library program promoting youth and family participation in outdoor learning called the Backyard Explorers (BYE) program. A design-based research (DBR) approach was utilized in designing an immersive learning space called Virtual Citizen Science Expo (VCSE) that contains BYE rooms. A DBR conceptual framework was used in designing BYE, the aim of the DBR conceptual framework was to make sense of the logical flow of the design of the virtual space, and to demonstrate the phase at which we identified our research problem within the educational context, and the theory we used in our research. This was followed by the design processes of DBR which includes

T. I. Asino (✉) · N. Colston · C. Abai
Oklahoma State University, Stillwater, OK, USA
e-mail: tutaleni.asino@okstate.edu

Y. Cai et al. (eds.), *Virtual and Augmented Reality Technology-Enhanced Learning*, Gaming Media and Social Effects, https://doi.org/10.1007/978-981-96-2332-7_10

designing the immersive learning space and finally, the design outcome after each iteration of the design process. By following this conceptual framework, we were able to understand how design choices can positively impact students' motivation, presence, and engagement.

1.1 What is Design-Based Research?

Anderson and Shattuck [4] defined DBR as a methodology designed by and for educators that provides possible ways to increase impact, transfer, and translate research in education to improve practices. Hung [5] stated that DBR is a methodology that is driven by theory to generate interventions that are complex that can be improved through empirical research that can contribute to the more basic understanding of the underlying theory. Wang and Hannafin [6] defined DBR as "a systemic but flexible methodology aimed to improve educational practices through iterative analysis, design, development, and implementation, based in collaboration among researchers and practitioners in real-world settings, and leading to contextually sensitive design principles and theories." Lyon and Magana [7] defined DBR as interdisciplinary mixed-method research approach conducted in the field that serves applied and theory building processes. In educational research, DBR is an accepted paradigm that forms a usual association between theory and innovation in classroom practices [8]. The approaches used in DBR are ideal for research projects that focus on theory and practice [9]. The five main characteristics that define DBR are (1) the methodology is pragmatic, (2) grounded, (3) iterative and flexible, (4) integrative, and (5) contextual [7]. DBR is pragmatic because it contributes to educational theories and practices that result in design principles that can be applied in educational settings such as a classroom [10]. The methodology is grounded; therefore, to generate interventions, DBR should be able to pull from current educational theories or evidence-based teaching methods and be iterative. It is iterative in nature because the design is refined continuously using multiple methods such as mix methods to ultimately answer the research question [6]. Lastly, DBR is easily embedded into a classroom context; therefore, it is contextual [7].

Figure 1 presents a DBR conceptual framework that consists of three phases when designing a learning environment. Phase 1 includes theory to use in the study, the problem that a researcher found that is context specific in a learning environment that is backed up by previous studies [11]. For example, Asino et al. [12] found in their study that there is a problem with promoting environmental science education in rural areas in the great plains of the US. This led to Asino et al. [12] designing an environmental science education learning space that is accessible anywhere at any time that learners can enter and learn about water, drought, and environmental monitoring using VR. In this study, the diffusion of innovation theory was used to explain how learners adopt the use of VR over time as a learning tool to learn about water, drought, and environmental monitoring. In phase 2, the DBR methodology is applied. The researcher analyzes the problem and explores possible interventions; the

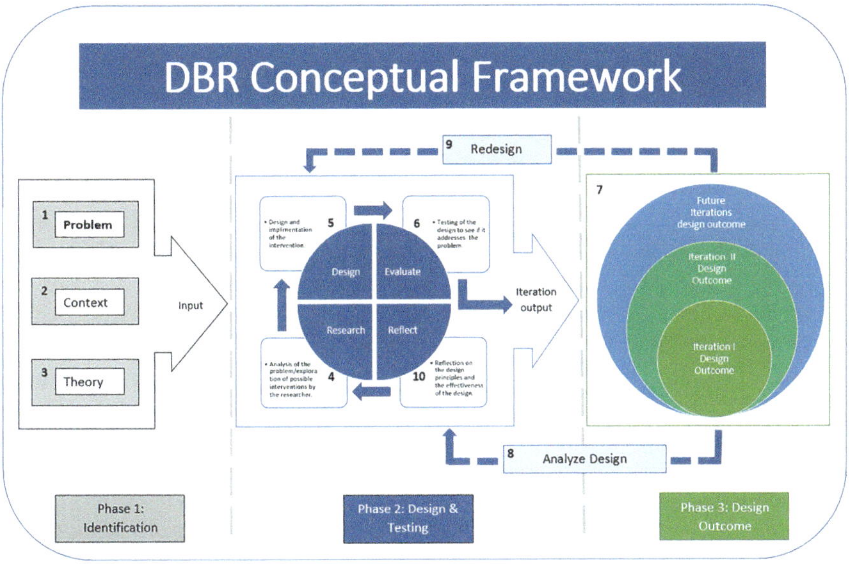

Fig. 1 Design-based research conceptual framework

learning environment then is designed, implemented, and tested to see if it addresses the learning problem. Finally, the researcher reflects on the design principles and the effectiveness of the design. Phase 3 is the design of the learning environment after each iterative cycle of the design process to create a learning environment that is more authentic and realistic. In this phase, after each iterative cycle, the redesigned learning environment is expected to be more authentic and realistic. For example, in this framework, the first design outcome (Iteration I Design Outcome) is represented by a small circle. As the design goes through each iteration, it becomes more authentic and realistic. The authenticity of the learning space in the second iteration is represented by the growth of the circle from small to medium (Iteration II Design Outcome). As some DBR projects end at the second iteration, others continue to apply multiple iterations after the second design. This is represented by the large circle (Future Design Outcome). This means that the more iterations a DBR project applies, the more authentic and realistic the design of the learning environment becomes.

2 Literature Review

Literature published on the use of DBR in education and how this methodology is utilized in different educational contexts demonstrated that this methodology has impacted learning positively. For example, the different educational contexts that utilized DBR such as designing virtual learning spaces for an environmental science education program [12], a medical emergency course [13], online flipped classrooms [5, 9], and a training solution for pre-service teachers [14] are some of the studies that demonstrate the application of DBR in designing authentic educational environments found that using DBR to design educational spaces led to positive learning outcomes for the learners. Furthermore, studies have shown that there has been an increased utilization of DBR in formal and informal educational settings in the United States (US) and globally to provide technological interventions. DBR has potential and has contributed in a significant manner in the quality and utilization of research in education. Following the DBR iterative approach to design can lead to interventions that can be implemented to create authentic and realistic educational scenarios that can result in positive learning outcomes [4]. The findings from the studies on use of DBR explained why this methodology should be used in designing different educational scenarios. In this chapter, four different educational scenarios demonstrated the use of DBR in designing different contexts in each of the scenarios. Scenario 1 explored the use of DBR to design an environmental science education learning space and a medical emergency course using virtual reality while in scenario 2, DBR was used to design a Math, and English flipped courses. In scenario 3, DBR was utilized in designing a teacher STEM education training solution and an engineering college-level course, and in scenario 4, DBR was used to design mobile learning courses. By discussing the findings, this literature review hopes to demonstrate the impact DBR has on education.

2.1 Scenario 1. Using DBR to Design Immersive Learning Environment

Emerging technologies such as virtual reality (VR) can be very engaging for learners and can allow them to learn about important subjects in different educational contexts [15], for example, using VR to learn about water and drought in an environmental science education [12]. VR enables new ways of interacting with information by allowing instant communication, real-time conversations, and collaboration within purpose-built environments [16–18]. In addition, the VR learning spaces or the immersive learning spaces are virtually available to learners. This means learners can enter these virtual spaces, immerse themselves and learn [12]. With the positives that VR brings, the design of these immersive learning spaces requires design skills and knowledge to build learning spaces that utilizes the DBR methodology. For example, the study by Asino et al. [12] on use of VR for an environmental science education used the DBR methodology to design an immersive learning space called VCSE that learners can enter and learn about drought, water, and environmental monitoring. This study applied three iterations of the design process to collect data for analysis and to develop interventions to improve the overall design of the learning space. The study found that after undergoing the first iterative cycle of the design approaches, initial design issues that were not known were discovered. For example, instructional materials were placed in areas that do not follow the logical flow of how learners move in space, navigational instructions to move around the learning space using different devices were hard to notice when entering the landing spot on the space. In addition, technological issues such as a need for high-speed Internet emerged as learners of mobile devices who have low Internet speed tried to access the learning space. With these issues identified, interventions were developed and implemented. In the second and the third cycle of the iterations design process, the need for additional instructional materials was needed in the space, such as video instructions, 3D model of the water cycle. Learners were then invited into the learning space and their responses and interactions were recorded. The learners found the space fun and engaging, a few found the immersive space exciting while others found the use of immersive spaces as a place to meet and teach virtually interesting.

In a separate study, Cochrane et al. [13] used DBR to design a mobile VR learning space. This study applied four iterations of the DBR design approach in designing two projects: a critical care education course using VR and a visual design course using augmented reality at a university in New Zealand. A four-phase iterative approach was used to develop a more authentic learner-centered educational experience for the students to prepare them for the real-world situations. VR was used to develop simulation scenarios and exercises in an enhanced paramedic education setting. For example, in their learning scenarios, students responded to emergency situations as paramedics, used mobile technologies to communicate and access remote medical databases, recorded their experiences of the scene, simulated environments that were hazardous, and used drones that are mobile piloted to explore inaccessible crash and disaster scenes. The study found that students in these classes felt that they were more prepared to respond to emergency situations in a real-world setting. Findings from [12] and [13] suggest that using DBR as a methodology to design immersive learning spaces leads to fun and engaging learning experiences that are more authentic and realistic for the learners.

2.2 Scenario 2. Use of DBR in Designing Flip Classrooms

Online learning came into existence in 1995 when the Blackboard Learning Management System (LMS), previously known as WebCT was developed. This technology allowed users to upload texts and pdf files. Since then, many distinct overlapping terms have emerged to reference online learning. This included blended learning, online education, online courses, etc. [19]. Increasing use of online education globally and the advancement in technology has led researchers and practitioners to explore innovative ways such as designing flip classrooms for learners. FLIP stands for flexible environment, learning culture, intentional content, and professional educator [5]. In this online learning environment, students have access to the learning materials that they can access anywhere and anytime and collaborate with their peers and teachers before class [9]. Similar approach to designing a virtual learning environment, the DBR methodology can be used to design flipped classrooms. For instance,

Lo and Hew [9] designed a flipped learning environment to examine if this method of teaching supports student engagement in math. The study was conducted in a secondary school in China where the DBR approach was utilized over two years to investigate the methods that support student engagement in a flip math class. Two iterations of the design approaches were used in this study based on reflections from teachers and student interviews to solve issues such as accessing online class materials and engagement in class. The students from flipped class were compared to the students in a traditional class who were provided similar instructions. Based on the findings from exploratory studies, randomized experiences conducted, and review of literature, interventions were developed. A prototype integrated with a flipped lesson that supported student engagement was developed as an intervention and implemented in an authentic setting. Solutions were then evaluated, and activities refined. Finally, the development of the design principles was applied to promote student engagement. The study found that the students in the flipped class performed significantly better than the students in the traditional class and were more engaged compared to the students in the traditional class.

In a separate study, Hung [5] applied the DBR design approaches to redesign an English language course in a flipped classroom environment at a university in Taiwan. The study examined a flipped classroom method of teaching to see if this pedagogy enhanced the traditional courses and contributed to positive attitudes toward learning and higher students' achievements. The flipped classroom was an 18 week-long course redesigned from a baseline course with comparable course requirements, content, and instructor. The goal of this course was to develop competency in English communication through various interactive activities. The study found that students in the flipped classroom outperformed the students in the non-flipped classroom. This finding suggests that by applying the design approaches of DBR to design or redesign educational learning spaces such as flipped classrooms, this will lead to students having positive attitudes toward their learning experiences and there will be improved performances in their learning process.

2.3 Scenario 3. Applying DBR Approaches in STEM Education

Investments in education such as science, technology, engineering, and math (STEM) education are implemented to encourage students to engage in STEM and to develop creativity and innovative skills in their learning process [20]. Research shows that pre-service teachers designing curriculum for STEM education have a great influence in developing students' interest in STEM-related fields [21]. This is in line with [22], on robotics integrated STEM education for in-service and pre-service teachers. The study found that STEM education exposes students to the processes of engineering design, programming, math contexts and connections, and problem-based learning. While these studies demonstrate the importance of integrating STEM education in schools to engage students and to develop their creativity and innovative learning skills, teachers, on the other hand, need to be trained and be prepared with skills to design STEM courses at the pre-service stage [14].

Wu et. al. [14] explored the use of DBR to design an online STEM learning unit training solution for the pre-service teachers in eastern China. The study used a prototype design as the STEM training system for the pre-service teachers to simulate an authentic design practice context which was a virtual internship at a junior school. Three rounds of the iterative design process were utilized in designing the system, with each round addressing practical issues identified in the implementation stage. The first round of designing the online training system was text-based, improvements were implemented in the second round, by converting the design into a visual tool, and finally, in the third round, learning analytics tools were implemented. The study found that after three iterative cycles of design, there was a very significant improvement in the pre-service teachers' perceived TPACK and design thinking, therefore, suggesting that DBR design approaches are an effective method in designing and developing well training platforms and online STEM design practices for pre-service teachers.

A different study by Lyon and Magana [7] used DBR to design a learning environment that implemented computational modeling activities that operationalized computational thinking in a capstone engineering course in a university setting to understand how a designed modeling-based learning environment can be integrated into a naturalistic setting. The findings from the first iteration of the DBR classroom intervention were used to improve the design, the learning materials and collaboration in the classroom to promote computational thinking. In this iterative cycle, a productive failure method of teaching was applied. This means various solutions to pathways to complex and ill-structured problems were explored by the students without any instructional support from the instructors, resulting in students failing before finding an acceptable solution [23]. In the first phase of the productive failures, a problem in a narrative form was given to the students in their learning environment that allowed for collaboration. In this study, a food sterilization intervention complex modeling problem was given to the students. The students did not get any instructional guidance from their instructors but were encouraged to explore different ways to explore the problem. In the second phase, students were given access to the solutions created by experts and to other students. After building their models, students were allowed to view others' solutions, discuss their models' operations with each other, take notes, and compare their models with their others. The students documented this process to help them think through the different decisions others made with their models and compare it to their own. This study demonstrated that using DBR interventions in an engineering course can lead to developing computational thinking in the students' learning process. In all, the findings from these studies suggest that DBR is an effective method in designing and developing a STEM-related learning environment.

2.4 Scenario 4. Utilizing DBR to Design Mobile Learning Environments.

The use of mobile devices such as laptops, personal digital assistants, and mobile phones as learning tools has a great potential in classroom learning environments and in outdoor learning settings [24]. With this technology's ubiquity, the increased use of mobile devices is seen in educational settings globally [25]. However, devices such as laptops and PDAs have their own limitations. These devices have 3–4 h of battery life and need to be recharged. In addition, they require an Internet connection at buildings or areas with wireless access points. Mobile phones on the hand have battery life that lasts longer making them more suitable for a school day [25]. A study by Goundar [25] found that in developed countries, mobile phones have almost 99.999% connectivity rate while developing countries have a 68% of connectivity rate and experience a rapid growth of 10% per year. Mobile phones have several advantages for education compared to laptops and netbooks. They are lightweight, easy to orient, and adjust for digital reading and accessing content, they are fast and have a capability of instant switching between applications that allows for learning activities to proceed with minimal delay, and they are highly interactive because of the touch screen interface and features. In addition, they are versatile. This means learners can carry them inside and outside of buildings without closing and reopening the screen or storing them in cases and can be used for collecting data and notes. Finally, the low cost in developing apps for mobile platforms has allowed many free and low-cost mobile phone apps that are suitable for education to be developed. The potential mobile phones bring to education has caused researchers and practitioners to integrate the use of mobile phones in their studies. One of the methods that have been explored in designing mobile learning environments is DBR. For example, Bressler et al. [26] in their School Scene Investigator, a forensic science game series utilizes mobile augmented reality (AR) to engage students through experiences of flow, a psychological state that is positive that learners experience while playing games that are well designed and trigger interest in science. This project was conducted in a middle school and went through three iteration processes where students played science-related mystery AR games, collected, and interpreted the data they collected using their mobile devices. In each iteration, adjustments were made in the design to slightly fit a different context of the project. The study found that students that experience a higher state of flow indicated interest in science. The study also confirmed that a learning game that is effective and well designed to promote flow can trigger interest. In Schmitz et al. [27] HeartRun project, a cardiopulmonary resuscitation (CPR) training approach that consists of authentic activities, game tasks, locations, and physical objects to support process-oriented learning for first responders, the project's aim was to enhance the psychological preparedness for rescuers and preparing them for more prompt response in unexpected emergencies situation. Data were collected from high school students and their knowledge building practices were analyzed to better understand how to design mobile game learning environments to support learning. Three cycles of the DBR iteration process were used in the testing of the

Table 1 Four dimensions of engagement by immersive technology feature

	AR logbooks	Make it rain gauge	Virtual Citizen Science Expo
Learner(s)	Youth and families	Librarians and patrons	Public, informal science
Pedagogical model	Heads down, heads up	Place-based	Expert speaker
Representation	AR/3D models of natural phenomena	AR/simulated rain, with links to videos and maps	3 drought exhibitor rooms, 2 BYE rooms, project posters
Context	Library program (pop-up or online)	Passive indoor/outdoor display at the library	Virtual exhibit hall in Mozilla Hubs, in avatar form

prototype, the first trial, and second trial of the HeartRun project. The study found that participants like using simulation games and positively rated the game. In addition, the study showed that playing games with a mobile device motivated the learners, and integrations of time-critical tasks such as providing CPR to a manikin contributed to the learners' motivation in this study.

3 Method

Our research project focuses on the design and use of new technologies to support library programs related to citizen science. We applied a four-dimensional (Table 1) framework for understanding engagement within immersive technology spaces: the learner, the pedagogic model, the representation, and the context for learning [28]. An inductive method of data collection over several design iterations included observation, screen/video captures, feedback from librarians, and participant surveys.

4 Results and Discussion

This nature inquiry program follows an experiential learning cycle: Learn (watch video)-Try (logbook challenge)- Act (volunteer for citizen science). The BYE program incorporates immersive technologies in three ways: (a) logbook worksheets enhanced with AR codes for 3D models (e.g., water cycle), (b) an AR code on a rain gauge to raise awareness about drought monitoring, and (c) a VR exhibit hall (accessed on mobile devices or computer) sharing resources for nature inquiry and citizen science. The BYE program is designed for easy implementation by librarians, who often have limited time and experience in science education or nature facilitation. The wider trend of immersive technologies whether in natural, virtual, or blended environments for outdoor education is still relatively new. The evaluative framework for engagement helped to clarify the role of AR/AR technologies in an experiential learning setting. Our goal is to develop a theoretical and practical understanding of how to use mobile technologies and virtual environments to enhance environmental education. Our future program implementation and design-based research will focus on learner experience and engagement.

In step 1 of the experimental learning cycle (Fig. 2) in the BYE program, learners in the library start by watching a video and learn about the weather. In step 2, the learners are provided an AR-enhanced logbook where they complete the given task and fill out the logbook challenge. In addition, they learn about the water cycle by interacting with the AR water cycle. The Zappar App downloaded on the mobile devices is used to scan the zapcode. Once scanned, the 3D model water cycle is activated and is placed in the learning space where learners can view and interact with the model. During the learning cycle, the learners also learn about different citizen science organizations and projects that they can volunteer in and apply what they have learned about weather, data collection, and reporting. And finally, in step 3, the learners can join the citizen science group of their choice and volunteer by conducting environmental monitoring activities such as water qualities checks or rainfall measurement and reporting.

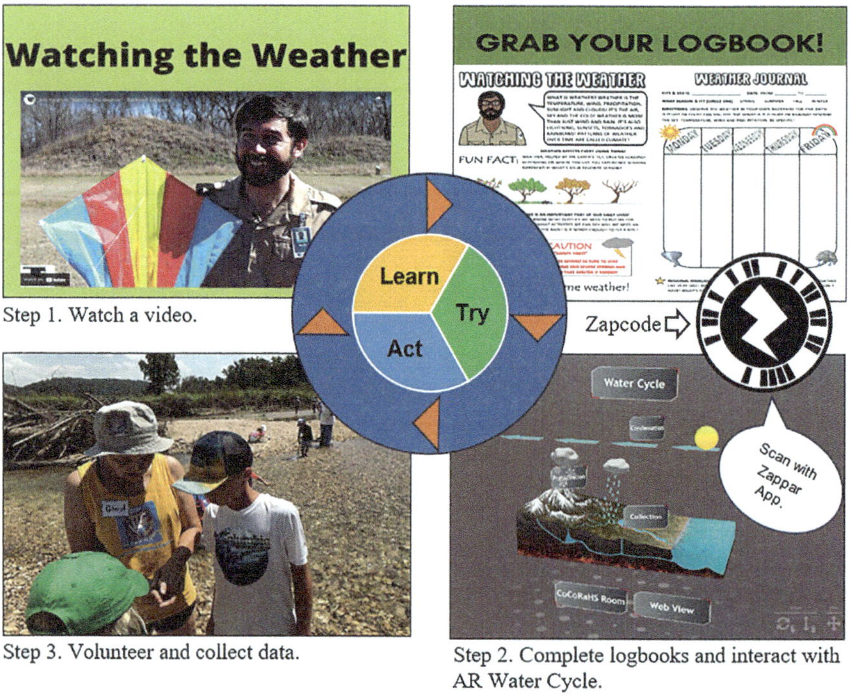

Fig. 2 Experimental learning cycle

5 BYE Immersive Technologies Integrated Learning

(a) Logbook worksheet with AR Code.
(b) AR rain gauge for drought awareness.
(c) VR exhibit hall with nature inquiry and citizen science resources.

The integration of immersive technologies in the BYE program included the logbook worksheet with AR code (Fig. 3) that activates a 3D water cycle (Fig. 4) that learners can place in their learning space and interact with while learning. AR was also used for augmenting a rain gauge use as part of the library display (Fig. 5) to create awareness for drought monitoring and installation, measuring, and reporting (Fig. 6). VR was incorporated to share information on nature enquiry and citizen science. An exhibit hall (Fig. 7) was designed with a lobby to host live events and has two water rooms (Fig. 8) and one drought room and two BYE rooms. The two water rooms are: (1) Community Collaborative Rain Hail and Snow Network (CoCoRaHS) and (2) BlueThumb, and the drought room consists of the National Drought Mitigation Center (NDMC) while the two BYE rooms consisted of the leaf and insect rooms. In this VR learning space, learners enter as avatars, attend live events, interact with each other, learn about water and drought, leaves, and insects, and about various citizen science projects that they can volunteer in as citizens to collect and report scientific data and conduct different environmental monitoring activities.

Fig. 3 AR-enhanced logbook

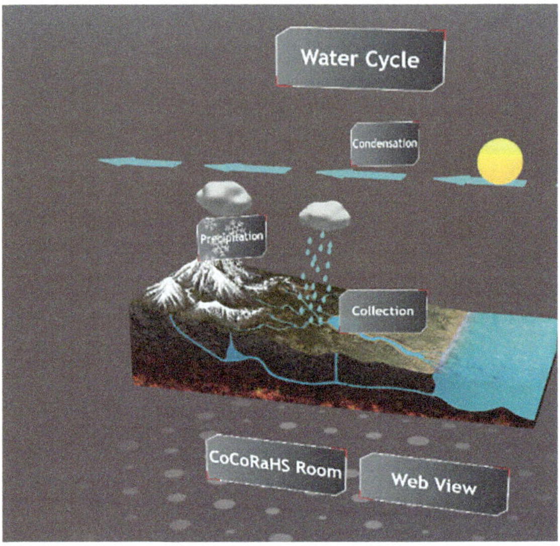

Fig. 4 AR water cycle

Fig. 5 AR rain gauge on display

6 Conclusions

Anderson and Shattuck [4] stated that interest in DBR is increasing even though the evidence does not show that the methodology is positively impacting education research. Questions on the involvement of the researcher in the design process have emerged. It was argued that similar to anthropological research, the researcher's involvement in the design process of DBR methodology may cause biases on the

Fig. 6 AR rain gauge

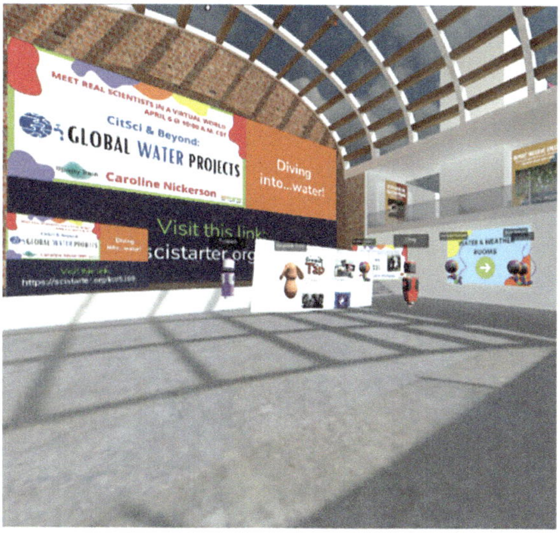

Fig. 7 VR exhibit hall

pedagogical approach, the credibility, and trustworthiness of the study. This can be seen as a negative to DBR; however, the positive impact that DBR has brought to education outweighs the negative. Finally, studies have shown that education requires more than just a traditional classroom where teachers teach, and students learn. Preparing students to be capable of working in the real world and solving problems has been a challenge that not only requires traditional ways of teaching methods or clinical simulation learning environment, but it requires a more authentic approach in teaching in education [13]. The use of DBR overtime in designing educational

Fig. 8 Live event in VR water room

scenarios has been a proven methodology in education to create authentic learning experiences for all learners [4]. Therefore, by using this proven methodology to design learning environments that use the AR and VR immersive technologies in library programs such as the BYE, AR rain gauge, and VR exhibits hall, and rural libraries will be able to provide learning opportunities and share information, that will allow learners to access the shared information and learn about citizen science using their mobile devices and personal computers.

References

1. Bodzin, A., Junior, R.A., Hammond, T., Anastasio, D.: Investigating engagement and flow with a placed-based immersive virtual reality game. J. Sci. Educ. Technol. **30**(3), 347–360 (2021)
2. Johnson-Glenberg, M.C., Ly, V., Su, M., Zavala, R.N., Bartolomeo, H., Kalina, E.: Embodied agentic STEM education: effects of 3D VR compared to 2D PC. In: 2020 6th International Conference of the Immersive Learning Research Network (iLRN), pp. 24–30. IEEE (2020, June)
3. Kljun, M., Geroimenko, V., Čopič Pucihar, K.: Augmented reality in education: current status and advancement of the field. In: Geroimenko, V. (ed.) Augmented Reality in Education. Springer, Cham (2020)
4. Anderson, T., Shattuck, J.: Design-based research: a decade of progress in education research? Educ. Res. **41**(1), 16–25 (2012)
5. Hung, H.T.: Design-based research: redesign of an English language course using a flipped classroom approach. TESOL Q. **51**(1), 180–192 (2017)
6. Wang, F., Hannafin, M.J.: Design-based research and technology-enhanced learning environments. Educ. Tech. Res. Dev. **53**(4), 5–23 (2005)
7. Lyon, J.A., Magana, A.J.: The use of engineering model-building activities to elicit computational thinking: a design-based research study. J. Eng. Educ. **110**(1), 184–206 (2021)

8. Ryu, S.: The role of mixed methods in conducting design-based research. Educ. Psychol. **55**(4), 232–243 (2020)
9. Lo, C.K., Hew, K.F.: Developing a flipped learning approach to support student engagement: a design-based research of secondary school mathematics teaching. J. Comput. Assist. Learn. **37**(1), 142–157 (2021)
10. Barab, S., Squire, K.: Design-based research: putting a stake in the ground. J. Learn. Sci. **13**(1), 1–14 (2004). https://doi.org/10.1207/s15327809jls1301_1
11. Armstrong, M., Dopp, C., Welsh, J.: Design-based research. Educ. Res. (2022)
12. Asino, T.I., Colston, N.M., Ibukun, A., Abai, C.: The Virtual citizen science expo hall: a case study of a design-based project for sustainability education. Sustainability **14**(8), 4671 (2022)
13. Cochrane, T.D., Cook, S., Aiello, S., Christie, D., Sinfield, D., Steagall, M., Aguayo, C.: A DBR framework for designing mobile virtual reality learning environments. Australas. J. Educ. Technol. **33**(6) (2017)
14. Wu, B., Peng, X., Hu, Y.: How to foster pre-service teachers' STEM learning design expertise through virtual internship: a design-based research. Educ. Tech. Res. Dev. **69**(6), 3307–3329 (2021)
15. Williams, S., Enatsky, R., Gillcash, H., Murphy, J.J., Gračanin, D.: Immersive technology in the public school classroom: when a class meets. In: 2021 7th International Conference of the Immersive Learning Research Network (iLRN), pp. 1–8. IEEE (2021)
16. Monahan, T., McArdle, G., Bertolotto, M.: Virtual reality for collaborative e-learning. Comput. Educ. **50**(4), 1339–1353 (2008)
17. Parong, J., Mayer, E.R.: Learning science in immersive virtual reality. J. Educ. Psychol. **110**(6), 785–797 (2018)
18. Stefan, L.: Immersive collaborative environments for teaching and learning traditional design. Procedia, Soc. Behav. Sci. **51**, 1056–1060 (2012)
19. Singh, V., Thurman, A.: How many ways can we define online learning? A systematic literature review of definitions of online learning (1988–2018). Am. J. Distance Educ. **33**(4), 289–306 (2019)
20. Afari, E., Khine, S.M.: Robotics as an educational tool: Impact of Lego Mindstorms. Int. J. Inform. Educ. Technol. **7**(6), 437–442 (2017)
21. Kim, C., Kim, D., Yuan, J., Hill, B.R., Doshi, P., Thai, N.C.: Robotics to promote elementary education pre-service teachers' STEM engagement, learning and teaching. Comput. Educ. **91**, 14–31 (2015)
22. Ortiz, M.A., Bos, B., Smith, S.: The power of educational robotics as an integrated STEM learning experience in teacher preparation programs. J. Coll. Sci. Teach. **44**(5), 42–47 (2015)
23. Kapur, M.: Productive failure. Cogn. Instr. **26**(3), 379–424 (2008)
24. Sung, Y.T., Chang, K.E., Liu, T.C.: The effects of integrating mobile devices with teaching and learning on students' learning performance: a meta-analysis and research synthesis. Comput. Educ. **94**, 252–275 (2016)
25. Goundar, S.: What is the potential impact of using mobile devices in education. In: Proceedings of SIG GlobDev Fourth Annual Workshop, vol. 3, pp. 1–30 (2011, December)
26. Bressler, D.M., Shane Tutwiler, M., Bodzin, A.M.: Promoting student flow and interest in a science learning game: a design-based research study of School Scene Investigators. Educ. Tech. Res. Dev. **69**, 2789–2811 (2021)
27. Schmitz, B., Klemke, R., Walhout, J., Specht, M.: Attuning a mobile simulation game for school children using a design-based research approach. Comput. Educ. **81**, 35–48 (2015)
28. De Freitas, S., Rebolledo-Mendez, G., Liarokapis, F., Magoulas, G., Poulovassilis, A.: Learning as immersive experiences: using the four-dimensional framework for designing and evaluating immersive learning experiences in a virtual world. Br. J. Edu. Technol. **41**(1), 69–85 (2010)

Using a Hierarchy of Data Sources to Represent Cultural Heritage Sites a Case Study of Sophiatown, Johannesburg in Virtual Reality

I. F. Potgieter, D. M. Hendricks, and M. Ntshona

Abstract The mediation of cultural heritage (CH) studies through immersive technologies is becoming more prevalent. Immersive technologies like virtual reality (VR) enable new opportunities for learning about CH through place illusion. The use of VR has led to new epistemological concepts and increased levels of engagement with the represented content. VR can expand our knowledge of CH sites by spatializing CH practices and their identity. However, many CH in VR experiences fail to create the cultural presence necessary for the representations of CH to embody the many layers that make up a CH site. Many experiences of CH in VR focus their efforts on representing the physical structures that make up a CH site, often neglecting the site's intangible qualities. Since CH in VR experiences inform our understanding of historical realities and imaginations, these experiences need to be representative of as many facets of a CH site as possible. When CH in VR experiences fail to represent the multiple facets and intangible qualities of CH, they abstract our understanding of CH sites, limiting them to their physical structure. This paper proposes a methodology for working through data sources in a systematic way to construct immersive experiences that represent multiple dimensions of CH. This methodology can be used to represent immersive experiences of CH that represent both their tangible and intangible aspects. This is especially pertinent when working with demolished or degraded CH sites, as is common in the global south where non-physical data sources often outweigh physical ones. The paper uses Sophiatown (Johannesburg, South Africa) as a case study to demonstrate the viability of this methodology.

Keywords Virtual reality · Cultural heritage · Sophiatown

I. F. Potgieter (✉) · D. M. Hendricks · M. Ntshona
University of Johannesburg, Johannesburg, South Africa
e-mail: ipotgieter@uj.ac.za

Y. Cai et al. (eds.), *Virtual and Augmented Reality Technology-Enhanced Learning*, Gaming Media and Social Effects, https://doi.org/10.1007/978-981-96-2332-7_11

1 Introduction

1.1 Representing Cultural Heritage in Virtual Reality

Zhao [22] defines VR as "a digital environment in which visual perception, sense of hearing, and sense of touch are highly similar to those of actual environment within a certain range". One of the benefits of VR is that it provides a platform for many different media to simultaneously be presented and experienced. As you walk through a 3D environment in VR you can interact and engage with still images, video clips, soundtracks, animated 3D objects, etc. Denard [3] notes the potential inherent in VR for CH because of its ability to stitch together many different mediums. They state that "heritage visualisation is, above all, a hypothesis machine". VR can enable more holistic representations of cultural heritage by presenting various media simultaneously, capturing both tangible and intangible facets of the site concurrently. The ability of digital, including VR, representations of CH to stitch together many disparate sources makes it a useful tool for the representation of CH sites.

The London Charter [1] and the Principles of Seville [2] are the de-facto guidelines on how computer-based visualisations of CH, including VR should be made [3]. They advocate for the representation of both the tangible/physical aspects of CH as well as the intangible practices associated with them. The Principles of Seville [2] go so far as to argue that CH sites cannot be represented as lifeless buildings and empty landscapes because these are historical falsehoods. However, according to a European Union (EU) study [4], computer-based visualisations of CH mainly focus on object's visual appearance when they should also be representing the stories and experiences that form part of a CH site.

Zarri [5] argues that the representation of events, actions, behaviours, attitudes and their impact on one another in VR is difficult to achieve with standard representation models. Because of this, representations of CH in VR often focus on representing only the physical characteristics of a CH site for the sake of simplicity. This situation causes two main problems. Firstly, there are cases where memories, narratives, oral accounts, etc. are the most significant, and sometimes only, existing data sources of a CH site. This is especially true when trying to represent physically lost or degraded CH sites, a common scenario in regions such as Sub-Saharan Africa due to historical inequalities such as cultural erasure imposed by colonialism [6]. Sophiatown in Johannesburg, South Africa is a case in point. Since the physical CH site of Sophiatown has almost entirely been destroyed, any representation of it in VR cannot rely on physical sources alone. In the prevailing paradigm favouring physical aspects and objective sources, a site such as Sophiatown is essentially unrepresentable.

Secondly, because it is difficult to represent the non-physical aspects of a CH site using current representational models, tangible aspects are favoured for representation over the intangible aspects of CH. Bakka [7] notes that there are no CH sites that do not also have intangible aspects to them. Working with CH inherently involves working with objects and places that represent many different stories, values, and perspectives. The meaning of a CH site to different populations changes through time, carrying the weight of many past lives. Bias towards physical sources of data in VR results in spatially accurate but incomplete depictions of CH sites lacking the representation of intangible rituals, activities, and belief systems that convey the values inherent to the heritage. This paper argues for the use of a hierarchy of sources from current sources such as direct site measurements to historical/theorised sources such as ethnographic data (see Fig. 2). By using a hierarchy of sources not only can we avoid unnecessary speculation, but we are also able to integrate both tangible and intangible aspects of a CH site using a wide variety of different sources. Sophiatown is used as a case study because it provides an example of the kind of CH site that is difficult to represent in VR when relying primarily on current data sources (see Fig. 1).

Introduction to Sophiatown: Sophiatown was established in 1905 as a neighbourhood in Johannesburg, South Africa exclusively for the white population [8] around 6 kms (4 miles) from the Johannesburg city centre. However, the sale of property was sluggish, this was attributed to its proximity to the municipal sewerage works. Following this, a blanket restriction preventing black individuals from owning land in new suburbs was lifted leading to a rapid increase in black residents. Sophiatown became one of only two places in Johannesburg where Africans were allowed to buy and own property [8]. This led to a surge in the population of the neighbourhood, exacerbated by the Urban Areas Act of 1923 which designated Johannesburg as a white-only area. This designation did not apply to the Western Areas, which included Sophiatown [9]. This designation led to a further influx of non-white residents from the rest of Johannesburg to Sophiatown creating a multicultural melting pot.

Fig. 1 3D model of "Sisanyana's Shebeen". By IF Potgieter, 2021. A snapshot from the VR experience based on a recorded first-hand account of "Sisanyana's Shebeen", capturing its immersive details

Fig. 2 Table showing the hierarchy of sources to sequentially consider for the representation of cultural heritage. By IF Potgieter, 2023. The table captures the proposed hierarchy for working through different kinds of data sources when constructing representations of CH from the most to the least valid sources

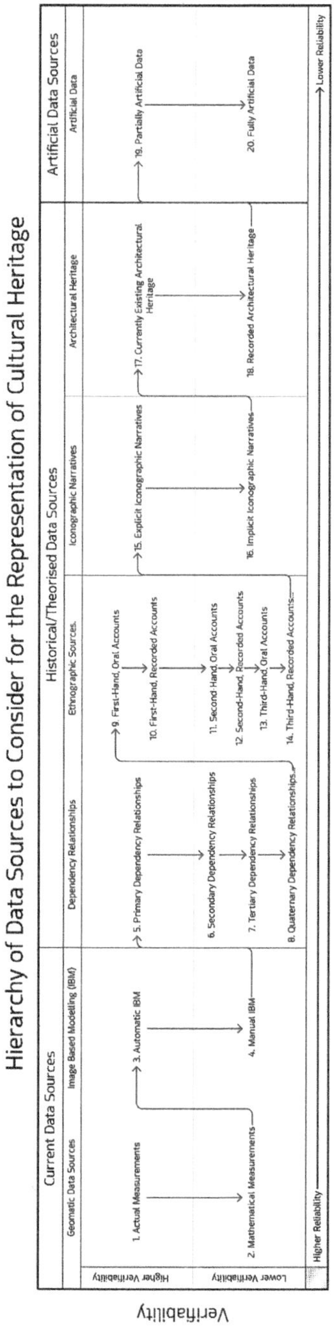

The result of these historical developments was that Sophiatown, in the 1940s and 1950s, experienced a dual identity. It was marked by economic hardship while also serving as a hub for political engagement and the creation of cultural expressions. It became synonymous with the evolution of black urban culture throughout South Africa during that era [10]. Sophiatown had a distinctive atmosphere of inclusivity and multiculturalism uncommon in urban South Africa [8]. Nevertheless, Sophiatown's reputation as a slum, along with the passing of the 1951 Group Areas Act was justification enough for its demolition and the forced removals of its residents by the apartheid government. They were scattered to multiple locations far away from the city centre, and the opportunities it provided, between 1955 and 1962 [10].

Mattera [11], a renowned South African poet, community activist, journalist, and musician from Sophiatown once noted that "Nobody can write the real story of Sophiatown". Despite this statement, Erlank and Morgan [10] have observed that:

"Through a range of writings and representations, the suburb has come to signify something distinctive in South African history", but it has only achieved this position through processes of isolation. Complex layers of context and relationships which might help to shed light on the phenomenon of Sophiatown have been trivialised through its construction by several hegemonic narratives.

Sophiatown has become a nostalgic signifier and is today presented as a quintessential apartheid space. In short, it has become a part of a grand historical narrative in the minds of many people. There are two main problems with the over-simplification of places and events produced through the construction of a grand narrative of history. Firstly, it forces us to exclude and isolate complexities that don't neatly fit into any hegemonic view of history. Secondly, it conclusively presents history as being about a few key events and a handful of heroic figures. This ignores the personal experiences of ordinary people whose stories are often more varied and complex than can be captured from a single top-down narrative of history. These two problems extend to the way we make discoveries and learn about CH. It is therefore important that representations of CH not contribute to a flattening or simplification of history, but rather that they present CH with all its complexity and contradictions.

2 Using Current and Historical/Theorised Data Sources for Representing CH

Principle 3.1 in the London Charter [1] defines research sources as "All information, digital and non-digital, considered during, or directly influencing, the creation of computer-based visualisation outcomes". Trinlae [12] states that CH in VR experiences will only empirically represent real events and places when they are made using diverse and valid data. Proving the validity of data sources is therefore critical. However, historical records are in many ways subjective and often incomplete. Selmanović et al. [13] have noted that working with historical records can lead to subjective misinterpretations of heritage, especially concerning intangible heritage. What then are the characteristics of valid sources?

Trinlae [12] defines the reliability of a data source as the confidence in the consistency between the initial description or observation of a CH site, event, or artefact and its subsequent recording. Not all sources have the same level of reliability; for instance, direct measurements taken at a site are generally more reliable than information gathered from oral accounts. The reliability of a source hinges on the extent to which the evidence assures that the data has remained unaltered from its original description to its recording. Günay [21] admits that there are limitations to the requirement for reliability, noting that in cases where CH sites are physically lost or where the amount of information that can be gathered from different sources is limited, data that has been verified by more than one source should be used. Here, Günay argues for the primacy of verifiability over reliability when faced with a source that has less reliability. In other words, if a source is less reliable, but is verified in multiple different sources, it can be considered to have some validity. The verifiability of a source is determined by how many sources agree on a data point. Günay sums this up by stating that "Verified data is data that could be found in more than one historic resource". Data can either be directly verifiable, meaning that the team doing the representation of CH can gather data directly from its source. Or the data may be indirectly verifiable meaning that data cannot be verified directly at its source but through non-first-hand accounts.

Günay has also noted that an in-depth examination of heritage involves multiple stages, each requiring the collection of diverse data types at every step. A successful representation of CH therefore relies on the use and synthesis of many different types of sources and shouldn't prefer one type over another. Rather any successful visualisation should attempt to systematically work with available sources from the most to least valid. Similarly, Bekele et al. [14] argue that VR experiences should blend current and historical (theorised) views of CH. Current data corresponds to geomatic and image-based sources while any other sources that cannot be physically measured can be considered historical/theorised sources. Artificial sources refer to sources that are entirely imagined and should be avoided. The complementary combination of current and historical data is referred to as "real-virtual data".

We propose that different kinds of current and historical/theorised data sources be considered and sequentially combined into real-virtual data using a hierarchy of data sources that run from the most to the least valid (see Fig. 2). This hierarchy will help determine what data to consider, what to do when conflicting data surfaces and how to deal with gaps in the available data. Very rarely will or should a VR representation of CH focus on one type of data source exclusively. Even in the case of a perfectly preserved heritage site, there will always be changes from the original and aspects that cannot fully be captured by one type of source only. What follows are data sources identified for use in VR representations of CH in order from the most to the least valid. Geomatic data should be considered first running along the gamut to architectural heritage and, eventually to artificial sources. It is important to note that the kinds of sources considered are dependent on the goal and end-use of the CH representation. For example, a representation attempting to physically document a CH site to determine what repairs need to be done to it need not move beyond geomatic data. Whereas CH representations attempting to provide an experience of CH to the public should consider all available data to provide a holistic experience of that CH. Ultimately the team making a CH in VR representation will have to decide for themselves what data is appropriate considering the representation's end goal.

2.1 Current Data Sources

Geomatic Data: Zhao [22] identifies two different types of geomatic data, namely actual measurements and mathematical measurements. He defines actual measurements as "The data acquired through 2D or 3D scanners, cameras, motion capture equipment, data clothes, data gloves, and various special data acquisition equipment (such as CT, B-ultrasound, nuclear magnetic resonance, satellite remote sensing.) or artificial measurement". In other words, any physical measurement of a CH site in its present state. Actual measurements should be considered first because they can be directly verified and thus have high reliability, assuming the person who took the measurements did so correctly and with accurate tools. Mathematical measurements differ slightly from actual measurements in that they are not direct measurements of the CH site but rather indirect measurements mediated through mathematical models, physical equations, and indirect measurements. Rodríguez-Gonzálvez et al. [15] elaborate on this, listing different techniques for mathematical measurements. Their list includes mobile laser systems (MLS), airborne laser scanners (ALS), terrestrial laser scanners (TLS), GPS measurements, and RGB/multispectral cameras. Mathematical measurements derive their accuracy from technology-mediated actual measurements. Consequently, their precision is contingent upon the reliability of the underlying technological mediation. Given this dependence on technology, it is advisable to prioritise data obtained directly from actual measurements before relying on mathematical calculations. This is because the verifiability of mathematical measurements tends to be lower in comparison, to directly measured data.

Geomatic data might be sufficient on its own when attempting to document the current, physical state of a CH site for further examination or in case something happens to the CH site as in Fassi et al.'s [16] 3D scan of the Milan Cathedral. However, geomatic sources are only capable of documenting the physical aspects of CH sites as they currently are. If the CH site has been physically lost, altered or degraded, or if it is necessary to document more than just the physical/tangible aspects of a CH site, then other sources need to be consulted as well.

Image-Based Modelling: Image-based modelling refers to the representation of a CH site based on reference images from the past. There are two kinds of image-based modelling (IBM), namely: Manual IBM [17], and automatic IBM [18]. Manual IBM involves the use of photographs, paintings, maps, films or other visual sources as reference material [17]. A 3D modeller would then take these reference materials to recreate the depicted scene. Automatic IBM involves the capture of multiple

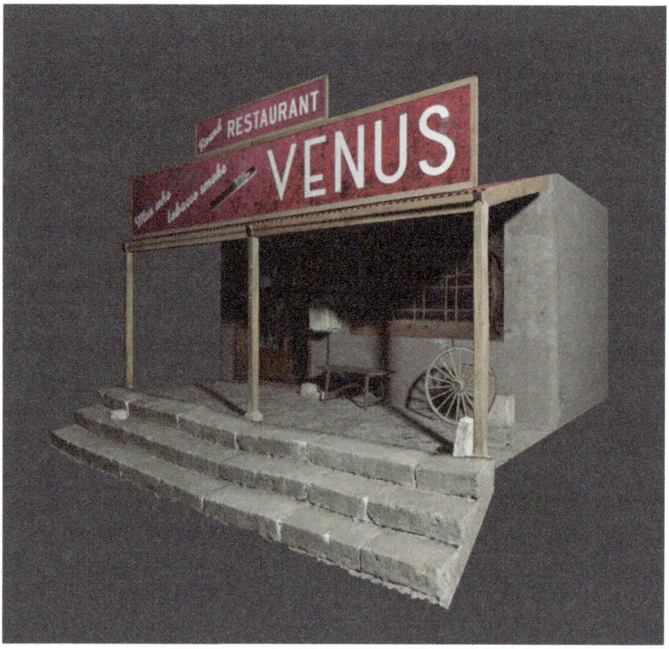

Fig. 3 3D model produced from manual IBM. By IF Potgieter, 2021, 3D representation of Venus café, Sophiatown

images on-site that are then digitally stitched together [18]. Examples of this include photogrammetry and LIDAR scanning.

Günay [21] believes that automatic IBM is preferable over manual IBM because its data is directly verifiable. This is because manual IBM can only reproduce the scene depicted in the visual reference, which is only indirectly verifiable. Visual references might also be from different periods, some parts of the CH site might not have been documented or, in the case of paintings, the reliability of the representation can be called into question. Nevertheless, Günay concedes that manual IBM may be required in cases where a CH site is inaccessible, no longer physically exists, or has been degraded and/or altered. The use of manual IBM from references was used in this paper's case study (see Fig. 3).

2.2 *Historical/Theorised Data Sources*

According to Denard [3], no virtual representation can be considered entirely "objectively" accurate. While there is an intention for a 3D model's geometry to faithfully represent the cultural artefact, this is especially challenging in computer-based representations. Digital representations, constrained by specific technologies, represent aspects of reality rather than reproducing the whole reality of a CH site.

Denard asserts that acknowledging interpretation's inevitable role in representing CH in VR provides validation for historical sources, which are less reliable than current data. Historical sources can effectively supplement data gaps based on knowledge of similar CH sites and contexts. It's crucial to recognise that a VR representation should never be mistaken for or replace the original heritage site, even in the absence or degradation of the physical site. He contends that understanding these representations grants freedom for interpretation and flexible interpolation of historical facts, reinforcing the idea that virtual representations are models of reality, not reality itself. He also suggests viewing CH representations in VR as memory structures rather than objective depictions of real sites. According to them, this aligns with the spirit of the London Charter. This approach accommodates situations where specific details about an object's placement within a larger structure are lacking. Documenting the visualisation process enhances the freedom to explore possibilities. Denard [3] notes that when a representation of CH, its reasoning, and supporting evidence are published, that representation of CH obtains methodological validity. Digital representations of CH, differing from static images or fixed models, enable dynamic testing of hypotheses by altering variables in a digital environment.

However, a representation of CH should not ignore current data in preference of historical data. Instead, historical sources should build on and complement current data. For example, the layout, materiality, and function of the ruins of the ancient Roman town of Baiae [19] were ascertained by gathering current data through direct measurement and analysis. Then the colour of the walls, height of the structures, roof tiles, window locations, and street furniture were all added into the VR representation through historical sources based on knowledge of similar architectural heritage. It is in the experimentation with historical sources where the real power of heritage visualisation as a tool for discovery and learning about CH lies. Each addition of historical data to a CH representation sets up a hypothesis of what that CH site might have looked like that can then be reviewed, examined and disputed. Representations of CH, therefore, have the potential to be dynamic containers of knowledge that could prompt debate and continuous adjustment as new knowledge surfaces or as old beliefs are refuted.

This research has identified four main ways in which historical data can be acquired including Induction and inference through dependency relationships, ethnographic sources, the conversion of iconoclastic narratives into 3D representations and finally, through a study of similar architectural heritage and/or CH sites (see Fig. 4).

Dependency Relationships: According to the London Charter, principle 4.10 "Computer-based visualisation outcomes should be disseminated in such a way that the nature and importance of significant, hypothetical dependency relationships between elements can be identified by users and the reasoning underlying such hypotheses understood". Here, the charter guides the use of dependency relationships as a historical data source. Denard [3] defines a dependency relationship as "A dependent relationship between the properties of elements within digital models,

Fig. 4 3D model of the Venus Cafe. By IF Potgieter, 2021. A snapshot from within the VR experience of the "Venus Restaurant" looking up Good Street. This scene contains elements sourced from geomatic, manual IBM, first-hand oral accounts and architectural heritage sources stitched together to create a holistic representation of CH

such that a change in one property will necessitate a change in the dependent properties". For example, if the height of a door is altered, a corresponding adjustment in the height of the doorframe will be required. In this way, representations of CH become intricate networks of interdependencies. It is this quality that makes representations of CH so valuable as a means of discovering new facets of CH sites. Representations of CH can be a way to piece together the implications of different data from multiple sources in relation to each other piece of data.

The reliability of a dependency relationship depends on the reliability of the data sources from which the dependency relationship has been interpolated. For example, if data about a door frame has been acquired through actual measurement, then inferring the size of the door as a dependency relationship of the door frame will give that data high reliability, the size of the door will be a primary dependency relationship. However, if we were now to try and ascertain what the door handle might have looked like based on our determination of the size of the door, that would be a much less reliable inference because it is indirectly verifiable, a secondary dependency relationship. The further away dependency relationships move from directly verifiable data measured on-site, the less reliable the data becomes and the less information they can provide us. In the above example where a theoretical doorframe at a CH site is known, information such as the size of the door, where the door hinges were, etc. can be inferred through a primary dependency relationship. However, if we want to infer anything about the door handle through a secondary dependency relationship, we may only be able to infer the existence of a door handle without being able to learn anything about its position, shape, material, etc. using dependency relationships.

Ethnographic Sources: Beyond the use of visual sources Martina et al. [20], have recognised that memories and life stories are often acknowledged as valuable historical resources. They can contribute diverse perspectives to the representation process generating new insights into the past, not only providing unique data but also often questioning established histories. They especially recognise the potential of memories and life stories to contribute to the representation of intangible CH, they note that the use of living memories and personal recordings as data in representations of CH serves not only to convey information but also to promote emotional engagement and cultural presence. Ethnographic data gathered through formal interviews with people who had first-hand experience with the CH site provide directly verifiable data. Secondly, recorded first-hand accounts should be considered such as novels, documentaries, radio recordings, recorded interviews, newspaper articles, etc. These sources were often mediated by a second party and may have been altered from what was originally said meaning that they are indirectly verifiable. Second and third-hand accounts may be considered next but offer much less reliability.

Because of this the more narratives that can be collected from different observers describing the same events or phenomena, the more reliable the data should be owing to its increased verifiability. Conversely, when many different narratives are considered the risk of conflicting accounts becomes greater. These conflicts are themselves of interest and are often indicative of underlying tensions between different viewpoints, hinting at conflicts and complexities present on a heritage site. In these cases, it is advisable to depict both accounts so that they may be compared to other relevant data.

Iconographic Narratives: Zarri [5] defines iconographic narratives as "those works of art whose aim is to tell, explicitly or implicitly, a story under visual form". They make a distinction between implicit and explicit narratives. Explicit iconographic narratives explicitly tell unambiguous stories. For example, a story is explicitly told in a picture like Géricault's "The Raft of the Medusa". Implicit iconographic narratives tell ambiguous stories that can be interpreted in multiple ways. An example of an implicit iconographic narrative is the Giaconda (Mona Lisa). Several possible stories are implicitly suggested when regarding the Gioconda or any possible still life. Explicit iconographic narratives should be considered before implicit narratives owing to their direct verifiability. In the context of cultural heritage, historical data could be extracted from the narratives told, either explicitly or implicitly, in paintings, drawings, frescoes, sculptures, bas-reliefs, high-reliefs, tapestries, mosaics, murals, photographs, comics, posters, advertising artworks, cartoons, movies, etc.

Architectural Heritage: Günay [21] has noted that architectural heritage serves as a significant source of data for the representation of CH. By examining architectural heritage from a time, style and location similar to the CH being represented, researchers can uncover valuable data. For example, while Sophiatown was demolished in the 1950s the neighbouring area of Westbury built at a similar time and in a similar style to Sophiatown has persevered to the present day (see Fig. 5). Some of the buildings in Westbury could therefore be used to help plausibly fill in the gaps in knowledge when representing Sophiatown in VR.

The London Charter states that "Computer-based visualisation should accurately convey to users the status of the knowledge that they represent, such as distinctions between evidence and hypothesis, and between different levels of probability". When using data from architectural heritage which has low reliability it is important to document and communicate where the data came from that has been used in the representation and to express that this data is not meant to provide information on what existed on a CH site. Rather data derived from studies of architectural heritage serves as an approximation that can be useful in certain cases to fill gaps in knowledge.

Artificial Data Sources: Zhao [22] refers to artificial construction, defining it as "Model data generated artificially or through hypothesis and imagination by artificial modelling software (CREATOR, 3D Studio, AutoCAD) such as 3D models of objects, cartoon, movie and television manufacture, etc". Similarly, Bekele et al. [14] define artificial construction as "model data generated by human imagination, where the world represented by the model data is completely fictitious". Artificial sources are completely speculative and should be avoided because they have no reliability or verifiability. Artificial sources are often used in films and video games and are not commonly used in serious representations of CH.

Artificial data sources can be used in representations of CH, by augmenting verifiable data with artificial data to fill small gaps in a representation, we can refer to these as partially artificial data sources. When entire sites, buildings, people and events are imagined, we can refer to them as fully artificial data sources. Exhaustive research

Fig. 5 Photograph of a shopfront in Westbury, built in a similar period and style to the buildings of the now demolished Sophiatown. By T Chapman, 2012: Retail Building: Steytler Str. Westbury. Used with permission from T Chapman

needs to first be done to find data sources with higher reliability before artificial sources may be considered. Partially artificial data may be helpful in cases where small details need to be filled in on a CH representation that cannot be ascertained by any other means. Similarly to architectural heritage, it needs to be made very clear in any representation of CH which elements of the representation are artificial and which parts are based on more verifiable evidence. This also helps other researchers to understand where further research and refinement may need to be done to improve the representation. Fully artificial data likely has no place in serious representations of CH though, as CH's function as a plausible representation of places and events that exist, or that existed in the past, is incompatible with fully artificial data sources. Artificial sources might be considered if they are necessary to a representation and compatible with the representation's end goal, for example, representations built for entertainment that explicitly are not attempting to represent historical facts. Even in use cases where the purpose of a CH in VR representation might allow for the use of artificial sources, any and all current and historical sources should be considered before artificial sources according to the hierarchy set out in this paper (see Fig. 2).

3 Technical Approach

Sophiatown was chosen as a case study because it belongs to the kind of CH sites that, as Zarri [5] pointed out, are difficult to represent using standard representation models (see Fig. 6). Sophiatown was seen as a problem to be dealt with by the apartheid government and was assumed to have little to no cultural value or significance in the eyes of that regime. Instead, the Western Areas, including Sophiatown, were seen as hotbeds of resistance—in the words of the Native Affairs Commission at the time, "for many years a source of great difficulty" [9]. This meant that in addition to most of the site having been demolished in the 1950s and 1960s leading to a dearth of geomatic sources, photographs and other sources for use in manual IBM were also insufficient to adequately represent the CH site. This lack of current sources is common to many CH sites, especially in the Global South [6].

Firstly, any available current data was obtained. For example, even though Sophiatown was demolished its street layout and plots remained consistent, these were measured as a base for the representation. Secondly, images and films were collected to be used as reference images for manual IBM. From this research, a specific site, the intersection of Good and Victoria streets, was chosen because many of the photographs, paintings and films of Sophiatown that could be found were of this area. This also pointed towards the significance of this area which among other things contained the Odin Cinema, Africa's largest cinema at the time [11: 76]. Next, primary dependency relationships were worked out through a process of inference based on the models produced from manual IBM (see Fig. 3).

Following this, interviews with a total of six former residents of Sophiatown were arranged via the Trevor Huddleston Centre (Sophiatown's primary museum). An interviewer experienced in these kinds of interviews was also recruited to assist

Fig. 6 3D model of a dwelling. By IF Potgieter, 2021. A snapshot from the VR experience based on a recorded first-hand account of a dwelling, capturing its immersive details

with the interview process. The interviews were intentionally conducted only after the representation was built to incorporate all previous steps on the hierarchy of sources to be considered for the representation of CH. This allowed us to include specific questions on places, details and areas that no data could hitherto be found on. Primarily though, the interviews were left open to allow the former residents to talk about the things that they remember about Sophiatown. This opened us to new and unexpected data we might not have been specifically looking for. Specific questions based on the gaps we identified were asked only if and when the former residents brought these places up themselves, or at the end of the interviews. Interviews were conducted in groups of three so that the interviewees could help stimulate each other's memories, discuss inconsistencies in their experiences and enjoy the comfort and openness of speaking that came with seeing old friends and acquaintances.

Recorded accounts were identified next in books, poems, etc. We specifically looked for recorded accounts that mentioned or elaborated on the places and events that the former residents mentioned in their interviews. Iconographic sources were considered next, these mostly consisted of analyses of paintings, especially those of renowned artist Gerard Sekoto who painted many scenes of life in Sophiatown. From these, details were gathered on specific events, big and small that he recorded in Sophiatown. Finally, to fill in some gaps in the model the architectural heritage of Westbury, an adjacent neighbourhood that was built in a similar time as Sophiatown and which still stood was analysed and used to flesh out the experience of Sophiatown (see Fig. 7). These copied buildings were intentionally left white and not textured to make it clear that they are not representative of the CH site in the same way as the buildings reproduced with reference images from manual IBM. Many of these buildings were also placed further away from the area than the user could navigate to.

This helped to flesh out Sophiatown and represent its density and size even though the specific buildings that existed in these areas were different from the ones created in the representation.

Since many different types of sources were stitched together. This project had to utilise many different technologies including 3D modelling, animation, film editing, game engine technology, coding, and virtual reality to translate the source material into an immersive and user-friendly experience of Sophiatown. The final product is an interactive VR environment. Users can navigate the 3D environment, interact with different elements and trigger different narratives depending on their actions.

Recordings of the interviews were used to provide context and elaborate on the areas of the representation described by the interviewed former residents. Additionally, a voice artist was used to record the narrative descriptions of life in Sophiatown found in recorded first-hand accounts, such as books and poems that were used to recreate it in 3D. This was to not only give more context on what the user is looking at but also to hint at the larger stories and events that comprise Sophiatown. Representing Sophiatown using multiple first-hand accounts meant that a potential student or museum patron was learning about Sophiatown not from the accounts of historians or professors but directly from the people who used to live there.

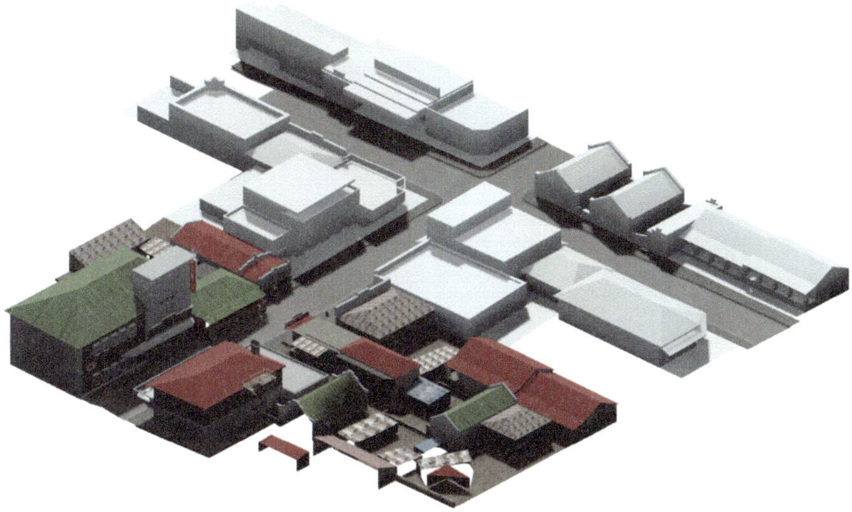

Fig. 7 3D model of the representation of Sophiatown circa 1955. By IF Potgieter, 2021. Overview of the VR representation of Good and Victoria Streets in Sophiatown including textured buildings produced through manual IBM and untextured buildings borrowed from architectural heritage

4 Conclusion

This paper underscores the significance of considering both tangible and intangible aspects of cultural heritage (CH) in virtual reality (VR) experiences of CH. While VR provides a platform to present various media simultaneously, representing both the physical structures and intangible qualities of CH sites concurrently is crucial. The existing guidelines, such as the London Charter and Principles of Seville, emphasise the need to capture both tangible and intangible aspects of CH, but the actual practice often falls short, focusing predominantly on visual appearances. This imbalance, favouring physical over intangible aspects, poses a challenge when dealing with historically marginalised regions such as Sub-Saharan Africa, where non-physical data sources often outweigh physical ones.

The chapter advocates for the use of a hierarchy of sources, ranging from current data to historical/theoretical sources and ultimately artificial sources if relevant, enabling the integration of tangible and intangible aspects of CH sites. The case study of Sophiatown in Johannesburg, South Africa, demonstrates the applicability of this methodology, particularly when dealing with physically lost or degraded CH sites. Sophiatown's complex history highlights the need for a nuanced approach that avoids the kind of oversimplification that often comes when representing CH from a limited set of data types. By acknowledging the multilayered nature of CH and embracing diverse sources, VR representations can contribute to a richer understanding of history, avoiding the pitfalls of flattening or simplifying narratives around CH sites. This approach becomes crucial when presenting CH sites that have been subject to difficult, recent history, preserving the multifaceted stories and experiences embedded in their heritage.

Acknowledgements This work is based on the research supported in part by the National Research Foundation of South Africa (Ref Numbers 0000-0003-1480-5750).

References

1. The London Charter for the Computer-based Visualisation of Cultural Heritage. http://www.londoncharter.Org. Last accessed 21 July 2022
2. ICOMOS: The Seville Principles (International Principles of Virtual Archaeology). Global Digital Heritage (2017). http://sevilleprinciples.com/. Last accessed 16 Oct 2023
3. Denard, H.: A new introduction to the London Charter'. In: Bentkowska-Kafel, A., Baker, D., Denard, H. (eds.) Paradata and Transparency in Virtual Heritage, pp. 57–71. Routledge, London (2012)
4. European Commission: Decision C (2020)6320 of 17 September 2020 (2020)
5. Zarri, G.P.: Formalization of the "immaterial features" conveyed by the iconographic cultural heritage entities. In: Sheade, M., Theopisti, S.L. (eds.) Emerging Technologies and the Digital Transformation of Museums and Heritage Sites, 2021, pp. 316–329. Springer Nature, Nicosia (2021)
6. Thiaw, I.: The Management of Cultural World Heritage Sites and Development in Africa, 1st edn. Springer, New York (2014)

7. Bakka, E.: Safeguarding of intangible cultural heritage—the spirit and the letter of the law. Musikk og Tradisjon **29**, 136–169 (2015)
8. Hart, D.M., Pirie, G.H.: The sight and soul of Sophiatown. Geogr. Rev. **74**(1), 38–47 (1984)
9. Lodge, T.: The destruction of Sophiatown. J. Mod. Afr. Stud. **19**(1), 107–132 (1981)
10. Erlank, N., Morgan, K.L.: Sophiatown. Afr. Stud. **74**(1), 1–9 (2015)
11. Mattera, D.: Memory Is the Weapon. African Perspectives Pub, Johannesburg (1987)
12. Trinlae, B.L.: Ethical considerations and methods for diversifying representations of cultural heritage: a case example of the Swayambhu UNESCO world heritage site in Nepal. In: Shehade, M., Stylianou-Lambert, T. (eds.) Emerging Technologies and the Digital Transformation of Museums and Heritage Sites, pp. 331–346. Springer International Publishing (Communications in Computer and Information Science), Cham (2021)
13. Selmanović, E., Rizvíc, S., Harvey, C., Boškovíc, D., Hulusić, V., Chahin, M., Šljivo, S.: VR video storytelling for intangible cultural heritage preservation. Sablatnig, R., Wimmer, M. (eds.) Eurographics Workshop on Graphics and Cultural Heritage. Eurographics Association, Vienna (2018)
14. Bekele, M.K., Pierdicca, R., Frontoni, E., Malinverni, E.S., Gain, J.: A survey of augmented, virtual, and mixed reality. Cultural heritage. J. Comput. Cult. Herit. **11**(2), 1–36 (2018)
15. Rodríguez-Gonzálvez, P., Campo, A., Muñoz-Nieto, A., Sánchez-Aparicio, L., González-Aguilera, D.: Diachronic reconstruction and visualization of lost cultural heritage sites. ISPRS Int. J. Geo Inf. **8**, 61–76 (2019)
16. Fassi, F., Mandelli, A., Teruggi, S., Rechichi, F., Fiorillo, F., Achille, C.: VR for Cultural Heritage: A VR Web BIM for the Future Maintenance of Milan's Cathedral. Augmented Reality, Virtual Reality, and Computer Graphics, pp. 139–157. Springer International Publishing, Berlin (2016)
17. Balletti, C., Dabrowski, M., Guevara, F., Vernier, P.: Digital reconstruction of a lost heritage: The San Geminiano's Church in San Marco's Square in Venice. In: International Conference on Metrology for Archaeology and Cultural Heritage, pp. 304–310. Metroarcheo, Trento (2020)
18. Thompson, E.L.: Legal and ethical considerations for digital recreations of cultural heritage. Chapman Law Rev. **20**(1), 153–176 (2017)
19. Bruno, F., Lagudi, A., Barbieri, L., Cozza, M., Peluso, R., Davidde Petriaggi, B., Petriaggi, R., Rizvic, S., Skarlatos, D.: Virtual tour in the Sunken "Villa con Ingresso a Protiro" within the underwater archaeological park of Baiae. Int. Arch. Photogramm. Remote. Sens. Spat. Inf. Sci. **42**(10), 45–51 (2019)
20. Martina, A., Bottino, A., Rubino, I., Daniel, C.: One day at the sands: exploring Las Vegas intangible heritage through virtual reality. Int. J. Herit. Digit. Era **4**(1), 1–19 (2015)
21. Günay, S.: Virtual reality for lost architectural heritage visualization utilizing limited data. The International Archives of the Photogrammetry, Remote Sensing and Spatial Information Sciences, vol. XLVI-2/W1-2022. 9th International Workshop 3D-ARCH "3D Virtual Reconstruction and Visualization of Complex Architectures", 2–4 March 2022, Mantua, Italy (2022)
22. Zhao, Q.: A survey on virtual reality, Science in China Series F Information Sciences **52**(3), 348–400 (2009). https://doi.org/10.1007/s11432-009-0066-0

An Interactive VR/Desktop-3D System for Nasal Sinus Anatomy Education

Yanlin Luo, Wenjing Li, Yuhui Wang, Weini Hu, Peidi Gu, and Jing Wang

Abstract Fully understanding the inner spatial structure of the nasal sinuses is necessary for mastering nasal sinus and skull base surgery. However, it is hard for medical students to have a whole picture of the sinuses using traditional educational methods. Immersive virtual reality (VR) systems have great potential as a supplement to anatomy education, yet few of them exist for theoretical anatomical knowledge of nasal sinuses. The purpose of this study was to investigate the usability of the immersive VR technology in nasal sinus anatomy education context among undergraduate medical students. Thus, an interactive nasal sinus anatomy learning system was developed and distributed in two versions: the immersive VR version and the non-immersive desktop-3D version. A total of 21 students were recruited for this study and were randomized into the VR group with HMD-based learning and the control group with desktop-based learning. The study gathered students' perceptions regarding usability and side effects including workload and cybersickness. In addition, objective data regarding interaction efficiency were gathered for analysis. Results showed no difference was found in the overall usability scores between the VR version and the desktop-3D version of the system. The VR group perceived a higher level of performance than the control group. However, the VR group needed more support from a technical person to use the system and were more likely to get lost in the virtual environment when rotating the models. Overall, this study suggested that more natural and intuitive interactions need to be explored to exploit the potential of immersive VR in nasal sinus anatomy learning.

Keywords Virtual reality · Anatomy education · Nasal sinuses · Immersive visualization · Volume rendering

---

Y. Luo · W. Li · J. Wang
School of Artificial Intelligence, Beijing Normal University, Beijing, China
e-mail: j.wang@bnu.edu.cn

Y. Wang · W. Hu
Peking University Third Hospital, Beijing, China
e-mail: j.wang@bnu.edu.cn

P. Gu (✉)
College of Future Education, Beijing Normal University, Beijing, China
e-mail: jarypig1983@gmail.com

Y. Cai et al. (eds.), *Virtual and Augmented Reality Technology-Enhanced Learning*, Gaming Media and Social Effects, https://doi.org/10.1007/978-981-96-2332-7_12

1 Introduction

The nasal sinuses (a.k.a. paranasal sinuses) are air-filled cavities within the bones of the skull. As shown in Fig. 1, there are four pairs of sinuses, including the frontal sinuses, the sphenoid sinuses, the maxillary sinuses, and the ethmoid sinuses, which are connected to the nasal cavity through the sinus openings. Skull base neoplasms are characterized by high risk and morbidity, such as pituitary adenomas and intracranial chordomas, for which endoscopic trans-nasal resection is the major option for treatments. The nasal sinuses are adjacent to many important blood vessels and nerves, so a mistake during surgery may cause irreparable health loss to the patient, which may lead to complications including severe bleeding, blindness, and damage to the central nervous system [1]. A full understanding of the complex sinus anatomy is a necessary prerequisite for medical students to master nasal sinus and skull base surgery, therefore it is necessary to develop effective and easy-to-understand approaches for nasal sinus anatomy education.

Immersive VR technology can offer high-resolution and interactive observation of anatomical structures [2, 3]. Its potential as a supplement to traditional anatomy education is supported by studies comparing traditional methods with immersive VR approaches [2–5]. Thus, an interactive nasal sinus anatomy learning system was developed and distributed in two versions: the immersive VR version and the non-immersive desktop-3D version. Our study focuses on whether immersive technology

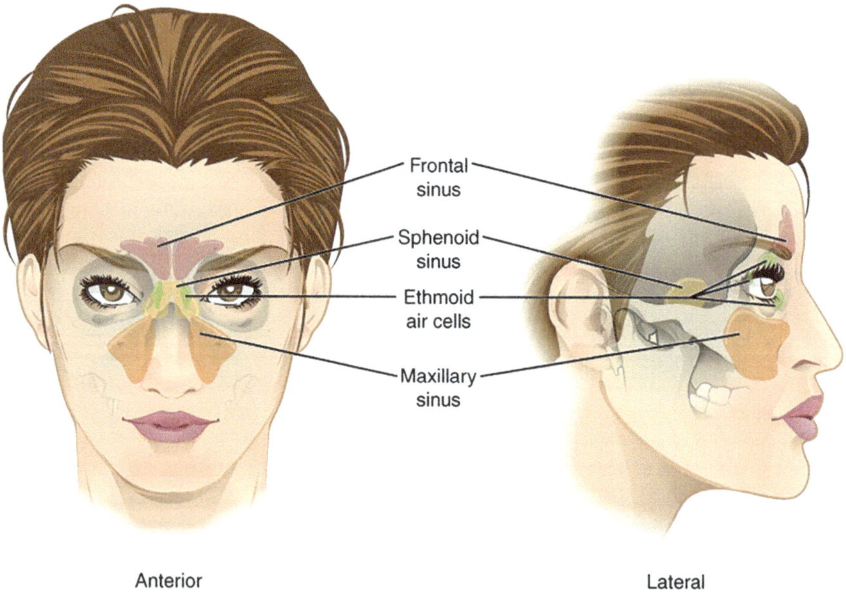

Fig. 1 Position of nasal sinuses (*Note*. From *Anatomy and Physiology 2e* (p. 247), by OpenStax, 2022, Livonia, MI: XanEdu Publishing Inc. CC BY 4.0.)

can enhance the usability of the learning system compared with the non-immersive method in the context of nasal sinus anatomy learning.

1.1 Traditional Anatomy Education

Traditional anatomy learning resources mainly include lecture slides, textbooks, atlases, cadavers, and physical models [6]. However, students encounter challenges in comprehending the spatial and physical characteristics of anatomy solely through the observation of 2D anatomical schematics from lecture slides, textbooks, or atlases, particularly among individuals with poor mental rotation ability [6, 7]. Cadaver dissection is considered the gold standard in anatomy learning, but cadavers have many limitations such as high cost, short supply, short duration of use, strong differences in color and texture from living body, and ethical and regulatory restrictions [8, 9]. The use of physical models is a more cost-effective alternative to cadaver dissection, as these models possess a longer shelf life and allow students to study a specimen repeatedly [10]. However, many of them lack the details necessary for a comprehensive understanding of anatomical structures [6].

1.2 Computer-Assisted Learning

A study by Johnson et al. [10] presented the thinking and evidence supporting the transition of anatomy education from the "passive, didactic, and highly detailed" anatomy course to a "more interactive, as well as functionally and clinically relevant" anatomy curriculum by using a hybrid of teaching methodologies. *Computer-assisted learning* (CAL) is a promising approach due to its interactive nature. With the rapid advances in three-dimensional (3D) visualization technology, CAL has become an effective approach in addition to more traditional methods of anatomy learning [11].

The technologies involved in CAL can range from fully immersive to non-immersive [12]. Fully immersive VR technology allows the user to be in a completely computer-generated space, in which users can have experiences similar to those in the real world [13]. This is usually achieved through the use of a head-mounted display (HMD) and trackers, which enables the user to get deeply immersed in the virtual world. Non-immersive technologies display virtual content through a computer screen and employ traditional media, such as keyboards and mice for user interaction.

1.3 Applications of Immersive VR Technology to Enhance Anatomy Education

Many studies comparing various methods of anatomy learning highlight the potential of VR-based learning. For example, a study by Moro et al. [8] enabled students to learn cranial anatomy using VR, AR, and a tablet-based application, respectively. The results show that the three modalities were equally effective, with students in the VR group being more prone to cybersickness but achieving similar scores to those in the AR group. However, cybersickness caused by limitations in visual smoothness and clarity can be mitigated with the continuous advancement of graphics cards and HMD devices, which implies that VR has the potential to serve as a more effective tool for learning. In a study by Stephan et al. [14], students who used a VR simulator to learn neuroanatomy had a more positive learning experience and were more motivated. In a study by Alfalah et al. [15], the utilization of a semi-immersive VR system employing stereo 3D glasses for heart anatomy learning resulted in higher student satisfaction when compared with the traditional physical model. These studies demonstrate the potential of VR technology in learning cranial anatomy, neuroanatomy, and heart anatomy.

To our knowledge, there is a scarcity of immersive VR systems specifically tailored for sinus anatomy learning. Marks et al. [16] implemented a VR nasal teaching system that bears some relevance to the context of this study, as it provides visualization of anatomical structures from an internal cavity perspective. They published the results of a quantitative evaluation of their system, which showed the advantages of VR over traditional teaching using lecture slides in terms of spatial understanding, ease of use, and motivation [17]. The obtained results encourage us to apply VR technology in the development of nasal sinus anatomy learning system, and to further investigate the evidence supporting VR applications as a usable tool for sinus anatomy learning.

While immersive VR applications have great potential to serve as complementary tools for anatomy learning, they also face unique challenges, such as cybersickness and the increased workload resulting from their high level of immersion [18]. To incorporate VR applications into nasal sinus anatomy education in a better way, it is necessary to investigate whether immersive VR technology can enhance the usability of the nasal sinus anatomy learning system. The purpose of this study is to assess the usability of immersive VR-based learning compared to the common non-immersive desktop-based learning in the context of nasal sinus anatomy education among undergraduate medical students. To this end, an interactive system for nasal sinus anatomy learning was developed and then delivered in two versions: the immersive VR version and the non-immersive desktop-3D version. Students' perceptions of system usability were collected, while side effects and log data regarding user interaction were also considered and gathered for usability analysis.

2 Design and Development of the System

An immersive system was designed and implemented to help medical students understand the anatomy of the nasal sinuses and their positions in relation to other structures (e.g., the skull). The system was implemented in three steps: firstly, CT scan data was used to extract the 3D meshes of the nasal cavity, nasal sinuses, skull, and skin; secondly, the virtual scene was constructed in two visualization modes, realistic rendering of the meshes and volume rendering of the volume model; finally, a series of human–computer interactions were designed to support anatomical structure observation.

2.1 *Mesh Model Extraction*

The medical image processing software 3D Slicer and the 3D modeling software 3DsMax were used to accomplish the 3D reconstruction. First, the human-head CT scan data was processed by the threshold tool using the segment editor module [19] of the 3D Slicer software. By adjusting different threshold ranges, different structures were covered to obtain the head (including skin and mucosa) model and the skull model, respectively. Second, the smoothing factor was set to make the model smoother. Then, the mesh files of the segmented models were exported in.obj format. The head mesh was further split into facial skin, nasal cavity, and nasal sinus meshes in 3DsMax software by using the "Detach" command of the editable poly. Finally, all the meshes were obtained, including the nasal cavity, nasal sinuses (frontal, sphenoid, maxillary, and ethmoid), and turbinates (upper, middle, and inferior).

2.2 *Virtual Scene Design*

The reconstructed meshes were imported into the Unity3D game engine to construct virtual scenes of nasal sinus and design 3D interactive functions. In addition to realistic rendering of the meshes with physically based rendering (PBR) materials, a volume rendering mode was extended using ray casting algorithm.

2.2.1 Realistic Rendering of Meshes

To achieve realistic rendering effect, the PBR materials of the meshes were created by applying Lit shader of high definition render pipeline (HDRP) template provided by Unity, and the final rendering effect is shown in Fig. 2. The materials of the meshes and their corresponding text markers were set to different colors to help distinguish each kind of anatomical structure. As shown in Fig. 1c, the material type of the skin

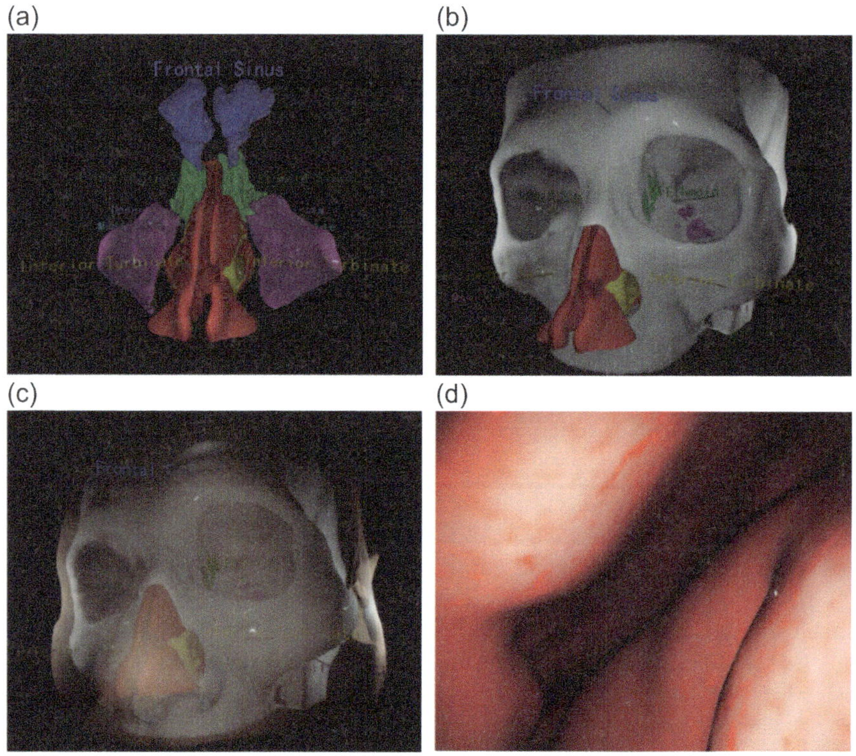

Fig. 2 Realistic rendering of the meshes with unity HDRP. **a** Nasal sinuses; **b** skull; **c** skin; **d** nose cavity

model was set to transparent to help the users observe the positions of the nasal sinuses relative to the skull and skin.

2.2.2 Volume Rendering

Volume rendering is a visualization technique in computer graphics and medical imaging to display 3D datasets [20]. It has a great advantage in presenting the internal structure of objects. Ray casting algorithm [21] is currently one of the most widely used volume rendering algorithms because of its high rendering quality. The semi-transparent effect of the human-head volume model was achieved by creating a Unity custom shader based on the ray casting algorithm. In the preparation stage, a 3D texture was first generated from the CT scan data for use in the sampling stage of the algorithm. Next, a custom shader was created in Unity, and the algorithm was implemented in the fragment shader. Adaptive sampling was performed on each ray, and the early ray termination method [22] was also used for acceleration. A segmented trapezoid transfer function [23] was applied to divide voxels into three

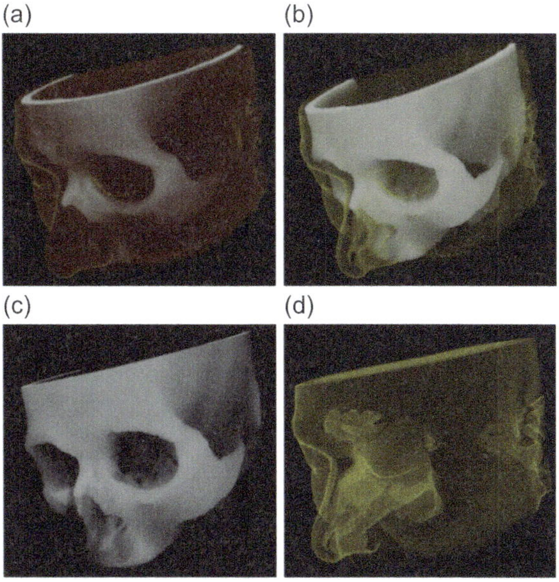

Fig. 3 Volume rendering under different parameters of the transfer function, **a** all the categories; **b** skin, mucosa and bone; **c** bone; **d** skin and mucous

categories: skin and mucosa, soft tissue, and bone, i.e., mapping them to different optical properties (including color and opacity). Finally, a cube was created in the Unity scene as the boundary for the volume model, and its shader was set to the custom shader described above. As shown in Fig. 3, the visualization of different categories can be achieved by adjusting the parameters of the transfer function.

2.3 Interactive Design

Basic interactions included zooming in/out and model rotating, model selecting, as well as model switching and splitting. Additionally, the internal nasal cavity roaming mode was designed to facilitate observation from an endoscopic perspective, and the layered display feature was designed to provide users with a more flexible representation, enabling them to extract the 3D information effectively. The interaction devices for the VR and desktop-3D versions were the HTC Vive controller and mouse, respectively.

Zooming in/out and Rotating: Users can observe the model at an appropriate distance by moving the camera back and forth with the controller button or the mouse wheel. The model can also be viewed from all angles by dragging it around using the controller or mouse buttons. Since the coordinate and orientation of the controller can be captured, a more intuitive way of interaction was achieved in the VR mode, just like dragging a virtual object in a realistic way.

Model Selecting: All anatomical structures were added with a collider component. The controller can emit a ray to identify the target model by collision detection, and then the model would be displayed in a highlighted color. When a model was selected through the controller trigger or the mouse button, the audio commentary would be played and the text commentary would be shown, so that the users can gain knowledge of the corresponding anatomical structures.

Model Switching and Splitting: By switching operations, the sinuses, nasal cavity, skin, and skull models can be displayed separately, which allowed users to observe the position of the nasal sinuses in the human head and their size in relation to other structures. It can also be used to split the nasal sinuses into smaller structures to see the details of the individual structures.

Internal Nasal Cavity Roaming: The roaming mode, as shown in Fig. 4 (bottom left), was designed to help users observe and recognize the anatomical structures that the endoscope may pass through during nasal sinus and skull base surgery. To obtain a wide field of view, the distance mapping method [24] was used to extract a centerline based on the volume model. The system was designed to provide the visualization effect under the endoscope by controlling the camera to move along the extracted centerline and simulating the lighting condition.

Layered Display of the Volume Model: The implementation of the layered display feature is based on the volume rendering technique discussed in Sect. 2.2.2. After switching to the volume rendering mode, the layers of interest (e.g., skin and mucosa, soft tissue, bone) can be displayed by changing the opacity of each category of the transfer function. The user was able to observe the shape and relative position of each structure easily, for all the structures can be displayed transparently.

Moreover, a logging module was implemented, which can provide user data for interaction efficiency analysis. If the logging feature was turned on, all user interactions (entering, exiting and interaction operations, etc.) would be recorded in the form of logs.

2.4 System Architecture and Technology

Our overall architecture is shown in Fig. 4. Both the VR and desktop-3D versions were developed based on the HDRP template of the Unity3D game engine (in version 2020.2.7). The VR version was developed based on the SteamVR Plugin (in version 2.7.3), where single pass instanced rendering (a.k.a. Stereo Instancing) was used to reduce the CPU and GPU workload.

For the VR version, the HTC Vive helmet was used as the display device, which has a resolution of 1080 * 1200 pixels for a single eye (a combined resolution of 2160 * 1200 pixels), a refresh rate of 90 Hz, and a field of view of 110°. And the interaction device was the HTC Vive controller. For the desktop-3D version, the display device was AOC D2352P monitor, which had a screen size of 23 inches and a resolution of 1920 * 1080 pixels. And the interaction device was the common mouse

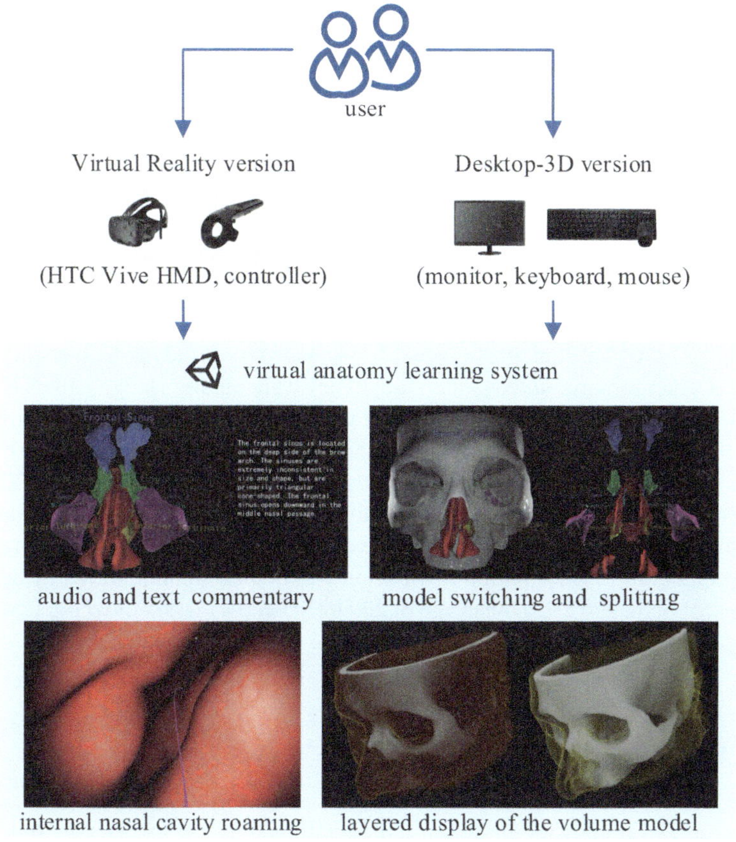

Fig. 4 Overall architecture of the system

and keyboard. Both versions used the same host computer with NVIDIA GeForce RTX 3090 GPU, Intel(R) Core (TM) i9-10900K CPU, and 16G RAM.

3 Method

The goal of this study is to compare the usability of the immersive VR version with the non-immersive desktop-3D version of the developed nasal sinus anatomy learning system. Thus, a random experimental design was chosen. Students' perceptions of system usability were collected, while side effects (workload and cybersickness) and objective data regarding interaction efficiency were also considered for usability analysis.

Table 1 Participants' demographic information

Age(years)		Group	Sex		Prior VR experience n (%)
Mean (SD)	Range		Male n (%)	Female n (%)	
20.89 (0.471)	20–22	3D (8)	3 (37.5)	5 (62.5)	3 (37.5)
		VR (10)	5 (50.0)	5 (50.0)	3 (30.0)

3.1 Participants

A total of 21 medical students ($n_{VR} = 11$, $n_{3D} = 10$) from Peking University Health Science Center were recruited in this study. They were randomized into the VR group with HMD-based learning and the control group with desktop-based learning. Two medical students in the desktop-3D group did not complete the required experimental tasks, as indicated by their interaction logs, and one medical student in the VR group did not complete the questionnaires, so the three were excluded from the data analysis. Thus, a total of 18 participants finished the complete experience and questionnaires, whose demographic information is shown in Table 1. The remaining 18 participants (8 males, 10 females) had a mean age of 20.89 (SD = 0.471), six of whom have prior VR experience. All the participants had normal or corrected visual acuity.

3.2 Measures

A high level of immersion in the virtual environment can bring various sicknesses and increased workload to users, which need to be considered to make VR technology gain "greater applicability, usability, precision, and control" [18]. Therefore, in addition to usability, the subjective workload and various symptoms induced by the system were also considered in our study. At the same time, the logs generated by the system served as source data for the calculation of interaction efficiency metrics.

Usability: The System Usability Scale (SUS) [25] was used in this study to assess users' subjective perceptions of system usability (including effectiveness, efficiency, and satisfaction). The questionnaire consists of 10 items, each in the form of a five-point Likert scale (1 = Strong disagree, 2 = Disagree, 3 = Neutral, 4 = Agree, 5 = Strongly Agree). The questionnaire has been proven to be valid and reliable [26]. In our study, the total usability score was converted to a range of 0 to 100 using the approach in [25].

Workload. NASA-TLX (Task Load Index) [27] was used to assess users' subjective perceived workload, which consists of six dimensions, namely effort, performance, frustration, as well as mental, physical, and temporal demands. The Raw Task Load Index (RTLX) was used this time, i.e., the weights on each subscale were removed and the total workload score was the average score of all dimensions.

Cybersickness. The virtual reality sickness questionnaire (VRSQ) [28] was used to assess the discomfort symptoms and side effects induced by the learning experience in the virtual environment. VRSQ consists of 9 symptoms, which can be divided into two factors: oculomotor and disorientation. It has a four-point Likert scale (0 = None, 1 = Slight, 2 = Moderate, and 3 = Severe) and has been proven to be a valid and reliable measure of cybersickness in consumer virtual environments [29]. In our study, the oculomotor scores, disorientation scores, and total scores of VRSQ were converted to the range of 0 to 100 as described by Kim et al. [28].

Interaction Efficiency. Based on the generated interaction logs, the time taken to complete the tasks and operation count for each kind of interaction were calculated. And they were considered as metrics of interaction efficiency.

3.3 Procedure

Participants were randomly divided into two groups to experience either the HMD-based VR version or the desktop-3D version of the developed system, as shown in Fig. 5. First, the participants were given a brief overview of the experimental procedure and were also informed of the purpose of the study. Then, all participants were given a practice session for about 2 min, to get familiar with the interface and the button control. Next, participants of both groups were asked to complete the three assigned experimental tasks shown in Table 2.

During the experiment, the researcher assisted the participants only when they voluntarily asked for help. In addition, the system recorded all interactions of the participants in the form of logs during their experience. After completing the experimental tasks, participants were asked to fill in their demographic information, three questionnaires (NASA-TLX, VRSQ, and SUS), and three open-ended questions

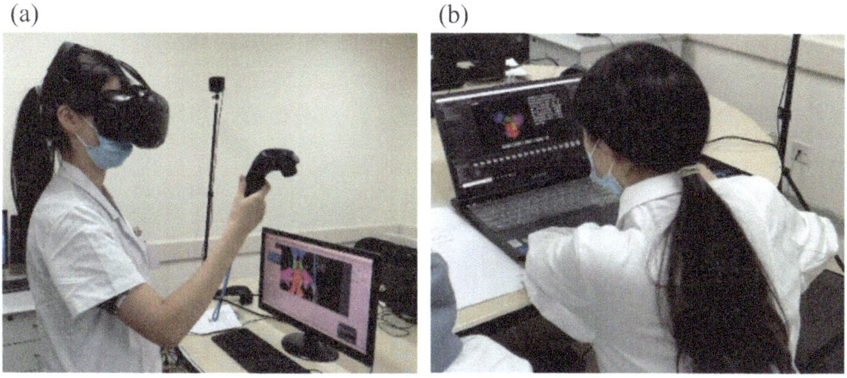

Fig. 5 Medical students experiencing different versions of the learning system. **a** HMD-based immersive VR learning; **b** desktop-based non-immersive learning

Table 2 Overview of experimental tasks

Item	Task	Corresponding modules
1	Learn all the anatomical structures and observe the relative positions between the sinuses and the skull/skin	• 3D mesh model exploration (zooming in/out, rotating, selecting, switching, splitting, and resetting) • Audio and text commentary
2	Observe the anatomical structures and relative positions to other parts in volume rendering mode	• 3D volume model exploration (layered display by adjusting the opacity)
3	Experience the automatic roaming in the internal nasal cavity	• Visualization of the internal nasal cavity

(feelings, likes, and dislikes of the system) on their mobile devices by scanning a QR code.

3.4 Data Analysis

A Mann–Whitney U test was used to compare scores of SUS, NASA-TLX, and VRSQ, as well as interaction efficiency between the VR group and the desktop-3D group. The metrics of interaction efficiency were calculated based on the interaction logs generated by the system, including the time taken to complete the tasks and operation count for each kind of interaction. Cronbach's alpha analysis was used to assess the reliability of SUS, NASA-TLX, and VRSQ. In addition, qualitative data of participants' subjective comments on the system were summarized into different themes and their counts.

4 Results

4.1 Comparison of VR and Desktop Conditions

Usability. The calculation of Cronbach's alpha for the SUS yielded a satisfactory level of internal consistency ($\alpha = 0.782$). The overall SUS scores of the desktop-3D and the VR versions were 65.94 (SD = 9.630, Med = 63.75) and 62.50 (SD = 14.577, Med = 60.00), respectively, with no significant difference found ($U = 29.500$, $Z = -0.939$, $p = 0.360$). The overall SUS scores of both versions were in the acceptability range of "Marginal High" and reached the "OK" adjective rating according to a study by Bangor et al. [30]. The results of each SUS item were converted to a scale ranging from 1 to 5, where 1 = negative perception, and 5 = positive perception. As shown in Fig. 6, item 4 was scored with a mean score of 2.40 by the VR group, and significant

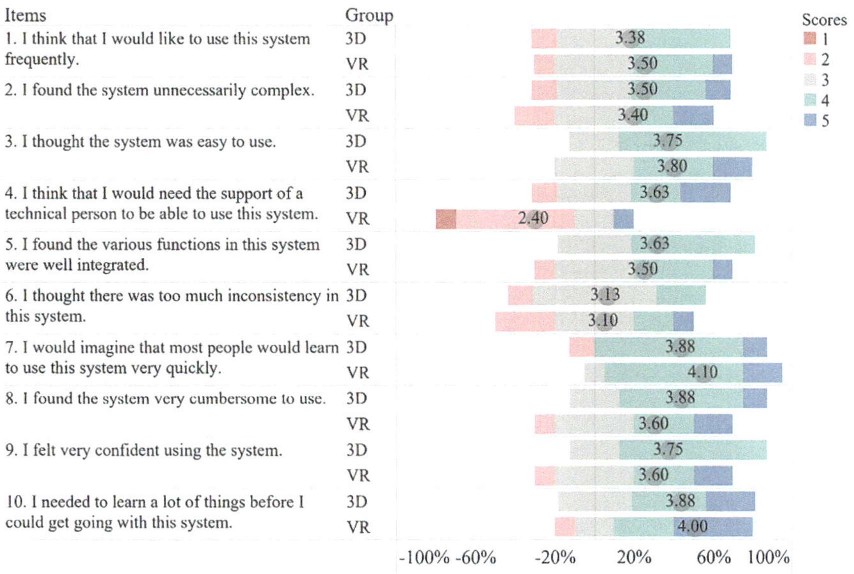

Fig. 6 Score distribution and the mean score of each SUS item

difference of item 4 (U = 15.000, Z = − 2.320, p = 0.027) was found between the VR group (M = 2.40, Med = 2.00, SD = 1.075) and the desktop-3D group (M = 3.63, Med = 3.50, SD = 1.061). Otherwise, all the remaining SUS items received positive mean scores (> 3) with 5 of them (50.0%) having the mean scores of no less than 3.60 in both groups.

Workload. The Cronbach's alpha of the NASA-TLX was 0.705. The results of the RTLX and all the subscales were expressed as scores between 0 and 100 (0 = low, 100 = high), which are shown in Fig. 7 ("Performance" is a reversed item, so the performance scores in the figure were transformed to a negative scale). There was no significant difference in RTLX (U = 32.00, Z = − 0.711, p = 0.515) between the VR group (M = 28.87, Med = 30.33, SD = 15.154) and the desktop-3D group (M = 33.21, Med = 34.17, SD = 8.230). However, participants in the VR group (M = 75.20, Med = 77.00, SD = 24.018) perceived a higher level of performance (U = 8.000, Z = − 2.846, p = 0.003) in completing the goals than those in the desktop-3D group (M = 37.50, Med = 38.50, SD = 19.398).

Cybersickness: Cronbach's alpha was calculated for the VRSQ (α = 0.894), as well as the oculomotor component (α = 0.831) and disorientation component (α = 0.755). As shown in Table 3, there was no significant difference found in scores of the total VRSQ, VRSQ oculomotor, and VRSQ disorientation between the desktop-3D group and the VR group.

Interaction Efficiency: The time to complete all three tasks and each task is shown in Fig. 8. The operation count for each kind of interaction was also compared, as

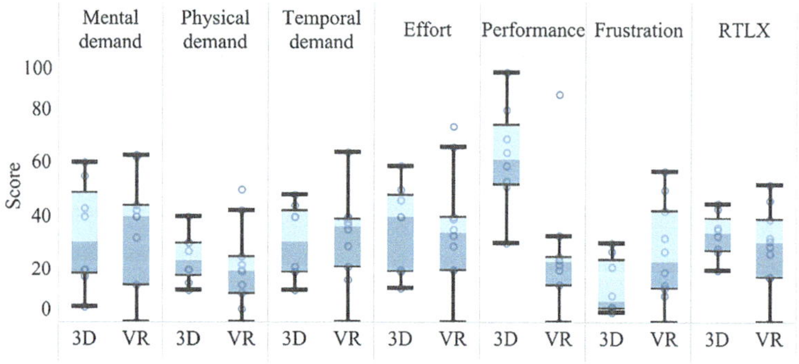

Fig. 7 Self-reported workload after exploring the VR version or the desktop-3D version of the system

Table 3 Comparison of the total, oculomotor, and disorientation VRSQ scores

Scores	Group	Mean (SD)	Med	U	Z	p
Total VRSQ	3D	6.46 (11.415)	0.00	28.000	− 1.119	0.263
	VR	14.67 (17.304)	13.33			
VRSQ oculomotor	3D	10.42 (17.107)	0.00	28.500	− 1.077	0.282
	VR	16.67 (18.002)	16.67			
VRSQ disorientation	3D	2.50 (7.071)	0.00	22.000	− 1.824	0.068
	VR	12.67 (16.763)	10.00			

shown in Table 4 (task 3 did not involve interaction operations). There was no significant difference in the number of rotate operations between the desktop-3D group and the VR group for both task 1 and task 2. However, in task 1, the VR group performed more reset operations ($U = 17.000$, $Z = -2.004$, $p = 0.043$) to put the nasal sinus model to its initial geometric state while performing fewer select operations ($U = 17.500$, $Z = -2.004$, $p = 0.043$) than the desktop-3D group.

4.2 *Qualitative Data*

Subjective comments from medical students are shown in Table 5, including their feelings, likes, and dislikes after exploring the system. Participants' subjective comments were categorized into four aspects: design, visualization, interaction, and device. The aspect that medical students liked about the system was its visualization of the spatial relationships of the anatomical structures, which could help them understand the 3D anatomy of the nasal sinuses. In addition, the roaming mode was popular among medical students due to the realistic visualization and the novel experience. However, the lack of identification of key anatomical structures in the roaming mode

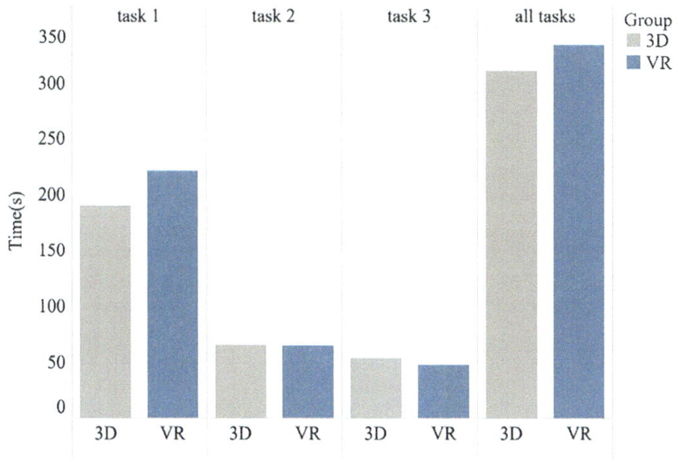

Fig. 8 Comparison of time spent on the tasks

Table 4 Comparison of interaction operation counts

Task	Operations	Group	M (SD)	Med	*U*	*Z*	*p*
1	Switch	3D	3.50 (2.390)	3.00	28.000	– 1.080	0.315
		VR	7.40 (7.619)	4.00			
	Split	3D	2.75 (1.488)	3.00	31.500	– 0.807	0.460
		VR	4.00 (2.667)	3.00			
	Rotate	3D	26.13 (10.535)	23.00	19.000	– 1.870	0.068
		VR	37.80 (14.390)	36.00			
	Reset	3D	0.50 (1.069)	0.00	17.000	– 2.153	0.043*
		VR	1.90 (1.524)	2.00			
	Select	3D	20.63 (7.230)	20.50	17.500	– 2.004	0.043*
		VR	14.00 (4.110)	12.50			
2	Adjust opacity	3D	6.12 (4.549)	5.50	27.000	– 1.162	0.274
		VR	8.10 (3.213)	8.00			
	Rotate	3D	7.62 (5.854)	7.50	34.500	– 0.492	0.633
		VR	7.30 (7.288)	6.50			
	Reset	3D	0.50 (1.069)	0.00	33.500	– 0.690	0.573
		VR	0.90 (1.287)	0.00			

became an aspect that some medical students disliked about the system, as it led to a feeling of getting lost. The disliked parts of the system also included the unnatural interaction and lack of detail in the nasal sinus model.

Table 5 Comments received from medical students

Both groups (n)	VR group (n)	Desktop-3D group (n)
Feelings		
"Good/nice" (5)	"Novel and fun" (2) "Feel dizzy in roaming mode" (1)	"Not much difference from the anatomical 3D software on the market" (2)
Likes		
Design "Excellent 3d functionality to help understand stereoscopic structures" (3) "Roaming mode" (5) **Visualization** "The spatial relationship of anatomical structures is shown in an intuitive way" (3)	**Design** "Splitting the nasal sinus model" (1) **Visualization** "Sense of reality" (2) "Stereoscopic" (1) **Device** "The VR display is clear" (1)	**Design** "Volume rendering mode" (1) **Visualization** "Clear structural distinction" (1) "Show the structure of the nose vividly" (1) **Interaction** "Easy to operate" (1) "See all sides of the model" (1)
Dislikes		
Design "The anatomy is not well understood during roaming for lack of structure labeling" (4) **Visualization** "Some models are slightly rough and need refinement" (4)	**Visualization** "The skull is not transparently displayed" (1) "The skull and skin model are not subdivided into anatomical structures" (1) **Interaction** "The use of the controller for rotation is somewhat counter-intuitive" (5) "Slightly difficult to operate" (1) **Device** "Lack of clarity in display" (2)	[No Data]

5 Discussion

This study evaluated the usability of the developed VR/desktop-3D nasal sinus anatomy learning system, which was designed with various human–computer interactions, including switching and splitting the virtual models, internal nasal cavity roaming, and layered display of the human-head volume model. The perceived usability scores of the two versions were compared, while side effects (cybersickness and workload) and interaction efficiency (time taken for the tasks and number of each kind of interaction operation) were also analyzed. The results showed that participants in the VR group required more support of a technical person, but they performed better than those in the desktop-3D group. The analysis of interaction efficiency suggested that designing more natural and intuitive interactions may be the key to improving usability of the VR version.

Both versions were easy to learn and the users were able to get a quick start, but participants with immersive VR learning significantly required more support of a technical person when using the system. This can be explained by the degree of user control and users' familiarity with the device.

Participants in the VR group were more likely to be disoriented due to the lack of convenience in certain interaction designs, which was supported by the calculation results of the participants' interaction logs. There was no difference in the time taken in task 1 between the two groups, but the VR group learned the nasal anatomy less often by select operations and performed more reset operations to put the geometric position of the model into place. This suggested that participants in the VR group wasted more time in adjusting the model, as five medical students commented on the VR version: "The use of the controller for rotation is somewhat counter-intuitive". The research of Birbara et al. [31] has shown that a certain degree of user control in virtual learning environment can reduce the likelihood of disorientation and contribute to a more beneficial learning experience. As the design of rotation operation in the VR version was not natural enough, participants in the VR group had a lower degree of control and were more likely to get lost, resulting in their perception that the VR version had less learnability compared to the desktop-3D version. This is also supported by the study of Huang et al. [32], in which interaction was found to be the predictor of perceived ease of use. Therefore, in the future, more natural and intuitive interactions need to be designed for the VR version to enhance the degree of user control and make it easier for participants to learn anatomy.

In addition, compared to the HTC Vive controller, the public is more familiar with the operation of the keyboard and mouse. The practice session for just about 2 min did not make up for this factual situation. Therefore, a lack of proficiency in the use of the HTC Vive controller may lead to the need for more support from a technical person, as reflected by the "slightly difficult to operate" comments of the VR version. As consumer-grade VR devices become more popular, the negative impact of the public's unfamiliarity with VR devices on usability will gradually diminish.

In this study, perceived workload and cybersickness were used as factors to explain the participants' perceived usability. The workload results showed that participants in the VR group perceived a higher degree of performance than those in the desktop-3D group, indicating that participants in the VR group felt they performed better. This may be associated with participant comments that the VR version was "novel and fun" while the desktop-3D version was "not much different from the anatomical 3D software on the market". This corresponds with the previous study suggesting that immersive VR can provide a more positive learning experience and enhance student motivation [14]. The delivery modality of HMD-based VR was more engaging and interesting for medical students than desktop-3D, potentially enhancing medical students' motivation and leading to better self-perceived performance.

Surprisingly, the difference in cybersickness between the two groups did not reach statistical significance. However, according to the study of Sharples et al. [33], the HMD induced more cybersickness among different VR display conditions. This may be due to the fact that most of the VR interactions designed in the developed system did not have properties that caused severe cybersickness, and the duration of the participants' experience was relatively short. Therefore, VR-induced cybersickness did not affect the participants much. Theories of the source of cybersickness are mainly sensory mismatch, postural instability, poison, and rest frame [34]. In this study, physical discomfort induced by the VR system may mainly originate from the sensory mismatch during roaming, as reflected by the comment of "feel dizzy in roaming mode" by the VR group. However, it should be noted that, as shown in Fig. 8, the average time spent by participants in roaming mode was only about 1 min. The short duration of roaming may be a factor contributing to the unrevealed differences in cybersickness between the two groups. In the future, more extensive roaming features need to be designed to increase the amount of time participants spend exploring independently in roaming mode to help them learn more about anatomy. At the same time, the side effects of cybersickness need to be avoided as much as possible. The study by Sharples et al. [33] reported that participants experienced higher levels of symptoms when they were unable to control their movement in the virtual environment. Therefore, a possible approach is to allow participants to move on their own in the virtual environment at a controlled speed in the roaming mode.

Limitations of the Study

This study has the following limitations. First, the sample size was small, so the Mann–Whitney U test, a nonparametric test, was used, and its test efficacy was not high enough. More medical students need to be recruited to participate in future experiments. Secondly, as the VR group only practiced the interaction for about 2 min, they were not familiar enough with the use of the VR devices. In the future study, it is necessary to ensure that participants are familiar with the VR devices. Third, the NASA-TLX and VRSQ for measuring perceived workload and cybersickness were filled out only once after completing all three tasks, which may be inappropriate for the three tasks that are not very similar. In the future study, it is more appropriate to ask participants to fill out the NASA-TLX and VRSQ once after each task. Finally, the system has a lot of room for improvement and is still in the early stages of development, so the learning effectiveness assessment has not been conducted yet. The total usability scores of the two versions both reached only an "ok" adjective rating. As reflected in the collected comments from participants, some of the models in the system were not fine enough, the prompting information of the roaming mode was not sufficient, and the model rotation of the VR version was difficult to operate. It is suggested that 2D display modes appear to be comparable to HMD-based immersive VR for spatial learning when the virtual learning environment lacks personalization cues during the user's movement [18]. Therefore, a better design of the prompting function is necessary. In addition, more accurate models and more natural ways of interaction need to be designed so that the system can be applied in the future to test the learning effectiveness of medical students using the system.

6 Conclusions and Future Work

In this study, a VR/desktop-3D system for nasal sinus anatomy education was designed and developed, and a pilot study focusing on usability was conducted to help improve the user experience before it was formally put into use for anatomy learning. The results show that no difference was found in the overall usability scores between the immersive VR version and the non-immersive desktop-3D version of the system. Participants with immersive VR-based learning perceived a higher level of performance but required more support from a technical person than those with non-immersive desktop-based learning. By considering interaction efficiency, these findings indicated that designing more intuitive and simple interactions and adding more prompts to help users get familiar with the system quickly may be the key to enhancing the usability of the immersive VR application. The results can provide a promising direction for enhancing the VR-based interaction functions, which encourages further exploration of the immersive VR-based interaction design in the context of nasal sinus anatomy education.

Acknowledgements This research has been supported by National Natural Science Foundation of China (No. 61977063 and 62377004).

References

1. Li, L., Yang, J., Chu, Y., Wu, W., Xue, J., Liang, P., Chen, L.: A novel augmented reality navigation system for endoscopic sinus and skull base surgery: a feasibility study. PLoS ONE **11**(1), e0146996 (2016)
2. Kurul, R., Ögün, M.N., Neriman, N.A., Avci, Ş, Yazgan, B.: An alternative method for anatomy training: immersive virtual reality. Anat. Sci. Educ. **13**(5), 648–656 (2020)
3. Bairamian, D., Liu, S., Eftekhar, B.: Virtual reality angiogram vs 3-dimensional printed angiogram as an educational tool—a comparative study. Neurosurgery **85**(2), E343–E349 (2019)
4. Jang, S., Vitale, J.M., Jyung, R.W., Black, J.B.: Direct manipulation is better than passive viewing for learning anatomy in a three-dimensional virtual reality environment. Comput. Educ. **106**, 150–165 (2017)
5. Maresky, H.S., Oikonomou, A., Ali, I., Ditkofsky, N., Pakkal, M., Ballyk, B.: Virtual reality and cardiac anatomy: exploring immersive three-dimensional cardiac imaging, a pilot study in undergraduate medical anatomy education. Clin. Anat. **32**(2), 238–243 (2019)
6. Ma, M., Fallavollita, P., Seelbach, I., Von Der Heide, A.M., Euler, E., Waschke, J., Navab, N.: Personalized augmented reality for anatomy education: personalized augmented reality. Clin. Anat. **29**(4), 446–453 (2016)
7. Brewer, D.N., Wilson, T.D., Eagleson, R., de Ribaupierre, S.: Evaluation of neuroanatomical training using a 3D visual reality model. Stud. Health Technol. Inform. **173**, 85–91 (2012)
8. Moro, C., Štromberga, Z., Raikos, A., Stirling, A.: The effectiveness of virtual and augmented reality in health sciences and medical anatomy: VR and AR in health sciences and medical anatomy. Anat. Sci. Educ. **10**(6), 549–559 (2017)
9. Preim, B., Saalfeld, P.: A survey of virtual human anatomy education systems. Comput. Graph. **71**, 132–153 (2018)

10. Johnson, E.O., Charchanti, A.V., Troupis, T.G.: Modernization of an anatomy class: From conceptualization to implementation. A case for integrated multimodal-multidisciplinary teaching. Anatom. Sci. Educ. **5**(6), 354–366 (2012)
11. Jamil, Z., Saeed, A.A., Madhani, S., Baig, S., Cheema, Z., Fatima, S.S.: Three-dimensional visualization software assists learning in students with diverse spatial intelligence in medical education. Anat. Sci. Educ. **12**(5), 550–560 (2019)
12. Suh, A., Prophet, J.: The state of immersive technology research: a literature analysis. Comput. Hum. Behav. **86**, 77–90 (2018)
13. Jerald, J.: The VR book: human-centered design for virtual reality. Morgan & Claypool Publishers (2015)
14. Stepan, K., Zeiger, J., Hanchuk, S., Del Signore, A., Shrivastava, R., Govindaraj, S., Iloreta, A.: Immersive virtual reality as a teaching tool for neuroanatomy. Int. Forum Allergy Rhinol. **7**(10), 1006–1013 (2017)
15. Alfalah, S.F.M., Falah, J.F.M., Alfalah, T., Elfalah, M., Muhaidat, N., Falah, O.: A comparative study between a virtual reality heart anatomy system and traditional medical teaching modalities. Virtual Reality **23**(3), 229–234 (2019)
16. Marks S., White, D., Singh, M.: Getting up your nose: a virtual reality education tool for nasal cavity anatomy. In: SIGGRAPH Asia 2017 Symposium on Education, pp. 1–7 (2017)
17. Marks, S., White, D., Mazdics, M.: Evaluation of a virtual reality nasal cavity education tool. In: 2018 IEEE International Conference on Teaching, Assessment, and Learning for Engineering (TALE), pp. 193–198 (2018)
18. Srivastava, P., Rimzhim, A., Vijay, P., Singh, S., Chandra, S.: Desktop VR is better than non-ambulatory HMD VR for spatial learning. Front. Robot. AI **6**, 50 (2019)
19. Pinter, C., Lasso, A., Fichtinger, G.: Polymorph segmentation representation for medical image computing. Comput. Methods Progr. Biomed. **171**, 19–26 (2019)
20. Kaufman A.E., Mueller K.: Overview of volume rendering. In: The Visualization Handbook, vol. 7, pp. 127–174 (2005)
21. Levoy, M.: Display of surfaces from volume data. IEEE Comput. Graphics Appl. **8**(3), 29–37 (1988)
22. Kruger, J., Westermann, R.: Acceleration techniques for GPU-based volume rendering. IEEE Visualization **2003**, 287–292 (2003)
23. Luo, Y., Gao, B., Deng, Y., Zhu, X., Jiang, T., Zhao, X., Yang, Z.: Automated brain extraction and immersive exploration of its layers in virtual reality for the rhesus macaque MRI data sets. Comp. Anim. Virtual Worlds **30**(1), e1841 (2019)
24. Wan, M., Liang, Z., Ke, Q., Hong, L., Bitter, I., Kaufman, A.: Automatic centerline extraction for virtual colonoscopy. IEEE Trans. Med. Imaging **21**(12), 1450–1460 (2002)
25. Brooke, J.: SUS: A quick and dirty usability scale. Usability Eval. Industry **189**(194), 4–7 (1996)
26. Brooke, J.: SUS: a retrospective. J. Usability Stud. **8**, 29–40 (2013)
27. Hart, S.G.: NASA-task load index (NASA-TLX); 20 years later. Proc. Human Factors Ergon. Soc. Ann. Meet. **50**(9), 904–908 (2006)
28. Kim, H.K., Park, J., Choi, Y., Choe, M.: Virtual reality sickness questionnaire (VRSQ): motion sickness measurement index in a virtual reality environment. Appl. Ergon. **69**, 66–73 (2018)
29. Sevinc, V., Berkman, M.I.: Psychometric evaluation of Simulator Sickness Questionnaire and its variants as a measure of cybersickness in consumer virtual environments. Appl. Ergon. **82**, 102958 (2020)
30. Bangor, A., Kortum, P., Miller, J.: Determining what individual SUS scores mean: adding an adjective rating scale. J. Usability Stud. **4**(3), 114–123 (2009)
31. Birbara, N.S., Sammut, C., Pather, N.: Virtual reality in anatomy: a pilot study evaluating different delivery modalities. Anat. Sci. Educ. **13**(4), 445–457 (2020)
32. Huang, H.M., Liaw, S.S., Lai, C.M.: Exploring learner acceptance of the use of virtual reality in medical education: a case study of desktop and projection-based display systems. Interact. Learn. Environ. **24**(1), 3–19 (2016)

33. Sharples, S., Cobb, S., Moody, A., Wilson, J.R.: Virtual reality induced symptoms and effects (VRISE): comparison of head mounted display (HMD), desktop and projection display systems. Displays **29**(2), 58–69 (2008)
34. Rebenitsch, L., Owen, C.: Review on cybersickness in applications and visual displays. Virtual Real. **20**(2), 101–125 (2016)

The Effectiveness of the Force Feedback Technology Application in the Immersive Physics Learning

Yanlin Luo, Xinyi Ye, Hailin Ji, Qifan Meng, Peidi Gu, and Jing Wang

Abstract Experiments are essential components of physics learning as students could have a deeper understanding of the domain topic through visual and haptic experiences in experiments. However, experimental activities in traditional educational settings face the challenge of a lack of guidance from experienced teachers and a shortage of facilities to support the diverse needs of physics experiments. Immersive learning supported by virtual reality technology could well-address current challenges in traditional classrooms. However, most immersive learning environments in literature focused on providing students with authentic visual and audial experiences, ignoring the importance of haptic experiences for learning physics in the immersive learning environment. As a result, students usually reported a lack of sense of presence and limited interactive experience. This study aims to evaluate the force feedback technology application's effectiveness in immersive physics learning. A pre-and post-test random experimental design was adopted in this study. Thirty-eight students were enrolled in this study and were randomly assigned into two groups, the experimental group and the control group. Data regarding students' domain knowledge, sense of presence, and efficacy of interaction were collected before and after the experiment. Results showed that the use of force feedback technology in immersive physics learning significantly improved students' sense of realism and efficacy of interaction, but did not affect the knowledge gains significantly. It is suggested to explore further the potential mechanisms on how to embed the force feedback technology in immersive physics learning environments.

Keywords Immersive learning · Force feedback technology · Virtual reality · Science education

Y. Luo · X. Ye · H. Ji · Q. Meng · J. Wang
Beijing Normal University, Beijing, China
e-mail: j.wang@bnu.edu.cn

P. Gu (✉)
Beijing Normal University, Guangdong, China
e-mail: jarypig1983@gmail.com

Y. Cai et al. (eds.), *Virtual and Augmented Reality Technology-Enhanced Learning*, Gaming Media and Social Effects, https://doi.org/10.1007/978-981-96-2332-7_13

1 Introduction

Laws and concepts of physics are difficult topics for students as there are many abstract, theoretical, and esoteric principles for learning. Physics experiments play important roles in testing theories and providing evidence for scientific principles, which help students to understand and remember the concepts. Due to limited resources, some experiments cannot be conducted in the classroom, which may result in students' misunderstanding and frustration. Experiments in school classrooms face the challenge that cannot be ignored, such as lack of experiment equipment, limited time for experiments, and concerns about experiment safety and costs, which impeded the effectiveness of students' physical experiments [1]. As computer simulation technology constantly evolves, new solutions for learning and experiments have been developed. Virtual reality (VR) technology has been recognized as one of the most prospective and important technologies in education due to its immersive, interactive, multi-sensory, and dynamic characteristics [2]. Most VR-based learning environments provide only visual and auditory experiences through VR devices, the feeling of touch with the virtual objects is seldom paid attention to in previous discussions. Force feedback technology can transmit the dynamic changes of the virtual objects to students in the form of force so that they can feel the sense of touching authentic objects. Adding force feedback for virtual learning environments may increase interaction and enhance immersion. Thus, the combination of VR and force feedback technology is becoming an important area of educational research and practical exploration, performing an important role in the field of distance learning, interactive teaching, and computer simulation [3]. Specifically, in some subjects, such as medicine, physics, and biology, the instruction may require hands-on practice for students to acquire experiences, construct knowledge, and master skills. Immersive virtual reality with force feedback has great potential in physics experiments and learning.

1.1 Immersive Virtual Reality

The impact of VR technology on learning has been widely explored. But the mechanisms of different types of VR in terms of learning content are yet to be further investigated [4]. In general, VR can be accessed through a variety of display devices, such as desktops, head-mounted displays (HMDs), or mobile devices, and matched interaction devices, such as keyboards, controllers, or styluses. Levels of immersion vary according to the adoption of different devices. Mandal et al. classify VR into three main categories: non-immersive desktop VR, semi-immersive VR, and immersive VR (IVR), where the IVR distinguishes from the other two by its enhanced immersive experience of users through technology-based multiple sensory stimulations [5].

When low-immersive VR is used in traditional classroom teaching, there are many problems such as lack of instruction interaction and timely feedback so that it is unable to provide students with satisfying experiences. IVR is advantageous in education as it renovates people's way of learning [6]. It expands the scope of learning by providing students the opportunities to interact with virtual objects that are hard to access in the real physical world. The combination of technology and instruction contributes to the application of IVR in K12 science education [7]. Students can further deepen their understanding of knowledge concepts by working with or interacting with virtual avatars and other peers in a 3D virtual environment [8]. Virtual laboratories developed by IVR can help students conduct experiments in physics, chemistry, geography, and biology to strengthen their practical skills [9].

In the IVR, the virtual environment offers purposeful control and adaptability, allowing for the creation of diverse scenarios and conditions. This capability enables students to observe and manipulate virtual items in a controlled manner. Moreover, IVR facilitates the simulation of scenarios that are inherently difficult to reproduce in a physical laboratory setting. For instance, students can explore the concept of friction in unconventional conditions, such as on rough or slippery surfaces, or even in microgravity environments. By immersing themselves in these simulated scenarios, students are able to gain valuable insights into the behavior of objects under unique circumstances, expanding their understanding of friction beyond the constraints of a traditional classroom setting. Additionally, the IVR offers an immersive and interactive learning experience that heightens student engagement and motivation. Overall, the IVR presents innovative avenues for enhancing science education and diversifying activity formats.

1.2 Force Feedback Technology

Force feedback technology is a new type of human–computer interaction that allows users to get a sense of force similar to the real world using haptic devices. Force feedback devices require the technologies to track the motion of the user, detect a collision, compute reaction forces, and exert forces on the user [10]. In a virtual environment, force feedback provides additional information that cannot be transferred by pictures or sound. Richard Mayer's cognitive theory of multimedia learning posits that presenting information through multiple sensory channels, such as visual, auditory, and haptic, can significantly enhance learning outcomes [11]. Mayer asserted that the human cognitive system operates distinct processing channels for visual and auditory information, and the utilization of multiple channels can facilitate learning by engaging different cognitive processes simultaneously. When considering haptic experiences, the incorporation of force feedback within a virtual environment introduces an additional sensory channel, enabling students to perceive and interact with the virtual world through touch and physical sensations. This integration of haptic feedback can effectively complement the visual and auditory channels, resulting in a more comprehensive and immersive learning experience for students.

Allan Paivio's dual coding theory postulates that information is processed and stored in two distinct cognitive codes: verbal and imaginal [12]. Verbal information relies on language processing, while imaginal information is based on non-verbal sensory experiences, including visual and haptic sensations. According to Paivio, when information is presented in both verbal and imaginal formats, it enhances learning by simultaneously activating and engaging both codes, facilitating information processing and retrieval. In the context of virtual learning environments, the incorporation of haptic experiences, such as force feedback, allows students to interact with physics concepts not only through verbal or visual representations but also through physical interactions. This integration activates both the verbal and imaginal codes, resulting in a more profound encoding and comprehension of the concepts at hand. By engaging multiple sensory channels, including haptic sensations, students can experience a richer and more holistic learning experience that aligns with Paivio's dual coding theory. The previous research has demonstrated that haptic elements can enhance students' sense of presence in virtual environments and balance the pressure on visual working memory [13, 14]. Adding force feedback may further enhance students' affective perceptions from a multisensory perspective. It is important for experiment-based learning.

Embodied cognitive theory is an integrative theory of mind, body, and environment, which believes that learning can be affected by the way students' minds and bodies interact with the environment [15]. Applying the theory of embodied cognition to the physics domain will produce multisensory interactions (visual, auditory, and haptics), which can serve as a cognitive foundation for K-12 students [16]. Therefore, adding simulated interactions to the virtual or authentic objects is a way to improve the learning effect for students. According to embodied cognition theory, the immersive physics learning environment supported by force feedback technology enables students to control the movement of virtual objects through interactive devices and receive force feedback similar to the physical environment, so that they have the practical experience to deepen their understanding of physics concepts and laws [17]. In a study conducted by Zheng et al. [18], immersive learning environments were employed to investigate the effects of gravity and friction experiments. The researchers discovered that the utilization of an immersive physics learning environment, coupled with force feedback, resulted in a heightened sense of immersion and more authentic interactions. These findings suggest that such an environment has the potential to enhance students' interest in learning and promote deeper memory retention.

1.3 Motivation

Force represents a fundamental concept in the realm of physics, playing a pivotal role in comprehending a wide range of related topics. By placing emphasis on the concepts associated with force, students can establish a robust groundwork, thereby facilitating their understanding of more intricate principles and laws in the field of physics. Force feedback technology has been applied to physics experiments and instruction in some virtual learning environments to diversify and improve students' experience [16, 19–22]. However, its effectiveness in various aspects of immersive physical learning is seldom discussed in previous research. Regarding the effects of force feedback on students' performance and attitudes in physics in a virtual learning environment (VLE), Civelek et al. [19] conducted an evaluation experiment with 215K-12 students on the topic of "Gravity in the Solar System". The results indicated that the use of force feedback in the VLE had a significant positive effect on students' performance and quality of learning compared to traditional classroom instruction. Research by Neri et al. [20] suggested that some physics mechanics concepts were not applicable in haptic scenarios without proper design. As they both focused on comparing traditional classroom instruction, they did not separately delve into the effects of force feedback in the VLE on immersive learning. Han and Black [16] investigated the effectiveness of force simulations in physics learning, which revealed that force feedback was necessary for helping students construct the multimodal representation. However, their research mainly conducted the knowledge test to assess students' recall, inference, and transfer abilities and did not address other aspects of force simulation effectiveness in a VLE for physics learning. Further exploration is required to determine the extent to which force feedback in immersive physical learning can enhance students' virtual presence, contribute to substantial knowledge gains in their learning outcomes, and improve their interaction efficiency.

1.4 Research Purpose

The aim of this study is to investigate the effectiveness of force feedback on students' sense of virtual presence and learning outcomes in physical learning, in order to explore deeply the design and development of VR with force feedback technology in the field of education and instruction. As this system is designed for students who are immersed in a simulated world, it has great potential to create a sense of presence in a VLE. It is necessary to explore whether force feedback technology can enhance students' sense of presence in a virtual learning environment, also improve the efficiency of interaction, and produce significant knowledge gains. This study attempts to explore the rationality and effectiveness of setting up force feedback in a physical learning environment.

2 Method

2.1 K12 Physics Teaching Case

Newton's laws of motion and sliding friction are pretty important in K12 physics teaching. Due to the abstract concept of forces, it is difficult for students to perceive the exact effects and laws of forces, both in traditional classroom teaching and in

Table 1 Physics concepts in two cases

Content	Knowledge point		Description
Newton's laws of motion	Newton's second law	Investigate the relationship between acceleration and force at a given mass	Acceleration is proportional to the force at a given mass
	Newton's third law	The relationship between action and reaction	Action and reaction are equal and opposite, acting in the same line
Kinetic friction coefficient	Gravity		Gravity is proportional to mass and directed vertically downward
	Elasticity		Elasticity is directed in the opposite direction to the deformation of the object
	Friction		Friction is directed in the opposite direction to the relative motion (tendency)
	Force analysis of an object on an incline		The object is subject to the combined action of gravity, elasticity, and friction
	Determinants of kinetic friction coefficient		Kinetic friction coefficient is related to the material and roughness of the contact surface
	Determinants of sliding friction		Sliding friction force is proportional to the kinetic coefficient and the positive pressure

most VR learning environments. This study applied force feedback technology to an immersive learning environment, teaching the topics "Newton's second laws", "Newton's third laws" and "Kinetic Friction Coefficient" to enhance students' understanding and application of physics knowledge. The description of the involved physics concepts is shown in Table 1.

2.2 Case Design

According to the physics teaching case, two virtual experiment cases were designed: "Newton's Laws of Motion" and "Kinetic Friction Coefficient". Students could feel the presence of three types of forces, including gravity, elasticity, and friction, through the stylus of the Geomagic Touch device. The description of the design is shown in Fig. 1a, b. As depicted in the left side of Fig. 1a, it could be observed that the same mass of the ball experienced diverse forces. Newton's second law states that if a user applies a substantial force to the ball, it will undergo significant acceleration and travel a considerable distance. Conversely, on the right side of Fig. 1a, as the ball moved forward, the cannon moved in the opposite direction, exemplifying Newton's third law. In Fig. 1b, it could be seen that the slider encountered the same elastic force on similar angles of the slopes. Consequently, in accordance with the friction formula, the sliding speed of the slider was solely influenced by the coefficient of kinetic friction. Users could directly perceive these distinct frictional experiences through the stylus, allowing for a tactile understanding of the differences.

Figure 2 shows the "Newton's Laws of Motion" case. The user first lifted the red ball into the cannon with the stylus and then pulled the spring tensioner at the bottom left. Depending on the magnitude of force pulled by the user, the cannon shot out

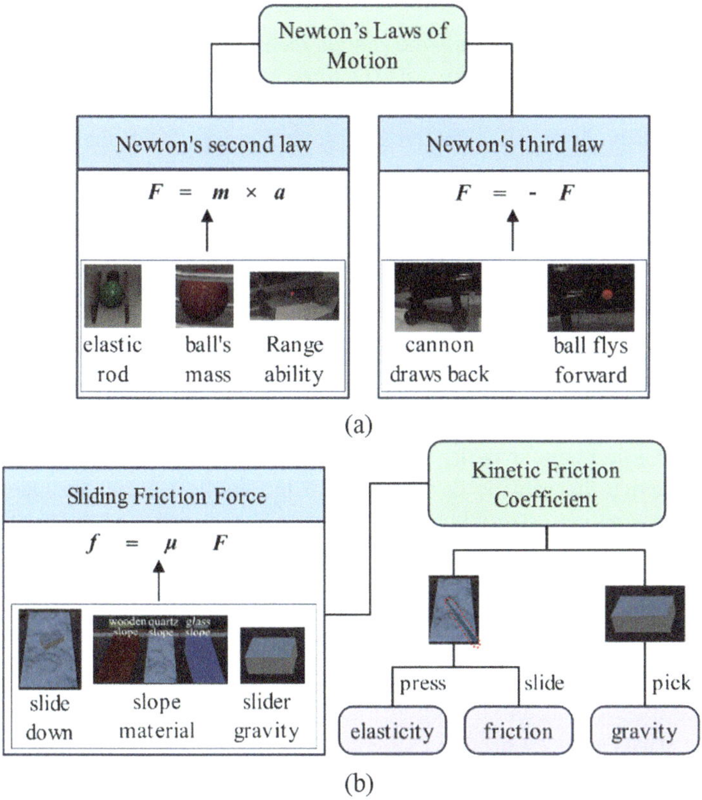

Fig. 1 **a** Design procedure of Newton's laws of motion. **b** Design procedure of kinetic friction coefficient

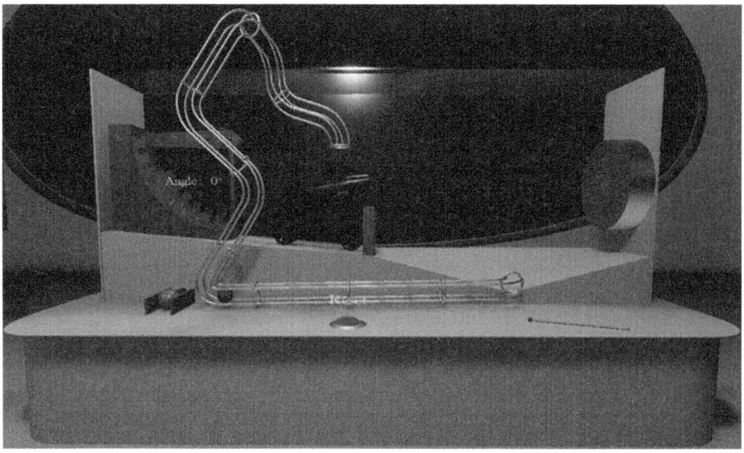

Fig. 2 Newton's laws of motion

the ball by the given force. While the cannon received the reaction force and moved back, it pushed the pointer backward. The force generated by the cannon on the ball varied in magnitude, as did the reaction force applied to it. For the ball, its initial acceleration is proportional to the force. With the force feedback, the user could learn Newton's second and third laws.

Figure 3 shows the "Kinetic Friction Coefficient" case. The user used the stylus to touch a slope while feeling the presence of elasticity and frictional force. Then they could use the stylus to pick up the cuboid and slide it down the slope. The three slopes had the same size and inclination but had different kinetic friction coefficients. The student controlled the cuboid to move through the stylus to feel the effect of the different friction coefficients.

Users were able to move 3D models (balls, cuboid) and pull the spring tensioner through a force feedback virtual model, which simulated the process of operation as if it were a first-hand experience. Geomagic Touch is a professional haptic force feedback device that can be developed with the application of the OpenHaptics toolkit. The user interacted with the stylus and receives received different kinds of force feedback feeling like touching the authentic objects. Based on the complicated setups and scenes, this device offered various controlling strategies that changed promptly. According to the predefined gravity and friction coefficients, the magnitude of the force would be computed and delivered to the stylus as feedback. The model could run with and without force feedback enabled. In the case of no-force feedback, the device functions as a tool to control the object's position and movement. When using the stylus to move the object, the user would feel the gravity and friction as the relative position of the object changes. In addition, when using the stylus to pull the spring tensioner, it was presented that the amount of elasticity was proportional to the length of the spring stretch.

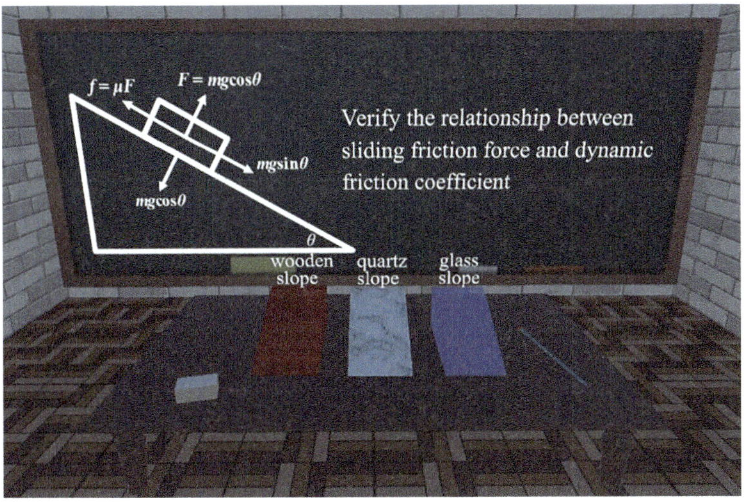

Fig. 3 Kinetic friction coefficient

2.3 Function Design

The function design included three parts: scene roaming, learning guide, and interactive experience. Scene roaming brought the immersion of the learning environment to the user and reinforced their memory. At the same time, it offered the user the space to move freely, enabling them to gain a deeper impression through observation from multiple perspectives. The learning guide included audio and text information to help users explore the environment. The audio would be played when the user stepped into the exhibit area. And there were some texts shown on the blackboard to guide the user. The interactive experience included controller interaction, haptic feedback interaction, and sound interaction. The user got the haptic feedback through the stylus. When an operation was completed, such as the cannon' shooting, an audio sound was played. Through the multisensory immersive interaction of visual, force, and auditory, the physical phenomenon and concepts were imparted to the user.

2.4 Force Feedback Implementation

2.4.1 Gravity

For an object of mass m, the vector of gravity to be applied to it is

$$\vec{F} = A * \left(0, -\frac{mg}{G_{\max}}, 0\right), \tag{1}$$

where G_{max} is the maximum value of the preset force; g is the acceleration of gravity; and A is the 3D transformation matrix from Unity world coordinate system to Geomagic Touch device space coordinate system. 3D space in Unity is represented by Cartesian coordinate system, i.e., the horizontal plane is the xz coordinate axis and the vertical direction is the y axis, which needs to be converted to the XYZ coordinates of the Geomagic Touch device.

2.4.2 Elasticity

The deformation of an object produces elasticity. Assume that the depth of deformation is H, and the vector of elastic forces on it is

$$\vec{F} = A * \left(\frac{TD}{D_{max}} \cdot \vec{D} \right), \tag{2}$$

where D_{max} is the maximum depth, $\vec{D}$ is the directional vector normal to the collision point, and A is the 3D transformation matrix, as defined previously.

2.4.3 Friction

Friction is based on relative motion and elasticity, so it is necessary to confirm that the stylus is in contact with the surface of the object and generates relative motion. The tangential partial velocity of the stylus at the contact surface is

$$\overrightarrow{v_t} = \vec{v} - \frac{\vec{v} \cdot \vec{D}}{\left|\vec{D}\right|} \cdot \frac{\vec{D}}{\left|\vec{D}\right|}, \tag{3}$$

where $\vec{v}$ is the velocity vector of the user manipulating the stylus, $\vec{D}$ is the normal direction vector at the point of collision. Finally, by vector subtraction, the tangential partial velocity of the stylus at the contact surface is obtained, i.e., the relative sliding velocity v_t with respect to the contact surface.

Therefore, the vector of frictional forces on the object is

$$\vec{f} = -\mu \cdot \left|\vec{F}\right| \cdot \frac{\overrightarrow{v_t}}{\left|\overrightarrow{v_t}\right|}, \tag{4}$$

where μ is the sliding friction coefficient of the contact surface and $\vec{F}$ is the magnitude of the elastic force on the object.

2.4.4 Combined Forces

In practical scenarios, objects are often not subject to a single force, but to a combination of forces. For example, in the "Newton's Law of Motion" case, the ball was subject to gravity; when the ball was picked up with the stylus, it was subject to tension, which was essentially elasticity; and if the ball was in contact with the pipe, it might also be subject to friction. The solution is to add the calculated forces together as F, which is

$$F = \sum f_i, \quad i = 1, 2, 3 \ldots \tag{5}$$

Finally, use the interface function provided by the Geomagic Touch device, setForce(deviceName, F, 0), to simulate the haptic feedback effect in the virtual environment.

2.5 *Experiment*

2.5.1 Equipment Settings

A pilot study was conducted with college students who were novices to force feedback devices to assess the effectiveness of the system. The equipment setting of the experiment includes (1) computer systems equipped with CPU as Intel Core™ i7-10875H, 16 GB RAM, NVIDIA GeForce RTX™ 2080 Graphics card, and 1 TB SSD hard disc; (2) HTC Vive HMD as VR device; (3) Geomagic Touch as a force feedback device. The physics VLE system was built on the Unity3D engine and used OpenHaptics and SteamVR plugins provided by the manufacturer. The parameters of force were set by the researchers according to the actual situation.

2.5.2 Measures

In order to assess the effectiveness of force simulation in this VLE for physics, academic pre-and post-test, and virtual presence questionnaires were used in this study.

(a) **Academic Test**

The items on the pre-test were selected from classical physics test questions in middle school, including 5 items of Newton's third law, 2 items of Newton's second law, and the last 3 items involving knowledge of friction. The knowledge examined in each item was closely related to the interaction task and was used to measure students' a priori knowledge base prior to completing the learning task. The post-test items were highly consistent with the items in the pre-test, with five items of Newton's third law, two items of Newton's second law, and the last three items of friction. However, the questions in the pre-test and post-test items were not identically the same. The post-test items also considered more of the transfer of knowledge and were closely related to real-life phenomena. All items were checked by experienced secondary school physics teachers to keep the pre-test and post-test items at the same level of difficulty as much as possible.

(b) **Virtual Presence Questionnaire**

The virtual presence questionnaire consisted of thirty-three 6-point Likert scale items from 1 as "strongly disagree" to 6 as "strongly agree" (with some items reversed), which was designed to assess the perceived presence of students in the experimental and control groups during the task by recording students' perceptions of control, sensory, distraction, and realism. Witmer and Singer [23] first used a questionnaire to identify aspects of presence in a virtual environment, which was subsequently described as containing the four factors of control, sensory, distraction, and realism. Their questionnaire has been widely used to assess virtual presence in various systems and its validity was verified by multiple studies. Later Hite et al. [24] adapted Witmer and Singer's questionnaire for use in evaluating presence in 3D, haptic-enabled, VR Science Instruction. The questionnaire used in this study was based on the questionnaire adapted from Hite et al. From the original questionnaire containing sixty-two six-point Likert scale items, some deletions were made around the aspects of force perception. Our questionnaire involved with control (eight items), sensory (seven items), distraction (nine items), and realism (nine items). Among them, the control factor referred to aspects of students' manipulation of virtual objects using the stylus and the transition of viewing through the HMD in the virtual environment. The sensory factor included visual effects from the HMD, auditory perception from the headset, and force feedback from the stylus. The distraction factor focused on the inconvenience of using the hardware interface (HMD and stylus) and the disturbance from the outside world, which reduced the student's perception of presence. The realism factor focused on the degree of similarity between the sensory perception brought about by the virtual environment and the actual phenomena in the real world.

2.5.3 Participants

There were 38 college students (24 females and 14 males, mean age of 19.9 years) who were enrolled and randomly assigned to the "experimental" and "control" groups. The student population N for each group was $N_E = 19$ and $N_C = 19$ for the experimental and control groups, respectively. Both groups used the stylus, but the experimental (force) group interacted with the model with force feedback technology activated, and the control (no-force) group received no-force feedback technology. All students participated voluntarily in the research and all had similar educational backgrounds.

2.5.4 Procedure

Prior experience with VR devices and force feedback devices was collected from students in addition to their demographic information. This was followed by a pre-test of knowledge of relevant physical concepts. Knowledge post-test needed to be tested after a certain interval to avoid instant memory. In the beginning, each student learned how to use the devices and watched a demo video that was made to introduce the whole interaction task. The researchers guided the students to do some informal exercises for about five to ten minutes. The experimental (force) group and control (no-force) group interacted with the model to finish the same task. There were two tasks in total:

(i) In "Newton's Law of Motion" case, each student lifted a small red ball along a track and threw it into a cannon, then pulled the spring tensioner to shoot the ball. This process was a complete interaction, and each student was required to successfully finish it twice.
(ii) In "Kinetic Friction Coefficient" case, it was required to touch three different materials of slope surfaces to feel the elasticity and friction (wood > quartz > glass) using the stylus. Then students picked up the cuboid and slid it down the three slopes to feel the difference in the sliding speed with different coefficients of kinetic friction.

All students would seat in a chair to finish the tasks by hand without excessive movement to avoid motion sickness. During the task, the interactions of students were recorded in videos, which would be analyzed afterward. The usability and effectiveness of the system were reflected in the interaction performance, while usability was measured by task-finishing time, and effectiveness was measured by the accuracy of performing the task.

All questionnaires were distributed in an online platform powered by www.wjx.cn. Students scanned the QR code to fill out the questionnaires. The interaction data was collected with the help of software, which included nine videos to record the complete interaction process.

2.5.5 Data Analysis

The purpose of this experiment was to evaluate the effectiveness of force feedback in the immersive physics learning environment, mean scores and standard deviations of the academic test and survey questions were used. And to compare the effectiveness of force feedback on students' knowledge of physics concepts, paired t-tests were performed to assess the statistical significance of average results' differences. In addition, Mann–Whitney's U test (2-tailed, $\alpha = 0.05$) was conducted to ascertain the significant differences of each group on presence survey questions and interaction data. And a reliability analysis (Cronbach's alpha) was calculated for each subscale of the questionnaire.

3 Results

A total of 38 students were recruited to fill out the survey and no one did not finish the tasks. Data collected through questionnaires were analyzed by SPSS 25. Table 2 presents the summary of the demographic information of the students. With regard to their prior experience with VR devices, 18 students (47.4%) reported they had never used VR devices before, and 19 students (50.0%) occasionally used VR devices, while the remaining one student (2.6%) reported having more prior VR experience. At the same time, only three students (7.9%) reported having some prior experience using force feedback devices, while the remaining 35 students (92.1%) had never used force feedback devices before.

A paired *t*-test was conducted on the data (Table 3). The paired *t*-test showed that the treatment conditions did not have a statistically significant increase in their post-test score. The mean score of the post-test minus the mean score of the pre-test was used as the knowledge gain of both groups, then calculating an effect size between the treatment and control groups. The result showed that the knowledge gains of the treatment and control groups were 0.26 and 0.89 ($t = -0.853$, $p = 0.399$), respectively.

A Mann–Whitney's *U* test (2-tailed, $\alpha = 0.05$) was used to compare the virtual presence survey scores of the experimental and control groups. To ensure the consistency of data, eight items were reverse-scored. The Cronbach's alpha of each group was calculated to check the internal consistency of each group. The Cronbach's alpha for the experimental group was 0.855, 0.771, 0.838, and 0.805 on the four factors, with a total value of 0.918. And that for the control group was 0.832, 0.901, 0.879,

Table 2 Demographic characteristics of students

Demographic category	Total (%)	Experimental group (%)	Control group (%)
Gender			
Male	14(36.8)	6(31.6)	8(42.1)
Female	24(63.2)	13(68.4)	11(57.9)
Age			
18 or younger	7(18.5)	6(31.5)	1(5.3)
19–22	30(78.9)	12(63.2)	18(94.7)
23 or older	1(2.6)	1(5.3)	0(0)
Prior VR experience			
Never	18(47.4)	8(42.1)	10(52.6)
Occasionally	19(50)	10(52.6)	9(47.4)
Usually	1(2.6)	1(5.3)	0(0)
Prior force feedback device experience			
Never	35(92.1)	18(94.7)	17(89.5)
Occasionally	3(7.9)	1(5.3)	2(10.5)

Table 3 *t*-test results of pre- and post-test between two groups

Group	N	Pre-test		Post-test		Paired t-test
		Mean	SD	Mean	SD	
Experimental group	19	8.00	1.972	8.26	1.284	$t = -0.545$ $p = 0.592$
Control group	19	7.79	2.250	8.68	0.885	$t = -1.594$ $p = 0.128$

Table 4 Reliability measures (Cronbach's alpha)

Presence category	Experimental group	Control group
Control ($N = 8$)	0.855	0.832
Sensory ($N = 7$)	0.771	0.901
Distraction ($N = 9$)	0.838	0.879
Realism ($N = 9$)	0.805	0.777
Whole test ($N = 33$)	0.918	0.937

Note "N" is the number of items

and 0.777 on the four factors, with a total value of 0.937. The reliability was excellent for alpha values greater than 0.9. As shown in Table 4, the values for the two groups reflect strong internal consistency of items' responses.

Figure 4 shows the mean ranks of four factors of virtual presence comparing the experimental and control groups. It was shown that the experimental group got a higher score in each factor.

Table 5 shows the differences in scores between the experimental and control groups on the four factors of virtual presence. The difference in whole presence

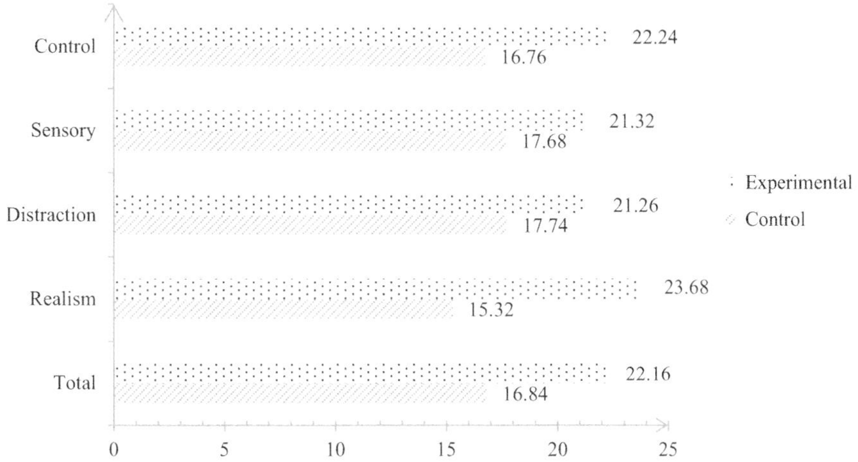

Fig. 4 Mean ranks of virtual presence for two groups

responses between the two groups was not significant. But the value of the realism factor ($p = 0.020$) indicated that the experimental group had a significantly higher sense of realism than the control group.

Table 6 shows the interaction time and accuracy for each task of the experimental and control groups, including the mean, maximum, and minimum values. In two cases, the experimental group had shorter interaction times and better accuracy rates compared to the control group. The shorter the interaction time, the better the usability of the system. And the higher the accuracy rate, the more effective the system is.

Table 7 shows the differences between the experimental and control groups in terms of interaction time and accuracy rate. There was a significant difference between the experimental group and control group in the interaction time and accuracy rate of "Newton's Laws of Motion" case. However, there was no significant difference between the two groups in the interaction time and accuracy rate of the "Kinetic Friction Coefficient" case.

Then ANOVA of multiple factors was used to analyze the influence of different experiment tasks (Table 8). In Table 8, Factor A referred to the different experiment

Table 5 Differences in two groups' presence responses

Presence category	Experimental group Mean rank	Control group Mean rank	Mann–Whitney *U*	*Z*	*p* value
Control item	22.24	16.76	128.50	− 1.522	0.128
Sensory item	21.32	17.68	146.00	− 1.015	0.310
Distraction item	21.26	17.74	147.00	− 0.980	0.327
Realism item	23.68	15.32	101.00	− 2.326	0.020*
Whole test	22.16	16.84	130.00	− 1.475	0.140

Note p-values were calculated using Mann–Whitney U (Alpha 2-tailed, 0.05)
$^*p < 0.05$

Table 6 Interaction descriptive statistics of the two groups

Group	*N*	Newton's laws of motion		Kinetic friction coefficient	
		Time (s)	Accuracy (%)	Time (s)	Accuracy (%)
Experimental group	Mean	108.5	83.5	85.8	86.1
	Lower bound	55.0	50.0	45.0	50.0
	Higher bound	229.0	100.0	164.0	100.0
Control group	Mean	142.4	63.8	97.8	80.7
	Lower bound	54.0	40.0	65.0	37.5
	Higher bound	296.0	100.0	164.0	100.0
Total	Mean	125.5	73.6	91.8	83.4
	Lower bound	54.0	40.0	45.0	37.5
	Higher bound	296.0	100.0	164.0	100.0

Table 7 Differences in two groups' interaction

Case	Item	Experimental group Mean rank	Control group Mean rank	Mann–Whitney *U*	*Z*	*p* value
Newton's laws of motion	Time (s)	15.76	23.24	109.50	– 2.073	0.038*
	Accuracy (%)	24.34	14.66	88.50	– 2.838	0.005*
Kinetic friction coefficient	Time (s)	16.00	23.00	114.00	– 1.942	0.052
	Accuracy (%)	20.61	18.39	159.50	– 0.677	0.498

Note p-values were calculated using Mann–Whitney U (Alpha 2-tailed, 0.05)
$^{*}p < 0.05$

Table 8 Results of multiple ANOVA

Factor	Time		Accuracy	
	F value	*p* value	*F* value	*p* value
Factor *A*	6.020	0.017*	7.183	0.009*
Factor *B*	12.882	0.001*	4.319	0.041*
Interaction *AB*	1.374	0.245	2.332	0.131

tasks and Factor B referred to whether force feedback was enabled. Both factors had a significant effect on the time and accuracy of the interaction, but neither interaction effect was significant.

4 Discussion

In this study, an immersive virtual environment with a force feedback application was designed and developed for physics learning. And the effectiveness of this application on students' learning outcomes, sense of presence, and efficacy of interaction were examined. Results from this study indicated some moderate evidence regarding the advantages of applying force feedback in an immersive physics learning environment. Specifically, our findings indicated that students' sense of realism and efficacy of interaction had been improved with the force feedback activated. The random assignment of the experimental and control groups revealed that the distribution of students in both groups was similar in terms of gender, age, and prior experience with the device.

Academic test results showed that the application of force feedback in our system did not have the positive effects of helping students learn the physics concepts in our system. No students spontaneously reported induced motion sickness, indicating that this was not disruptive to the students. The analysis revealed that the student's prior knowledge level was generally high and had a significant effect on knowledge gain. This may be the critical influencing factor for the learning effect. Given the complexity of the scenario and the potential variations between college and high school students, it is essential to consider the target audience's prior knowledge and learning capabilities. In future studies, greater attention will be given to including high school students as participants, allowing for a more comprehensive understanding of the effects on different age groups. Besides, the system also suffered from jitter, limited operating range, and locator jams caused by the hardware itself during operation, which might cause lots of distraction to students. Our researchers found that most of the students were focused on finishing the interaction task rather than on learning knowledge. Moreover, conceptual knowledge could be learned by visually observing the phenomena occurring in the virtual environment. A previous study by Webb et al. showed that after turning off force feedback, students were able to compensate for the lack of force feedback through visual cues [25]. However, Lontschar et al. [26] argued that haptic feedback for simple tasks (e.g., throwing heavy objects) brought higher stress to students and required more consideration of task design and technological improvements. Studies had shown that unintuitive interactions might increase students' cognitive load and decrease their attention to interactive tasks, thus affecting learning [27]. This result indicated that it was necessary to improve case design in order to take better advantage of force feedback devices and impart knowledge more effectively. In addition, these cases can be improved by adding some prompt cues and exercises to better illustrate the relationships between force and abstract concepts [20]. As previously mentioned, the level of prior knowledge of students, sample size, and learning time should be taken into account for future research. Considering that the current pre- and post-tests were limited to measure students' knowledge, new items should be designed, such as adding more types of questions. This result should be qualified as tentative until the experiment can be replicated with an appropriate sample.

According to the virtual presence survey, force feedback technology had an advantage in enhancing the sense of realism. Since students in the experiment used the same hardware interface for interaction, force feedback did not affect the students' sense of control, sensory, and distraction of the virtual environment. Realism factors included the similarity between the virtual environment and the real world and the consistency of simulation [23]. Results indicated that force simulation in this system had worked successfully in terms of realism. Previous experience has shown that providing better sensory feedback in a virtual environment led to better realism of the students' perception [28]. This was consistent with the findings of Jones et al. [29] that force perception has improved the student's sense of realism in a virtual learning environment.

The analysis of interaction data revealed that both accurate force perception and well-designed tasks contributed to the system's usability and effectiveness, thereby enhancing overall interaction effectiveness. With force feedback enable, students took less time to interact with the model while finishing the task, which was consistent with the findings of Schönborn et al. [13]. However, Vélaz et al. [27] argued that the actual task-finishing time for students in the force group was not significantly different from that of the video group. Therefore, it is necessary to consider the factors that influence students' interaction time and interaction behavior using force feedback devices in different learning tasks and disciplinary contexts. Also, with force feedback activated the accuracy of students finishing tasks was also significantly better. The research by Schönborn et al. [13] also showed that the activating force feedback resulted in more accurate placement of docking between the final models. Compared to the virtual environment without force feedback technology application, force simulation provided a haptic "space" outside of vision that was more constrained to the accessible position of the model, thus saving time in controlling the model to move and reducing the number of incorrect moves. In "Newton's Laws of Motion" case, this haptic "space" constraint was more precise, while the "Kinetic Friction Coefficient" case did not provide much haptic "space" constraint. This explained why the interaction time and accuracy of the latter did not differ significantly between the two groups. Therefore, when designing learning tasks with force feedback technology, more consideration should be taken to how to provide additional information beyond visuals to improve student interaction efficiency.

5 Conclusion

The purpose of this study was to explore the impact of force feedback technology in immersive physics learning on students' learning effectiveness, presence perception, and interaction efficiency in order to deeply explore the design and development of VR combined with force feedback technology in the field of K-12 physics education and instruction. First, for learning effectiveness, force feedback application did not have a significant effect on learning these physical concepts for the college students in this study, which might be due to the high level of prior knowledge they already had. Secondly, for the perception of presence, activating force feedback helped to enhance the student's sense of realism in the virtual environment. And students tended to find this kind of force simulation very realistic and extremely similar to the real world. Finally, in terms of interaction efficiency, activating force feedback helps students locate feasible movement positions and thus perform the task of moving the model more efficiently.

Students did not achieve the expected significant knowledge gains with force feedback enabled. This may be due to the fact that the relation between force feedback application and physical concepts was not highlighted in a high cognitive load environment, where students' attention was focused on using the device to complete interactive tasks and visual stimuli dominated their attentional system. Therefore, it is necessary to design systems that are more supportive of force feedback technology for learning physics concepts through technological improvements and the selection of appropriate tasks. Implementing visual cues and instructional aids in the learning environment can help students establish meaningful connections between force feedback sensations and conceptual knowledge. In the further, if force feedback experiments will be conducted in real classrooms, the role of the teacher also can be used to let students pay more attention to the knowledge learning. Future research is to validate the impact that prior knowledge may have on students' sensory perception and learning outcomes in virtual environments.

Acknowledgements This work is supported by the National Natural Science Foundation of China (No. 61977063, No. 62377004).

References

1. Yang, K.Y., Heh, J.S.: The impact of internet virtual physics laboratory instruction on the achievement in physics, science process skills and computer attitudes of 10th-grade students. J. Sci. Educ. Technol. **16**(5), 451–461 (2007)
2. Pelletier, K., Brown, M., Brooks, D.C., McCormack, M., Reeves, J., Arbino, N., Mondelli, V., et al.: EDUCAUSE horizon report teaching and learning edition. EDUCAUSE (2021)
3. Themelis, C., Sime, J.A.: From video-conferencing to holoportation and haptics: how emerging technologies can enhance presence in online education? In: Emerging Technologies and Pedagogies in the Curriculum, pp. 261–276. Springer, Singapore (2020)
4. Parong, J., Mayer, R.E.: Learning science in immersive virtual reality. J. Educ. Psychol. **110**(6), 785 (2018)
5. Mandal, S.: Brief introduction of virtual reality & its challenges. Int. J. Sci. Eng. Res. **4**(4), 304–309 (2013)
6. Freina, L., Ott, M.: A literature review on immersive virtual reality in education: state of the art and perspectives. In: The International Scientific Conference E-Learning and Software for Education, vol. 1, no. 133, pp. 10-1007 (2015)
7. Georgiou, Y., Tsivitanidou, O., Ioannou, A.: Learning experience design with immersive virtual reality in physics education. Educ. Tech. Res. Dev. **69**(6), 3051–3080 (2021)
8. Hew, K.F., Cheung, W.S.: Use of three-dimensional (3-D) immersive virtual worlds in K-12 and higher education settings: a review of the research. Br. J. Edu. Technol. **41**(1), 33–55 (2010)
9. Sanfilippo, F., Blazauskas, T., Salvietti, G., Ramos, I., Vert, S., Radianti, J., Oliveira, D., et al.: A perspective review on integrating VR/AR with haptics into STEM education for multi-sensory learning. Robotics **11**(2), 41 (2022)
10. Massie, T.H., Salisbury, J.K.: The phantom haptic interface: A device for probing virtual objects. In: Proceedings of the ASME Winter Annual Meeting, Symposium on Haptic Interfaces for Virtual Environment and Teleoperator Systems, vol. 55, no. 1, pp. 295–300 (1994)
11. Mayer, R.E.: Multimedia learning. In: Psychology of Learning and Motivation, vol. 41, pp. 85–139. Academic Press (2002)
12. Paivio, A.: Imagery and verbal processes. Psychology Press (2013)
13. Schönborn, K.J., Bivall, P., Tibell, L.A.: Exploring relationships between students' interaction and learning with a haptic virtual biomolecular model. Comput. Educ. **57**(3), 2095–2105 (2011)
14. Hite, R.L.: Perceptions of Virtual Presence in 3-D, Haptic-Enabled, Virtual Reality Science Instruction. North Carolina State University (2016)
15. Wilson, M.: Six views of embodied cognition. Psychon. Bull. Rev. **9**(4), 625–636 (2002)
16. Han, I., Black, J.B.: Incorporating haptic feedback in simulation for learning physics. Comput. Educ. **57**(4), 2281–2290 (2011)
17. Minaker, G., Schneider, O., Davis, R., MacLean, K.E.: HandsOn: Enabling embodied, creative STEM e-learning with programming-free force feedback. In: International Conference on Human Haptic Sensing and Touch Enabled Computer Applications, pp. 427–437. Springer, Cham (2016)

18. Zheng, M.Y., Li, J.H., Zhang, H., Luo, Y.L., Shen, J.L., Zhu, X.M.: Immersive physics learning environment with force feedback. J. Graphics **42**(1), 79–86 (2021)
19. Civelek, T., Ucar, E., Ustunel, H., Aydın, M.K.: Effects of a haptic augmented simulation on K-12 students' achievement and their attitudes towards physics. Eurasia J. Math. Sci. Technol. Educ. **10**(6), 565–574 (2014)
20. Neri, L., Noguez, J., Robledo-Rella, V., Escobar-Castillejos, D., Gonzalez-Nucamendi, A.: Teaching classical mechanics concepts using visuo-haptic simulators. J. Educ. Technol. Soc. **21**(2), 85–97 (2018)
21. Magana, A.J., Serrano, M.I., Rebello, N.S.: A sequenced multimodal learning approach to support students' development of conceptual learning. J. Comput. Assist. Learn. **35**(4), 516–528 (2019)
22. Yuksel, T., Walsh, Y., Magana, A.J., Nova, N., Krs, V., Ngambeki, I., Benes, B., et al.: Visuo-haptic experiments: exploring the effects of visual and haptic feedback on students' learning of friction concepts. Comput. Appl. Eng. Educ. **27**(6), 1376–1401 (2019)
23. Witmer, B.G., Singer, M.J.: Measuring presence in virtual environments: a presence questionnaire. Presence **7**(3), 225–240 (1998)
24. Hite, R.L., Jones, M.G., Childers, G.M., Ennes, M., Chesnutt, K., Pereyra, M., Cayton, E.: Investigating potential relationships between adolescents' cognitive development and perceptions of presence in 3-D, haptic-enabled, virtual reality science instruction. J. Sci. Educ. Technol. **28**(3), 265–284 (2019)
25. Webb, M., Tracey, M., Harwin, W., Tokatli, O., Hwang, F., Johnson, R., Jones, C., et al.: Haptic-enabled collaborative learning in virtual reality for schools. Educ. Inf. Technol. **27**(1), 937–960 (2022)
26. Lontschar, S., Deegan, D., Humer, I., Pietroszek, K., Eckhardt, C.: Analysis of haptic feedback and its influences in virtual reality learning environments. In: 2020 6th International Conference of the Immersive Learning Research Network (iLRN), pp. 171–177. IEEE (2020)
27. Vélaz, Y., Rodríguez Arce, J., Gutiérrez, T., Lozano-Rodero, A., Suescun, A.: The influence of interaction technology on the learning of assembly tasks using virtual reality. J. Comput. Inf. Sci. Eng. **14**(4), 041007 (2014)
28. Steuer, J.: Defining virtual reality: dimensions determining telepresence. J. Commun. **42**(4), 73–93 (1992)
29. Jones, M.G., Hite, R., Childers, G., Corin, E., Pereyra, M., Chesnutt, K.: Perceptions of presence in 3-D, haptic-enabled, virtual reality instruction. Int. J. Educ. Inf. Technol. **10**(1), 73–81 (2016)

Toward Universally Accessible Interactive Augmented Reality in Education: A Case Study of Developing a Photoelectric Effect Experiment in Interactive Augmented Reality

P. H. Myburgh, I. F. Potgieter, and Umesh Ramnarain

Abstract This chapter explores the creation of an approachable, affordable, and engaging augmented reality (AR) experience for instructing secondary pupils about the photoelectric effect experiment and provides a suggested workflow to guide educators and scholars in the creation of AR experiences. The project follows the Accessible, Interactive, Mobile, and Simple (AIMS) philosophy and makes use of free software, readily available materials, and low-end mobile devices that are not connected to the internet. The AR experience was created by a group of self-taught programmers and 3D modelers and includes custom 3D models, image recognition, and interactive virtual buttons. The experience, which was presented at the 2022 AI Expo Africa, received insightful comments that resulted in iterative improvements. In 2024, it will be put to the test in high schools in South Africa. The project shows that small research teams and dedicated teachers could create educational AR experiences using freely available platforms and 3D content creation tools streamlining the process further.

Keywords AR · STEM education · Natural science education · Access to education · Inclusive education

P. H. Myburgh (✉)
Metaverse Research Unit, Institute for Artificial Intelligent Systems, University of Johannesburg, JBS Business Park Campus, Auckland Park, Johannesburg, South Africa
e-mail: hermanm@uj.ac.za

I. F. Potgieter · U. Ramnarain
Centre for Advanced Learning Technologies in Science Technology Engineering Arts and Mathematics (CALTSTEAM), University of Johannesburg, Auckland Park Campus, Johannesburg, South Africa

Y. Cai et al. (eds.), *Virtual and Augmented Reality Technology-Enhanced Learning*, Gaming Media and Social Effects, https://doi.org/10.1007/978-981-96-2332-7_14

1 Introduction

The United Nations' Sustainable Development Goals (SDGs) emphasizes the importance of education in achieving a sustainable future, with the 4th Goal dedicated specifically to "Ensure inclusive and equitable quality and promote lifelong learning opportunities for all" [24]. Education is a key driver of social development, poverty reduction, and economic growth. Through quality education, individuals are often able to improve their livelihoods, reducing poverty (SDG1) [26] and through poverty reduction reducing hunger (SDG2). Education has also been linked to the attainment of SDG 3—good health and well-being [14]. This in turn results in societies that are more inclusive (SDG10), with decent work opportunities resulting in economic growth (SDG8), resulting in a larger tax-payer contingent which could be used to address the remaining SDGs.

However, many countries are struggling to ensure that every child has access to a proper education. Inequality in access to education [12], ensuring quality education for all [19], and other challenges often hamper governments' intentions to provide proper schooling. This is particularly true for governments in the Global South, for which societal pressures such as widespread poverty, challenging socio-economic circumstances, and political instability may result in inadequate schooling of children and adults [2]. Furthermore, the effects of the COVID-19 pandemic, and the measures implemented to curb the spread of the virus, highlighted the fragility of the educational system [21]. This was especially true for education in the science, technology, engineering, and mathematics (STEM) subjects, which are heavily reliant on a combination of theoretical and practical application of the subject matter [8, 16].

In the process of transferring STEM knowledge through teaching, instructors typically begin by presenting the theoretical concepts of a topic, followed by practical demonstrations and experiments to illustrate their applications. Through these experiments, the theoretical knowledge is applied in a real-world scenario, showcasing the practical application of said knowledge. Many Southern African schools, however, do not have access to the necessary equipment, reagents and technical expertise to provide learners with the opportunity to experience the practical aspects of study [9, 10, 22]. Moreover, in times of disturbed educational activities (such as the COVID pandemic and social unrest), learners are unable to access laboratories—resulting in many STEM learners only garnering theoretical knowledge within STEM subjects. This, in turn, results in students with only an abstract understanding of the subject matter [20].

A further challenge exists in STEM education, apart from the lack of access to educational laboratories. With increasing demand for STEM education, impacting all spheres of the educational offering, the physical infrastructure needs are hard to keep up with [7]. For high-student number schools, the solution is often to split classes, having multiple sessions with different groups demonstrating the practical application of theoretical knowledge. Although a valid interim solution, it does open the pedagogy to a few further challenges. For one, the instructor may not give their full attention to all groups equally—resulting in some inequalities in the quality of training. Secondly, for many schools, especially in developing countries, educators are overburdened with multiple subjects and high numbers of students in classes resulting in teaching-fatigue—also impacting the quality of education [15, 23, 25]. Lastly, setting the experiment up is often a timeous activity, that can result in challenges in ensuring that each class starts on time, and does not impact the lesson duration.

To overcome these challenges, some educators rely on the use of virtual laboratories (VLs). Often such laboratories or computer-aided simulations, accessible via the internet, are two-dimensional. In 2019, the majority (92%) of virtual laboratory experiences were 2D desktop based [17]. In their comparison of 2D and 3D virtual chemistry laboratories, Ali and Ullah [1] note that 3D VLs allow for more immersive and realistic experiences and affords for spatial depth in chemical images, while 2D VLs had the advantages of being less strenuous on the eyes and required less bandwidth to complete online. While virtual laboratories (VLs) in chemistry practices have demonstrated superior cognitive learning outcomes compared to traditional passive media and, in some cases, equal or greater learning outcomes than hands-on laboratories, it is important to note that the existing research has primarily focused on cognitive aspects, and there is a need for more investigations into the fine motor and practical skills associated with hands-on laboratory tasks [4].

Based on the effectiveness of VLs to overcome challenges in STEM education, we set forth to develop a virtual experiment focusing on the photoelectric effect. From practical experience and teacher feedback, the photoelectric effect was one of the topics that high school students in South Africa especially struggled with. The photoelectric effect is a phenomenon in which electrons are emitted from a metal surface when light of a certain frequency is shone on it. This effect was first observed by Heinrich Hertz in the late nineteenth century. It was later explained by Albert Einstein in 1905, who proposed that light was not just a wave but also consisted of individual packets of energy, known as photons. Einstein won the Noble prize for his work on the photoelectric effect in 1921. The discovery of the photoelectric effect provided the first experimental evidence of the wave-particle duality of light, a cornerstone of quantum mechanics. In schools, the photoelectric effect is typically taught as part of the physics curriculum, usually in the context of wave-particle duality and quantum mechanics. Students learn about the basic principles of the effect and the experimental setup used to demonstrate it. They also learn about the relationship between the intensity of the light and the number of electrons emitted and how it differs from the classical wave theory. Additionally, students may also learn about the applications of the photoelectric effect, such as in photocells, which are used in devices such as automatic door openers, light meters, and burglar alarms.

There exist other virtual photoelectric effect experiments, ranging in quality and cost to access. Some of these virtual experiments are included in Table 1.

There is a trend toward the development of VLs for extended reality (XR)-enabled devices. Mobile devices are especially important in an South African context [11], since these are often more accessible in resource limited educational institutions. With modern smartphones and tablets capable of allowing users to experience AR,

Table 1 Examples of virtual laboratory experiments depicting the photoelectric effect

Experience	2D/3D	Hardware	Cost	Description
PhET Interactive simulations[a]	2D	PC	Free	PhET provides a virtual experiment that allows students to explore the photoelectric effect. Users can adjust the frequency and intensity of the light, and observe the effect on the number of electrons emitted
Virtual Labs[b]	2D	PC	Free	The Indian Institute of Technology provides a virtual experiment on the photoelectric effect, where students can adjust the frequency and intensity of light and observe the effect on the number of electrons emitted
Go-Lab[c]	2D	PC	Free	The Go-Lab provides a virtual experiment on the photoelectric effect, in both English and Dutch and developed by Jakob Sikken and Anjoy Anjewierden. Students are able to change several parameters, to attempt to ellicit the photoelectric effect
Labster[d]	3D	PC/VR	Paid	Labster provides a virtual experiment that allows students to explore the photoelectric effect, where students can adjust the frequency and intensity of the light and observe the effect on the number of electrons emitted
Riyanda et al. [18]	3D	AR	N/A	An AR-based, preliminary report on a self-developed application that can run on mobile devices with more than 2 GB RAM
AROSE [3]	3D	AR	N/A	An AR-based wave-particle duality learning tool called "AROSE" to examine how AR technology affects students' perceptions of learning physics and feelings of self-efficacy

Abbreviations *AROSE*—Augmented Reality Optical Simulation Experiments; *PHeT*—Physics education technology

[a]https://phet.colorado.edu/en/simulations/photoelectric

[b]https://vlab.amrita.edu/?sub=3&brch=195&sim=840&cnt=4

[c]https://www.golabz.eu/lab/photoelectric-effect-lab

[d]https://www.labster.com/simulations/photoelectric-effect

and in some instances true mixed reality (MR), this opens a novel avenue for educational resource developers to explore. Work by other groups, including Riyanda et al. [18] and Cai et al. [3], expands on the available VL experiences for teachers in the near future, and may either become available as commercial endeavors or as free tools in the pedagogical arsenal of educators. AR shows promise as a tool to support and improve students' learning motivation and interest [5], allowing for VLs to become more accessible to students and practical facilitators alike.

2 Concept and Design

Planning of an immersive experience is essential to the success of the project. During the planning phase, our team identified the core features of the interactive AR experience, the design feel, and the level of immersion. A serious oversight during the planning phase of the project was to not include peer-reviewed journal articles as part of the investigation into already existing AR solutions focused on the photoelectric effect. Although the team did searches on the internet, application stores and known teaching resources (such as PHET), our initial findings did not include the work of Riyanda et al. [18], published in the Jurnal Teknologi Informasi Dan Pendidikan. As such, it is advised that extensive searches should be conducted prior to the development of new experiences, to avoid the duplication of work.

2.1 Project Planning and Learning Experience Design

During project planning, the final experience was set to adhere to the following criteria:

- **The experience had to be created by members of the team with no commercial contracting occurring**. Since the team members have no formal training in XR experience creation, apart from publically available online resources (i.e., Youtube and Stackoverflow, *inter alia*), this mimics the possibility of teachers and educators developing their own experiences for students.
- **The final experience may not use an active internet connection during play time**. Mobile connectivity remains a challenge in many resources limited settings. Mobile data, as an example, is an expensive commodity in some sub-Sarahan countries [6], resulting in experiences requiring an active internet connection to often limit the practical usability of such an experience.
- **The experience had to run on an entry-level mobile device (either a smartphone or a tablet)**. For development and testing of the experience, a Lenovo Tab M8 HD (ZA5H) device was acquired, with an quad-core processor, 2 GB RAM

and 32 GB storage.[1] The device has a 5MP back camera, and a relative cost of US$ 150 (ZAR 2500) at the time of purchase.

- **To engage students, the experience had to be interactive, allowing for experimentation with different parameters**. Interaction is an important in practical learning of STEM subjects. As such, we required the experience to guide learners in their quest for knowledge.
- **Any input, such as images for recognition, should be printable from the user's own printer**. Although high-end AR experience use custom created 3D printed models, we decided to opt for a paper based input modality, allowing teachers to download and copy the input without additional waiting times for deliveries. This makes it easier to reach rural regions.
- **Since no capital was available for the creation of the experience, this interactive AR experience should be built using free software**. There are several challenges facing educators, one of which is access to funds to create their own XR teaching experiences. With the University of Johannesburg's recently launched Master and Doctoral of Philosophy Degrees in Virtual and Augmented Reality in STEM Education focused on equipping teachers with the necessary skills to develop their own XR VLs, the creation of such experiences on a nearly free basis is an essential design component.

Incorporating the criteria above, we developed a simplified strategy for the development of VLs for the South African market. Such experiences should be Accessible, Interactive, Mobile and Simple (AIMS). Furthermore, our previous experience and evaluations of different software to implement in the creation of similar experiences [13] resulted in a predefined flow for the creation of this experience.

The following software was used for the creation of our interactive experience:

- **Adobe Illustrator 2022**: Adobe Illustrator is a vector graphics editor widely used for creating digital illustrations, logos, graphics, and other visual content. It offers a comprehensive set of tools and features for creating and manipulating vector graphics, including support for working with multiple layers, shapes, paths, and more. Although Illustrator is a premium product, Adobe does offer a free trial period with access to the full software package. Alternative free solutions such as Inkscape, Gravit Designer, Figma, Vectr, and GIMP also exist.
- **Blender v3.4.x**: Blender is a free and open-source 3D creation software that has gained popularity in the fields of animation, film, architecture, and more. Its wide range of features and tools make it a versatile tool for creating high-quality 3D models, animations, and visual effects. This article investigates the use of Blender in various scientific applications and evaluates its potential as a tool for scientific visualization, simulation, and analysis.
- **Unity3D v2021.3.x**: Unity is a powerful game engine and development platform that has been widely adopted for the creation of interactive and immersive experiences in fields such as architecture, medicine, education, and more. The versatility

[1] According to the supplier website: https://www.takealot.com/lenovo-tab-m8-hd-za5h-8-32gb-lte-tablet/PLID68964385

of Unity has led to its use in a variety of scientific domains, including simulations, virtual reality, and interactive visualization. This article explores the use of Unity in scientific research and highlights the benefits and challenges of using Unity as a tool for scientific investigation. Installation of Unity on a personal computer is described in Appendix.

- **Vuforia Software Development Kit for Unity**: The Vuforia SDK for Unity is a plugin that integrates the Vuforia AR engine into the Unity game engine, allowing developers to create AR experiences for a variety of platforms. With its intuitive visual interface and support for features such as image and object recognition, Vuforia for Unity offers a streamlined and efficient way for developers to create interactive AR experiences that can enhance learning, communication, and engagement.

3 Technical Implementation

The photoelectric effect experiment involves shining light onto a metal surface, which releases electrons from the metal surface in a process called photoemission. The current of the released electrons can be measured using an ammeter, which provides a measure of the number of electrons released per unit time. An image that would be used to identify the experiment that students would experience was designed on Adobe Illustrator (Fig. 1). Using a vector-based image editor is advised, as this will allow for printing of the image on any size of media. This allows teacher to print individual experiments, in classes where students have their own AR enabled devices, or a single large format should teamwork be required.

Next, this image was uploaded to the Vuforia developer platform for image recognition, and the subsequent database was downloaded. After importing the Vuforia SDK into Unity (Appendix), the database was added to Unity. A process of setting up an AR camera and the Vuforia license followed, after which any 3D models placed on the image recognition asset will be presented once the image is recognized. This presentation occurs in 3D space, with tracking of the image, allowing for visualization of the 3D model from any viewing angle while the image is view of the physical camera of the mobile device. 3D models used in this experience were custom created by the research team using Blender modeling software. A digital current meter, battery, metal plates, and light source are some of the 3D generated assets.

Once the AR experience was set up and tested to recognize the created image in Fig. 1, interactivity was added into the experience. Interactivity was mainly achieved by programming virtual buttons, another feature of the Vuforia SDK for Unity. This allows the program to easily overlay virtual buttons, unseen to the user of the experience, that are triggered in the event of occlusion of predefined regions of the recognized image. Our users were allowed to change the light wavelength, the metal plates, and the light intensity. The combinations of these events were programmed in a C# script, and updates to the various 3D models (i.e., the metal plates and light color)

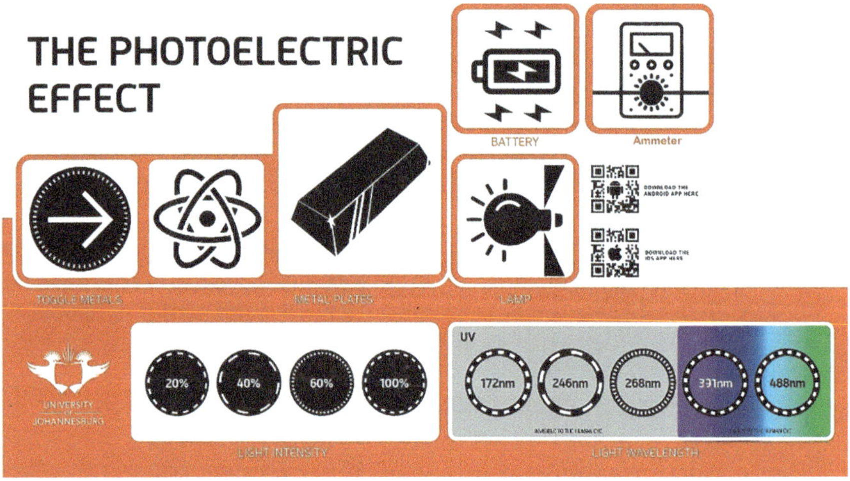

Fig. 1 A vector generated image used for recognition by the Vuforia engine for Unity. The image is optimized for printing on international "A" series paper sizes defined by the InternationaOrganisation for Standarisation (ISO), with an aspect ratio of 1: $\sqrt{2}$ allowing for easy scaling

Fig. 2 An image of the created immersive AR experience. The scannable quick response code links to a video hosted on Youtube.com showcasing the prototype of the experience

were scripted based upon user input. We opted for a digital ammeter, as this allows for the viewer to easily read the change in potential occurring during input given. The completed experience can be viewed from Youtube by scanning the quick response code presented in Fig. 2.

4 Testing and User Feedback

The created immersive AR experience was presented at the 2022 AI Expo Africa trade show during September 2022. Based on industry feedback addressing a few minor bugs in the coding, the experience is currently being improved by addressing these found issues. One such issue was drifting—as the experience was running for a whole day, the 3D models would shift out of position of the tracked image. This resulted in the virtual buttons not being "on top" of their dedicated spaces on the printed image used for recognition. As a result, users would have to search for the virtual button on random spaces. This has largely been overcome by adding a reset option—allowing the user to reset the image recognition without having to exit the application. It is thus recommended to add a reposition button, that re-recognizes the image and overlays the 3D models and virtual buttons on the correct spaces. Other than this, our experience performed well on the entry-level tablet, with no noticeable stress on the hardware (i.e., overheating or poor response times). Battery life was also good; however, it should be noted that it was a new device and that battery life is highly dependent on the number of charging cycles that the battery experienced in its lifetime.

The created project will be tested in South African high schools in 2024 and the results published in a peer-reviewed open-access journal. It is not currently envisioned that the project will become commercial, as the need for innovative, accessible teaching aids far outweighs the financial gains that could be made from such an endeavor.

Future XR experiences should always be created following an in-depth review of available literature, including scholarly output. The combination of free(mium) software used here, including Unity and Blender, provides educators with the ability to create custom made experiences for their pupils, that can be reused and standardized in subsequent teaching engagements.

5 Conclusion

The creation of an interactive augmented reality application is within the reach of small research teams, and even single teachers with a passion for creating novel pedagogical tools. Our team were self-taught programmers and 3D modelers, using free online resources and tutorials to create an AR experience that could be implemented in resource limited schools. The creation process will become easier and quicker with the advances made in utilizing online, artificial intelligence enabled platforms such as ChatGPT to assist with coding challenges, and 3D content creation platforms similar to those developed by Nvidia. Although we inadvertently used the same approach as [18], our experience allows for more experimentation to occur, as students are able to change the wavelength of the light, the intensity of the light source, as well as the metals used in the metal plates.

Appendix: Installation Guides

This appendix contains installation guides for software that was implemented in the creation of the presented interactive AR experience. Please note that the guides were accurate at the time of going to press, however changes may occur to the installation process in future. These guides were generated using ChatGPT.

Blender

Installing Blender is a straightforward process that can be completed in a few simple steps. Here is a basic guide on how to install Blender:

1. Download Blender (https://www.blender.org/download/): The first step is to download Blender from the Blender website. You can choose between the latest stable release or the latest daily build, depending on your needs.
2. Install Blender: Once the Blender package is downloaded, run the installer and follow the instructions to install the software.
3. Launch Blender: Once the installation is complete, you can launch Blender by clicking on the Blender icon in your start menu or on the desktop.
4. Set up Blender: On the first launch, Blender will prompt you to set up your preferences, including the interface, input devices, and add-ons. You can configure these settings to meet your needs.
5. Get started: Once the setup is complete, you can start using Blender. You can explore the interface, create new projects, and begin working with the various tools and features available in Blender.

This is a basic outline of the steps required to install Blender. The exact process may vary depending on your operating system and specific version of Blender.

Unity3D

1. Download Unity Hub (https://unity.com/download): The first step is to download the Unity Hub, which is a centralized tool for managing your Unity installations and projects. You can download the Unity Hub from the Unity website.
2. Install Unity Hub: Once the Unity Hub is downloaded, run the installer and follow the instructions to install the software.
3. Log in to Unity Hub: Open the Unity Hub and log in with your Unity ID. If you do not have a Unity ID, you can create one from within the Unity Hub.
4. Add a new Unity version: Once you are logged in, you can add a new Unity version by clicking the "+" icon in the "Installs" section and selecting the version you want to install.

5. Download and Install Unity: Follow the on-screen instructions to download and install the selected Unity version. This process may take several minutes, depending on the size of the installation package and your internet speed.
6. Launch Unity: Once the installation is complete, you can launch Unity from the Unity Hub.

This is a basic outline of the steps required to install Unity3D. The exact process may vary depending on your operating system and specific version of Unity.

Vuforia SDK for Unity3D

1. Download Unity: The first step is to make sure you have Unity installed on your computer. If you don't have Unity, see guide above.
2. Create a Vuforia account (https://developer.vuforia.com/vui/auth/register): To use the Vuforia SDK, you need to create a Vuforia account. You can do this by visiting the Vuforia website and clicking the "Sign Up" button.
3. Download the Vuforia SDK: Once you have created a Vuforia account, you can download the Vuforia SDK for Unity by visiting the Vuforia Developer Portal and clicking the "Download" button.
4. Import the Vuforia SDK into Unity: Once the Vuforia SDK is downloaded, open Unity and import the SDK into your project by selecting "Assets" > "Import Package" > "Custom Package" from the top menu and selecting the downloaded SDK package.
5. Activate the Vuforia license: To use the Vuforia SDK in your Unity project, you need to activate a Vuforia license. You can do this by selecting "Window" > "Vuforia Configuration" from the top menu, and entering your Vuforia license key.
6. Build and Run: Once the Vuforia license is activated, you can build and run your Unity project with the Vuforia SDK. You can use the Vuforia SDK to create AR experiences and applications in Unity.

Image Recognition Setup

Setting up Vuforia in Unity for image recognition involves the following steps:

1. Create a Unity Project: Open Unity and create a new project.
2. Import Vuforia Engine: From the Unity Asset Store, import the Vuforia Engine package into your project. You can do this by selecting "Assets" > "Import Package" > "Custom Package" from the top menu and selecting the Vuforia Engine package.

3. Activate Vuforia License: Create a Vuforia Developer Account and activate your Vuforia license. You can do this by visiting the Vuforia Developer Portal and entering your license key into the Vuforia Configuration in Unity.
4. Create an Image Target Database: Create a new image target database by selecting "Vuforia" > "Image" > "Add Image Target." You will need to provide a name for the database, and then add the target images to it.
5. Create an Image Target: Create a new image target by selecting the target image from the database, and then dragging it into the scene.
6. Attach a Vuforia Behavior: Add the Vuforia Behavior component to the Image Target by selecting "Add Component" > "Vuforia Engine" > "Vuforia Behavior."
7. Add an AR Camera: Add an AR camera to the scene by selecting "Game Object" > "Vuforia Engine" > "AR Camera."
8. Add 3D Model: Add a 3D model to the scene that will appear when the target image is recognized. You can do this by dragging a prefab or model from the Assets panel into the scene.
9. Configure Image Target: Configure the Image Target by setting the "Image Target Behavior" component to recognize the target image, and then set the "Virtual Button Behavior" component to recognize the virtual button.
10. Test the AR Experience: Test the AR experience by building and running the Unity project on a device that supports AR. When the target image is recognized, the 3D model will appear.

This is a basic outline of the steps required to set up Vuforia in Unity for image recognition. The exact process may vary depending on your specific project requirements and the version of Vuforia and Unity you are using.

References

1. Ali, N., Ullah, S.: Review to analyze and compare virtual chemistry laboratories for their use in education. J. Chem. Educ. **97**, 3563–3574 (2020)
2. Benavot, A., Antoninis, M., Bella, N., Delprato, M., Härmä, J., Jere, C., Joshi, P., Koseleci, N., Longlands, H., Mcwilliam, A., Zubairi, A.: Education for All 2000–2015: review and perspectives. Zeitschrift für internationale Bildungsforschung und Entwicklungspädagogik, 38 (2015)
3. Cai, S., Liu, C., Wang, T., Liu, E., Liang, J.-C.: Effects of learning physics using Augmented Reality on students' self-efficacy and conceptions of learning. Br. J. Edu. Technol. **52**, 235–251 (2021)
4. Chan, P., van Gerven, T., Dubois, J.-L., Bernaerts, K.: Virtual chemical laboratories: a systematic literature review of research, technologies and instructional design. Comp. Educ. Open **2**, 100053 (2021)
5. Chang, R.-C., Yu, Z.-S.: Using augmented reality technologies to enhance students' engagement and achievement in science laboratories. Int. J. Dist. Educ. Technol. (IJDET) **16**, 54–72 (2018)
6. Chinembiri, T.: Despite reduction in mobile data tariffs, data still expensive in South Africa (2020)

7. Chowdhury, H., Alam, F., Mustary, I.: Development of an innovative technique for teaching and learning of laboratory experiments for engineering courses. Energy Procedia **160**, 806–811 (2019)
8. Gamage, K.A.A., Wijesuriya, D.I., Ekanayake, S.Y., Rennie, A.E.W., Lambert, C.G., Gunawardhana, N.: Online delivery of teaching and laboratory practices: continuity of university programmes during COVID-19 pandemic. Educ. Sci. **10**, 291 (2020)
9. George, M.J.: Assessing the level of laboratory resources for teaching and learning of chemistry at advanced level in Lesotho secondary schools. S. Afr. J. Chem. **70**, 154–162 (2017)
10. Kaupitwa, E., Amuthenu, S.H.: Exploring the status and the utilisation of biology laboratories in a secondary school in Namibia. Open Access Lib. J. **9**, 1–13 (2022)
11. Mdlalose, N., Ramnarain, U., Penn, M.: Science teachers' perceptions on using mobile-based formative assessment for inquiry-based teaching: benefits and constraints (2022)
12. Mutisya, M., Muchira, J.M., Abuya, B.A.: Understanding wealth inequalities in education access in urbanizing Sub-Saharan Africa. Front. Educ. **6** (2021)
13. Myburgh, P.H.: Reflecting on the creation of virtual laboratory experiences for biology students. Front. Educ. (Lausanne), 7 (2022)
14. Myburgh, P.H., Nienaber-Rousseau, C., Kruger, I.M., Towers, G.W.: Education, smoking and CRP genetics in relation to C-reactive protein concentrations in black South Africans. Int. J. Environ. Res. Public Health **17**, 6646 (2020)
15. Rabe-Steynberg, A.: MEdPsych Masters, Stellenbosch University (2021)
16. Radhamani, R., Kumar, D., Nizar, N., Achuthan, K., Nair, B., Diwakar, S.: What virtual laboratory usage tells us about laboratory skill education pre- and post-COVID-19: focus on usage, behavior, intention and adoption. Educ. Inf. Technol. (Dordr.) **26**, 7477–7495 (2021)
17. Reeves, S.M., Crippen, K.J.: Virtual laboratories in undergraduate science and engineering courses: a systematic review, 2009–2019. J. Sci. Educ. Technol. **30**, 16–30 (2021)
18. Riyanda, A.R., Jalinus, N., Sukardi, S., Waskito, W., Ranuharja, F., Samala, A.D., Adi, N.H., Dewi, I.P.: Augmented reality technology for 3D photoelectric simulation. Jurnal Teknologi Informasi dan Pendidikan **14**, 217–222 (2021)
19. Roser, M., Ortiz-Ospina, E.: Global Education. Published online at OurWorldInData.org. (2016)
20. Shana, Z., Abulibdeh, E.S.: Science practical work and its impact on students' science achievement. J. Technol. Sci. Educ. **10**, 17 (2020)
21. Soudien, C., Reddy, V., Harvey, J.: The impact of COVID-19 on a fragile education system: the case of South Africa (2021)
22. South African Institute For Race Relations.: Only 15% of public schools have laboratories (2012)
23. Springer, A., Oleksa-Marewska, K., Basińska-Zych, A., Werner, I., Białowąs, S.: Occupational burnout and chronic fatigue in the work of academic teachers–moderating role of selected health behaviours. PLoS ONE **18**, e0280080 (2023)
24. United Nations General Assembly.: Transforming our world : the 2030 Agenda for Sustainable Development (2015)
25. van Tonder, C.L., Williams, C.: Exploring the origins of burnout among secondary educators. SA J. Ind. Psychol. **35**, 204–218 (2009)
26. Yuan, L., Ding, Y.: Poverty alleviation through education: China's approach and its significance to the world. In: Guo, D. (ed.) The Frontier of Education Reform and Development in China: Articles from Educational Research. Springer Nature Singapore, Singapore (2023)

Phantom Touch in VR and Its Potential Uses in the Education Field

Sasha Alexdottir

Abstract Phantom Touch in VR is an emerging phenomenon where a user can experience another person's touch in virtual reality without any real-life contact or haptic devices. It is a pseudo-haptic brain trick which allows interpersonal touch interactions to seem real to an individual. There is ongoing research on how Phantom Touch can be developed in people, as well as understanding what psychophysiological effect the pseudo-haptic has on users. An effective method has been developed through which users can generate Phantom Touch—the Head-to-Toe method (HTT). The technique was tested on participants, with heart rate and qualitative data being collected during the experiments to analyze Phantom Touch's effect. This chapter presents findings from the ongoing study and how the discoveries on Phantom Touch in VR can be implemented into the education field—to enhance teaching, student development, and support. The topics which are covered include—immersive education, student mental health, studying abroad, isolation (COVID-19), and cultural knowledge (including travel, language, and student exchange). There is discussion on possible virtual simulations and how they could be applied in academia.

Keywords Phantom Touch in VR · Pseudo-haptics · Education · Social touch

1 Introduction

When picturing education, most commonly people would think of student academic achievement, whether in the form of exam results, grades throughout the school year, or in-depth understanding of a particular subject. However, when looking closely at the education field, there are many other important aspects which make up a student's educational journey, such as learning discipline, empathy, confidence, personal growth, enhancing communication skills, improving qualifications, gaining connections and friendships, taking part in innovation, and contributing to building a better society. With these tasks and studies, there are also troubles which students

S. Alexdottir (✉)
Bournemouth University, Bournemouth, UK
e-mail: sa@phantomtouchvr.com

Y. Cai et al. (eds.), *Virtual and Augmented Reality Technology-Enhanced Learning*, Gaming Media and Social Effects, https://doi.org/10.1007/978-981-96-2332-7_15

face, such as coping with large amounts of work, stress, fear of failure, and other. Such occurrences and emotions can lead to poor mental and physical wellbeing of students.

In the present day, academic institutions are progressively incorporating technology, virtual reality, and augmented reality into the teaching curriculum [1] to aid students in learning, increase engagement, improve cognitive skills, enhance understanding in a subject, and make studying more enjoyable. In a survey by Kamińska et al. [2], there is great support for the use of VR, not only in the medical and engineering subjects, but in academia overall. Moreover, a study by Philippe et al. [3] looked at learning and teaching in VR, specifically with regard to serious games, simulations, and collaborative work. The authors wrote about the benefits of such learning mode, which included soft skills training when using serious games, simulations being useful for STEM subjects, and collaborative VR being practical for the immersion with innovative content and facilitating in student interactions. Herrera-Aliaga and Estrada [4] also discuss how technology is used today to replicate real-life scenarios in the medical field, pointing out the existence of virtual hospitals with accurate equipment. The study also mentions that the COVID-19 pandemic has increased the need for virtual learning and can have successful implementation in academia. A study by Oman et al. [5] explains the importance of VR simulations in medical training, stating that such technology will become a solution to the scalability of traditional simulation training as it will improve the accessibility, time, and cost. Graham and McAleer [6] also support such technology in education, explaining that students who took part in simulation-based learning reported that the activity has assisted them in the transition from students to newly qualified doctors. The research also mentioned the importance of 'design' in simulations, especially in the medical field, as they need to be as close as possible to real-life scenarios.

As mentioned in these studies, VR can be successfully implemented into education, especially when it comes to training, simulation, and visualization. Although learning interest, engagement, subject understanding, and cognitive skills are increased with such learning medium, some other aspects of education are abandoned, or not studied in such detail, for example, mental health and culture learning. With the use of VR, other aspects of a student's academic journey can be targeted to benefit them not only in their studies, but also wellbeing. Please refer to Sect. 5 of this chapter—Phantom Touch in the Education field.

Furthermore, 'touch' is not explored greatly for virtual reality in education. Hannaford greatly emphasizes in her books the importance of tactile elements in learning [7]. Hannaford wrote that: when touch in implemented into learning, much more of the brain is activated, building more complex nerve networks which aid in heightening the learning potential. Although tactile elements are much more widely used in primary education (via activities, textures, etc.), touch can be applied in higher education for various aspects of a student's learning—to benefit wellbeing, those studying abroad, in cases of a pandemic/epidemic and other. If touch were to be utilized in VR and simultaneously applied in education, students could have an interactive and engaging medium which can assist them in their academic journey, targeting the diverse sides of learning and student life.

2 Importance of Social Touch and Touch in Education

Since birth, humans require social touch to grow emotionally and improve their wellbeing and overall health. Cascio et al. [8] outline how crucial interpersonal touch is, from birth to old age. There is also great emphasis on the importance of social touch to build relationships, connections, and bonds between people. In a study by Henry Dwight Chapin [9], it was observed that most infants under the age of two had died due to failure to thrive, related to the lack of touch and affection. Cohen et al. [10] study also supports the claim that hugging can provide stress-buffering social support and can benefit human wellbeing—both in mental health and physical health. It has also been observed that interpersonal touch can relieve pain, diminish stress [11], and boost the immune system [12, 13]. Moreover, interpersonal touch can enhance cognitive and social capabilities in people; meanwhile, deprivation of touch may create greater cognitive deficit [14]. This means that implementation of social touch (in this case, virtual touch) in education can be beneficial for learners in improving their social and cognitive skills.

With such fruitful and informative research on social touch, the question arises on how can it be applied in education, in combination with virtual reality? Since touch can bring many benefits to the learners, it is most definitely worth considering implementing into the higher education sector. There have been studies looking at the use of interpersonal touch in schools—such as observing the effects of a teacher's brief touch on a student's forearm prior to demonstrating an exercise in front the peers [15]. The results showed an increase in volunteer rate and a greater sense of encouragement in learners. Another research surveyed teachers regarding social touch for young students [16]—the results from the tutors stated that interpersonal touch received by the student could promote their on-task behaviors, improve general behaviors, and promote cognitive development. Although this study was conducted on primary students, social touch can be just as beneficial to learners in the higher education sector. In another work, an experiment was conducted where college students were touched by their instructors during a conference, while the other half of students were not [17]. The findings showed that students who were touched during the conference had a better performance on the next course examination, scoring 0.58 standard deviations higher on the examination compared to the untouched students. The study concluded that interpersonal touch could improve class performance and student learning. Further literature on the benefits of social touch in education and other social contexts is shown in Table 1.

As observed in the table above—social touch can bring benefits in various contexts, some of which can also be used in the academic sector (such as relationship building, stress reduction, dopamine increase). With the substantial growth in virtual reality technology and its userbase [21], there can be great potential in implementing 'touch' for various aspects of education and learning. In the present day, VR research is aiming to improve user experience and enhance its hardware. There is also an increasing amount of software for virtual reality, for recreational purposes, education, medical, and other. Moreover, various 'Metaverses' are being

Table 1 Studies on the different types of touch, their benefits, and whether they can be applied in education

Study	Social context	Benefits	Applicable in higher education?
Cascio et al. [8]	Research on effects of social touch at different stages in life	Assists in social relationships, increase on-task behavior in students	Yes
Cohen et al. [10]	Experiments with ill participants—testing effect of hugging	Reduces stress	Yes
Fishman et al. [11]	Controlled experiments on students	Reduces stress and pain	Yes
Light et al. [13]	Experiments between couples	Increases oxytocin and decreases blood pressure in women	Yes
Guéguen et al. [15]	Observing effect of a teacher's touch prior to presenting in class	Improves engagement rate and encouragement levels in students	Yes
Owen et al. [16]	Surveying teachers on use of interpersonal touch in class	Promote on-task behaviour, improve general behaviors, and encourages cognitive development	Yes
Steward et al. [17]	College students touched by their instructors during a conference	Better performance on the next course examination	Yes
Wheldall et al. [18]	Studying the effects of contingent teacher touch behavior upon children's classroom behavior	Increases on-task behaviour and decreases disruptive behavior	Yes
von Mohr et al. [19]	Study on the effect of lack of social touch during COVID-19	Interpersonal touch is necessary in daily social life, promotes mental health and reduces anxiety	Yes
Mitchell et al. [20]	Reviewing research on cognitive and neural effects of positive and negative mood on executive function	Dopamine promotes positive mood	Yes

developed for virtual reality, such as VRChat [22], Meta Horizons, many of which have an expanding userbase [21].

With the technological advancements and the growing innovation in the field of virtual reality, it is worth considering combining with tactile sensations and applying it in higher education. With the fusion of VR technology and tactile sensations—much more of a student's educational aspects can be covered, not only in improving their academic experience, but also in advancing their learning, as well as improving their wellbeing, mental health, social skills, cognitive skills, and relationship building. This book chapter explores the potential uses of tactile synesthesia in virtual reality for higher education—looking at the current virtual platforms and technologies which can be used for the learner's benefit, especially in improving wellbeing, social connections, and immersive learning.

3 Haptic Devices and Pseudo-haptics for Interpersonal Touch

With the expanding research and innovation in virtual reality hardware, there is great interest in haptic devices—such as gloves, vests, and even full body suits. Haptic technology can offer various stimuli, such as vibrotactile [23], electro-tactile [24], ultrasound [25], or thermal [26]. However, most of the time haptic devices offer singular stimuli, or restricted feedback palette. This means that the 'touch' experience is very limited and identical for every user. In other words, the sensations received from the haptic device are exactly the same for every person. Moreover, the mentioned tactile feedback (thermal, vibrotactile, ultrasound, etc.) recreates unrealistic touch sensations—interpersonal touch does not necessarily feel like a vibration, or solely a thermal response. Every touch interaction varies between people and situation [27]. With the use of haptic devices in virtual reality, it can be said that the sensations are forced and could potentially desensitize individuals. This can lead to lack of interpersonal connection in social virtual environments, making people feel isolated and not grow as individuals, nor sustain bonds and relationships. Additionally, when we look at haptic devices, they often can only bring tactile sensations to only a specific body part, for example a glove [28]—which will only give feedback to the hands. This makes the tangible experience very limited to the user, not grasping the true potential of interpersonal touch. Moreover, not all haptics focus on social touch, or 'receiving' touch sensation—some devices are meant to represent object matter, such as sharpness, texture, or weight [28, 29]. Although these devices could potentially provide a more immersive virtual experience, questions arise on how these technologies can be implemented in a more social matter—such as students meeting in virtual reality. Price et al.[27] wrote about the conceptualization of touch in VR—their discussion focused on how touch itself is difficult to define, making touch-based interactions in VR hard to design and develop from both haptic and pseudo-haptic perspective. Altinsoya et al. [30] also wrote about haptic feedback and how it is not convincing and can obstruct the user. It is challenging to improve the devices and often causes deterioration.

Moreover, haptic technology is still robust and hefty—making it difficult to be made widely available to the public [31]. Not to mention that most of such haptics would need time to be set up and troubleshooted. Furthermore, such hardware can be expensive, especially if it were to be purchased for a large group—such as a whole class. Overall, for an institute to implement haptic technology, a lot of considerations must be made regarding the cost, its utility, and benefit to the students. There is great risk that such technology may not be very fruitful in the long term for a learner's educational and wellbeing purpose—the downsides which have been mentioned potentially can over-complicate the VR experiences, making it not worth implementing in schools and universities.

On the other hand, Phantom Touch in VR is a pseudo-haptic phenomenon where a person can experience sensations of touch without any physical contact or haptic influence [32]. It could be described as a 'brain trick' or an 'illusion of touch'. The phenomenon emerged from a social virtual game platform called VRChat [22], and a fair number of users have claimed to experience these phantom sensations (Fig. 1). Pseudo-haptic touch has been studied by other researchers, such as Sykownik et al. [33], and Kato et al. [34]. Although many of them do not offer an explanation as to why Phantom Touch occurs, nor how it can be developed in users. Therefore, there is an ongoing in-depth study on Phantom Touch in VR to answer these questions (Please refer to Chap. 4 of this work).

Compared to haptic devices, Phantom Touch in VR can generate a larger variety of sensation, such as tingling, heat, pressure, brushing, and many others (Fig. 2) [31]. This means that Phantom Touch can recreate sensations of interpersonal touch much more like the ones in real life—compared to a singular or limited stimuli of a haptic device. Moreover, Phantom Touch makes the virtual interactions much more varied, depending on a person's virtual setting and who they are with in the virtual world—rather than having the same singular feedback from a haptic device. This way, Phantom Touch can make people/students have stronger bonds and connections,

Fig. 1 Example of Phantom Touch interaction in a virtual Metaverse (VRChat). Here the researcher is testing Phantom Touch on a participant's arm. The virtual environment and avatars do not belong to the author and are for illustrative purposes only

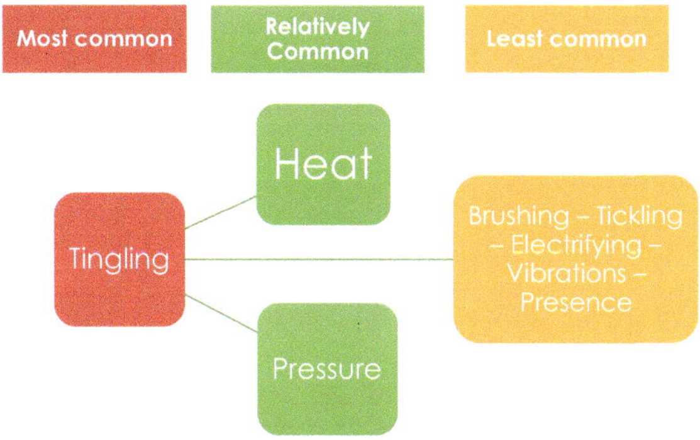

Fig. 2 Phantom Touch sensation development. Figure taken from Alexdottir and Yang [35]

as well as reinforce existing relationships. For the education field, this means that students can have much more broad and meaningful touch sensations when using VR for academic support, wellbeing, or socializing. Furthermore, this pseudo-haptic method is cheaper compared to haptic devices, as it does not require any additional equipment or software.

Neither does it need any plug-ins nor modifications to existing software for users to feel the sensations. The lack of additional hardware also makes Phantom Touch more accessible and easier to set up, or rather, develop. For institutions, such method can significantly reduce cost and set up time, allowing teachers and students to apply Phantom Touch in their VR experiences efficiently and quickly. Please refer to Table 2 which summarizes the benefits and downsides of haptic devices and Phantom Touch.

Overall, Phantom Touch in VR can be a more beneficial tool for academic institutes if they were to implement tactile sensations into virtual reality experiences for students. Comparing haptic devices and pseudo-haptic touch—it can be deduced that Phantom Touch would be a better option to implement in academia, particularly when it comes so interactions between learners, offering a greater sensation palette, which is more cost-efficient, accessible, and requires very minimal set up—making it faster for teachers and students to start their virtual experience.

4 Phantom Touch in VR Research

Phantom Touch in VR has been in research since 2021 and the current study aims include: creating a method for anyone to develop Phantom Touch across their body, learning what aspects increase Phantom Touch sensitivity and development, and understanding what psychophysiological effects Phantom Touch has on people. Some of these questions have already been answered and fruitful discoveries have been made [35].

Overall, the research to date has determined that: Phantom Touch can be generated in a person regardless of their VR experience, as there was no relation found between how long, and how often, individuals use their VR headsets, and how that connects to

Table 2 Summary of benefits and downsides of haptic devices and pseudo-haptics

Haptic type	Advantages	Disadvantages
Haptic devices—gloves, vests	Provides definite feedback	Provides singular feedback type or has a very limited sensation palette
	Supports feedback for 'giving' and 'receiving' touch	Requires setup and troubleshooting
	Can be used for social touch and sensing object matter in VR	Expensive or not widely available for purchase
		Can desensitize users
		Robust
		Unrealistic recreation of touch
		Not very accessible
		Provides sensations only to limited parts of the body, e.g., hands
		Not ecologically sustainable
Pseudo-haptics—Phantom Touch	Provides multiple feedback sensations	Requires a development program
	No extra software is needed	
	More accessible	
	No physical setup required, apart from VR headset	Uncertain on whether it can be used to sense objects
	Sensations are unique to every user	
	Cheaper than haptic devices	
	More parts of the body can experience tactile sensations	More research is currently carried out to see if sensations are similar across different days of usage
	Very ecologically sustainable	
	Better for interpersonal interactions in VR	

their experience with Phantom Touch. Furthermore, it was detected that a humanoid avatar is not necessary to experience Phantom Touch. The user just needs to find a virtual avatar that they can relate to on an emotional level and enjoy wearing in the virtual world. Moreover, it was discovered how Phantom Touch is felt, and which areas of the body are more susceptible to feeling these phantom sensations.

Furthermore, a working method was created for individuals to evoke Phantom Touch—The Head-to-Toe Method (HTT). The technique was tested on participants along with a control group, and it was discovered that HTT was 70% effective (Fig. 3). Additionally, heart rate was monitored in those who generated Phantom Touch, and it was found that the heart rate significantly dropped during the virtual interaction.

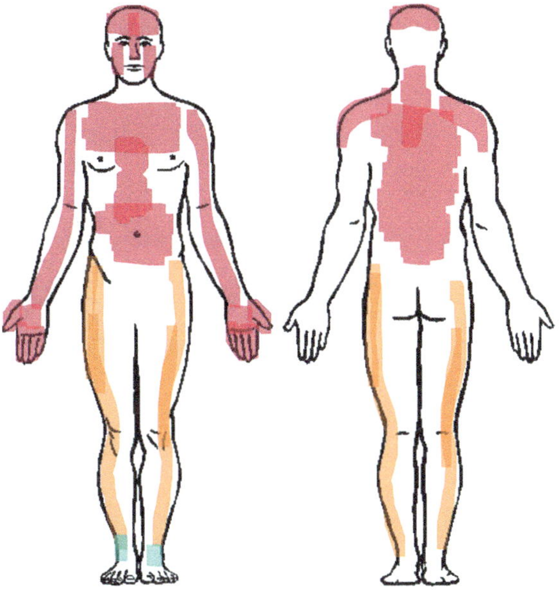

Fig. 3 Example of a participant's 'sensation map' after 7 trials of Phantom Touch through the HTT Method: red—very strong sensations, orange—prominent sensations, blue—something is felt ever so slightly. Figure taken from Alexdottir and Yang [35]

These results can be compared to touch in real life (where a relaxed heart rate is 60–100 bpm) [36], showing that Phantom Touch in VR has a similar effect on people (Fig. 4). The participants claimed that the phantom sensations during the interactions made them feel calm and tranquil. These results show potential in applying Phantom Touch in VR to aid in student wellbeing—where a learner's heart rate can be eased if they are feeling stressed or overwhelmed due to workload and other matters. Comparing these findings with other literature on interpersonal touch, such as Cohen et al. [10], supports the idea that social touch can diminish stress, even when used in virtual reality.

At present, the research is looking at speeding up Phantom Touch development and improving the method to generate the sensations. There is also ongoing work on increasing immersivity in VR, studying social context and behavior tests for participants, as well as having a more in-depth exploration of avatars and pseudo-haptic touch (looking at the sense of agency, sense of embodiment, and the sense of ownership).

Furthermore, current experiments on Phantom Touch are observing the phenomenon's effect on students, studying their Galvanic Skin Response (GSR), heart rate, and arousal rate during the virtual touch interaction. As the research is in progress, currently 64% students who underwent Phantom Touch development sessions have found the interactions enjoyable and helped them relax. Their sensations' descriptions have also matched with the earlier study [35], as well as the areas where they felt Phantom Touch the most.

With the findings that have already been made, Phantom Touch can be used in several fields—from social VR gaming platforms to medical care, and of course,

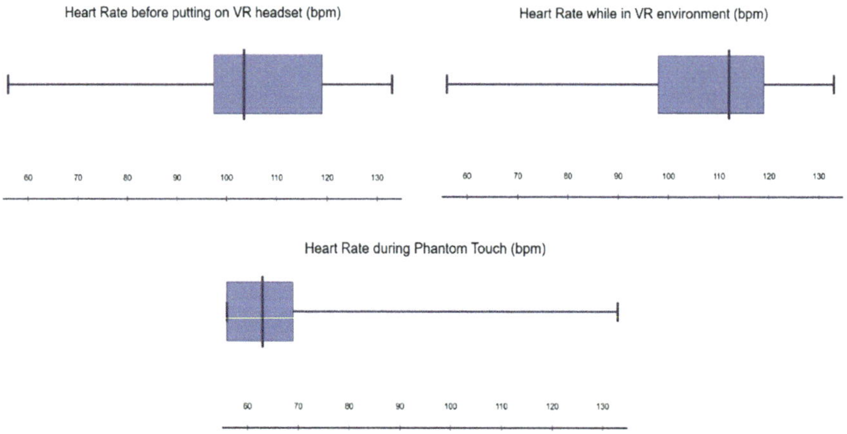

Fig. 4 Heart rate average comparison before putting on the VR headset, while in VR, and during Phantom Touch interaction

education. Phantom Touch in VR can offer academic institutions a unique virtual experience for their students—enabling them to feel tactile sensations, making learners more engaged in the VR simulations, whether it is for immersive education, wellbeing improvement, or academic support.

5 Phantom Touch in the Education Field

As mentioned earlier in this chapter, interpersonal touch and social touch play a grand role in human development and have the potential to offer students a multitude of support in their academic journey. With the rise of technology and virtual reality being used in education, Phantom Touch can be implemented to give students wellbeing support, provide greater immersive education, aid in abroad studies, assist in cultural learning, and be used in cases of a pandemic or distance learning. This section offers an insight on how social touch can be incorporated with the current available VR platforms and technology to aid in various aspects of academia. Moreover, there is also observation on the potential challenges in implementing Phantom Touch in VR in the academic field.

5.1 Mental Health

Over the past decade, much more attention has been paid to mental health and mindfulness. With that, research has shown that mental health in students is declining, partly due to being overworked and overwhelmed because of their studies, as well as feeling stressed, anxious, and having pressure to do as well as possible. In 2021, it was observed that 37% of first-year university students surveyed in England have symptoms of depression and anxiety [37]. A Mental Health study in Singapore estimated that 76.8% of 2568 students surveyed by the Community Health Assessment Team in 2021 were reported to have had mental health concerns [38]. In another study, "50 Current Student Stress Statistics: 2023 Data, Analysis and Predictions" shows disturbing statistics on student stress around the world [39]. In America, there was a 30% increase in the number of students who visited campus counseling centers between the school years 2009–2015, even though enrollment only increased by 5%. In the UK, only 23% of students are satisfied with the mental health resources available at their university. These are only a few of the many concerning data about student mental health, yet it is clear that mental health and wellbeing support must improve in academic institutions. Without sufficient aid, there is a great risk to student wellbeing, which could potentially make them fall ill, and develop depression, as well as many other issues.

So, how can virtual reality and Phantom Touch benefit student mental health? As mentioned earlier in this work, interpersonal touch between people can relieve anxieties and stresses [11], and instead be replaced with emotions of tranquility and calmness. There are existing VR meditation and relaxations applications, such as TRIPP [40], which offer guided meditation with visuals. Although such VR applications are available, they lack tactile experience and 'touch' interaction. One of the ways Phantom Touch can be applied into virtual reality and education is by creating touch-centered VR experiences for students—made to release dopamine and trigger positive response [20]. An appealing environment with a cute character could approach the student and speak to them about their studies, mental health, and wellbeing. The character could caress them and give hugs to make the student feel better. This persona could potentially be an agent or mental health professional who could work in real time with the student. In the foundation study on Phantom Touch in VR, it was discovered that a realistic virtual avatar is not necessary to evoke Phantom Touch [35], instead, the user, or in this case, a student, just need a virtual embodiment that they like and enjoy wearing. The study has shown that not only will this develop Phantom Touch faster, but also make people feel more at ease in the virtual environment. The described VR mental health simulation can be easily implemented into campuses and be cost-efficient, as haptic devices will not be needed and an economical headset like Oculus Quest 2 can suffice very well.

Furthermore, Phantom Touch can be used in virtual mental health sessions for students. Although pseudo-haptic touch may not be used directly in such setting, Phantom Touch highly relates to the sense of immersivity in virtual reality. Meaning that: those who have developed Phantom Touch are much more immersed into the virtual environment. Therefore, if students are more immersed into the virtual world, they are more likely to benefit from the sessions. With virtual reality, a much more comfortable environment can be created for students to share their concerns and worries. Moreover, the professional therapist or mental health supporter can work remotely and does not necessarily need to be on campus during working hours. Likewise for the students, virtual mental health sessions can be done from the comfort of their home or outside of study hours. This creates greater flexibility for all and aids learners in managing their worries and problems during their academic years.

On the other hand, it is important to test such virtual mental health programs and simulations and compare them to physical face-to-face sessions. Potentially, some students may feel dissociated in virtual reality and prefer to speak to a wellbeing specialist in-person. Inversely, there could also be a large group of students who would prefer virtual wellbeing help, with digital aspects which may not be available in real life (such as calming environments, cute characters, relaxing games, and more).

5.2 *Studying Abroad*

Over the years, more students have had the opportunity to study abroad. Statistics show that in 2017, there were over 5.3 million students studying abroad, and the number is increasing [41]. Studying abroad allows students to explore the world, discover different cultures, learn new languages, and get greater opportunities in their careers. With all the benefits that come with such an experience, there are also some downsides—one of them is homesickness. For those students who are away from home for months on end, sometimes even years, it is only natural for them to experience homesickness, and at times loneliness. Phantom Touch in VR can assist students who are away from home—allowing them to hug their family, friends, and loved ones, even though there is physical distance. Positive interpersonal touch interactions as such can boost the mood of the students, that way improving their mental health, energy levels, and happiness [12]. The family of the student also benefits in a similar manner—enabling them to have comforting interactions with their child who they miss.

Furthermore, since Phantom Touch interactions could take place on a social virtual platform with an interactive environment, the student and their family can have activities there too. This could include playing games, watching movies, exploring virtual worlds, and more—such activities can sustain friends and family relations when the student is abroad. It is a much more interactive and fun way of spending time with loved ones, compared to voice calls, video calls, or playing PC games—as in VR individuals have a sense of proximity with others in the virtual world, making them feel closer to one another. Communication through VR can be a good temporary fix when there are physical boundaries for a prolonged period.

On the other hand, the price for such an experience must be calculated—how many VR headsets would be needed? How many family members would require the VR headset? Would the school or university fund this? Would older family members want to participate? These are the questions which must be considered when it comes to integration of VR into education and such student aid. Please refer to the Discussion section of this chapter to find out more.

5.3 Isolation (COVID-19)

Since the 2020 pandemic with COVID-19, remote learning and E-learning became much more widely used and researched. Applications such as Zoom [42], Teams [43], Prezi [44], as well as other multimodal programs, have become more integrated into the academia. Although such applications are highly useful in cases where physical learning is not possible, they do not have a very high engagement number nor attract the students to study and learn [45]. One of the possible solutions to improve learning during a pandemic or epidemic is to implement VR, and potentially, Phantom Touch. With this medium, students can have more engaging lessons, since they will 'feel' as if they are in a classroom, surrounded by classmates and teachers. Several studies which were mentioned at the start of the chapter support this idea, that VR can bring great benefit to the learners, from soft and cognitive skills, simulation training, and more [2–4]. Moreover, as observed earlier in this work, many studies show the benefits of touch in education, which can boost student engagement, performance, learning, encouragement, and more [15–17]—this means that Phantom Touch could be applied in virtual remote learning. Virtual reality can also offer schools or universities to cater virtual environments for specific needs or subjects. With VR technology it is possible to recreate a university building, with designated spaces for learning and recreation. University of British Columbia in Canada has tested remote learning with the VRChat software and has received a positive response from the students—stating that they felt more engaged in the lecture and successfully learned the necessary subject material [46]. On the other hand, the virtual environment was quite basic in this demonstration (Fig. 5)—it would be worth exploring deeper into the abilities of existing virtual platforms and Metaverses to study how the virtual environment and avatars could enhance student learning.

Fig. 5 Virtual Lecture created with VRChat. James [46] (Accessed: February 1, 2023)

Furthermore, such VR technology can allow universities to have simulation training, so that the students would not miss out on any practical lessons because of a pandemic. Moreover, with virtual reality, students can collaborate and work on projects with great immersivity and proximity—rather than have video meetings which aren't as effective when it comes to engagement. Students can also socialize, make new connections, and sustain existing friendships with VR and Phantom Touch. Such pseudo-haptic can create the illusion of physical closeness, which can strengthen empathy and understanding between students and their peers. Phantom Touch can also be used for students to give each other a hug, high-five, handshake, and other friendly gestures. On the other hand, questions arise on whether Phantom Touch can be misused between learners, potentially enabling them to virtually bully and harass other students in a virtual space. In such instances, it is possible to report such people easily and quickly in VR, as well as muting them, and creating a personal space for every student (meaning no one can enter their 'personal circle'). Such practices are frequently used in social virtual platforms such as VRChat and can also be efficiently used in a virtual educational setting—making students feel safer and allowing them to focus on their studies.

Furthermore, although such use of virtual reality and Phantom Touch look good in theory, it may be tricky to recreate for large scale universities and schools. In such a case, it would be important to calculate how funds should be planned and what can be cut down. For example, what applications or software might not be useful anymore if the academic establishment purchases VR headsets for students and tutors? Are there any other funds which can be cut down if the physical buildings of a school are not used as much? If the university were to implement VR-based classrooms, how would this affect the students who prefer to work and study at the physical university building? These are some of questions which are vital to consider for virtual reality-based learning in case of a pandemic or epidemic.

5.4 Immersive Education

Several educational fields including STEM use VR to aid in student learning [47], whether it is through virtual training simulations or other skills training programs. There are also studies which support virtual simulation for training, STEM subjects, engineering, medical training, and more [2, 3]. There are already existing virtual applications for medical simulations, health and safety training, soft skills training, and others. Some of such programs include Acadicus by Arch Virtual, which offers various VR healthcare simulations and training programs. There is also BODYSWAPS [48], a VR application which aids people in communication, teamwork, confidence boosting, and leadership skills (Fig. 6). Phantom Touch in VR can be used in immersive learning, especially where human interaction and touch are required. It can provide more moving and realistic experiences for students. As mentioned previously, even though touch may not be used directly, Phantom Touch and immersivity are related—in order to develop Phantom Touch, immersivity is

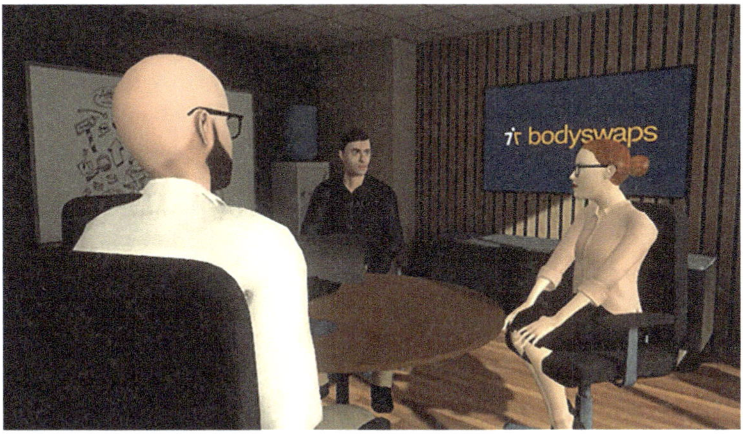

Fig. 6 Example of soft skills training from BODYSWAPS. Bodyswaps®, 2022 Available at: https://bodyswaps.co/mobile-app-gender-inclusion/ (Accessed: February 2, 2023)

key. If students perceive their virtual simulation more like reality, they will feel more engaged in the training program and have greater motivation to successfully complete the tasks. If students do not feel immersed in virtual reality, they will perceive their training as 'unrealistic' and will believe that: even if they fail or make mistakes, there is no 'real' negative outcome, and the simulation can be repeated. Hence, combining pseudo-haptic sensations can increase a student's feeling of presence in the immersive education program, meaning they are more encouraged to do well. On the other hand, it is important to study whether such virtual simulations with Phantom Touch can properly prepare students for implementation of the skills they learned in real life—simulations may be repetitive and predictable in some instances, making the knowledge hard to apply in varying, stressful situations in reality. Moreover, Phantom Touch may not be useful in all immersive education programs (such as surgical training or engineering), as the focus of the virtual simulation is more toward achieving a goal, such as completing a surgery correctly—meaning that pseudo-haptic sensations will not be crucial for the learner. Hence, each academic need must be worked out on an individual basis with the university or school to understand whether Phantom Touch would be useful for the students.

5.5 Cultural Knowledge and Empathy

To build a better, more knowledgeable, and harmonious society, academic institutions need to teach students cultural knowledge, awareness, and empathy. Discovering various cultures, languages, and traditions can be done in many different forms—reading from textbooks, watching documentaries, traveling, learning a language from a mobile application, and many others. Some institutions have annual trips abroad for their students (depending on the student's studied subject), so that they can have a better understanding of a particular nation's culture, language, and more. Probably the biggest factor as to why such trips are done quite rarely is due to the cost. Taking tens of students abroad can be quite expensive for some academic institutions, possibly also for the guardians who are partly paying for the travels.

Virtual reality and Phantom Touch can provide students with an immersive simulation or application to discover more about a country or a culture. Such virtual experiences can be done with agents, or perhaps with other students from a different country. This medium can also allow learners to 'visit' the country or place much more frequently and feel 'present' in the space. Virtual reality can also make the students much more engaged and willing to find out more about different traditions and countries. Virtual Reality may be not only be cost-efficient, but also safer and accessible compared to real-life traveling. Nonetheless, physical traveling should not be abandoned altogether and is still an important aspect of development and learning.

Moreover, when studying different societies and traditions, students can also learn empathy and understanding toward people of different cultures and customs. With the immersivity of virtual reality and pseudo-haptic touch, the experience is improved, and students are more likely to feel 'moved' and be 'touched' on an emotional level. Additionally, if there is human interaction in the virtual space, Phantom Touch can assist learners in being more submerged into the cultural experience. The use of virtual reality and Phantom Touch can be applied in different humanitarian subjects to benefit student learning, and as stated previously, some virtual simulations may need to be catered depending on a university's teaching subjects and interests—making such medium of learning possibly more complex and time-consuming if aspects of the virtual experience need to be altered or updated.

Moreover, it is vital to consider whether pseudo-haptic touch is necessary in some virtual experiences—possibly some students may not want to be touched and would like to keep their personal space in VR. Furthermore, not all cultures and nations have frequent social touch interactions among acquaintances/peers [49, 50], meaning that Phantom Touch may not be very applicable in some virtual cultural simulations between learners.

6 Discussion

Virtual reality and Phantom Touch can be successfully implemented into various aspects of education—allowing students to grow as individuals, learn, enhance their knowledge, and care for their wellbeing. Phantom Touch in VR can provide unique educational experiences, making the most out of (virtual) technological advancements. Virtual reality learning can also potentially reduce costs, for example, traveling and training (please refer to *Immersive Education* and *Cultural Knowledge and Empathy*). Furthermore, Phantom Touch and VR can be used as a great advantage for distance learning, especially in cases of a pandemic or epidemic. With all the benefits that have been mentioned in this work, questions arise on the necessity of virtual reality and Phantom Touch in different educational scenarios. It can be difficult to define in which instances pseudo-haptic touch is imperative, and in which it is not. In other words, for immersive education or simulation training, 'touch', may not be a key factor. For example, a surgery simulation—the focus is not on 'touch' specifically, but rather on completing the surgery successfully. This means that the student's attention in the virtual simulation will be on the procedure rather than tactile sensations. On the other hand, researchers such as Hannaford [7] and Shams et al. [51] support the idea that multisensory learning, where touch should be implemented into education and academia, so that students could grasp subject understanding faster. In such case, perhaps pseudo-haptic touch should be highly considered by academic institutes and allow students to experience tactile sensations in their learning, benefiting their progress. Nonetheless, for virtual reality and Phantom Touch to be implemented well into different aspects of education—such plans would mostly be done on an individual basis—depending on an institute's needs, courses, subjects, and student needs. It may be quite hard to cater for every single university or academic institution—perhaps, it may be worth exploring Phantom Touch and virtual reality simulations which could provide for several universities at once. One of the most promising aspects could be mental health—as all institutions are likely to face students who would like to improve their mental wellbeing during their studies. Implementation of Phantom Touch and VR for student mental health could create a new way of supporting people's wellbeing, perhaps not only in academic institutions. In such case, it is possible to create a 'one-fits-all' experience for students to improve their wellbeing with Phantom Touch in VR. Comparably to other aspects which have been mentioned, such as abroad studies, immersive education, and others—much more research would need to be conducted to understand how to supply ready-made simulations for a greater number of universities and schools.

Furthermore, in this chapter Phantom Touch was explored from the perspective of social touch, where one person interacts with the other, considering both parties are willing to initiate and receive touch. In reality, not all people enjoy interpersonal touch interactions—as this depends on their personality type [52], culture, upbringing, and the relationship they have with the person they are receiving and initiating touch with. This means that not all students may want to experience Phantom Touch in VR, as it could potentially bring them more downsides than positives. Interestingly, those partakers from the Phantom Touch research who did not enjoy virtual interaction, never developed any pseudo-haptic sensations [35]. This indicates that those students who would rather not experience Phantom Touch while being in VR, will simply not experience it, keeping them within their own comfortable space. Moreover, this implies that those learners who enjoy touch will greatly benefit from tactile interactions in VR and are much more likely to evoke the sensations.

7 Conclusions and Future Work

This book chapter explores the potential of virtual pseudo-haptic touch in the academic field—offering an insight on the importance of social touch, specifically in the field of education. The chapter sheds light on the existing virtual reality platforms and how they can be altered/improved, and applied in academia, targeting various aspects of a student's educational journey, including mental health, wellbeing, training simulations, and more. Phantom Touch in VR can be successfully used in the education field based on the existing, and ongoing experiments on the pseudo-haptic, although, for this to be fully proven, it is necessary to test it in institutes and schools. Phantom Touch in VR needs to be tested in different scenarios and simulations (such as mental health, remote learning, and training) to fully understand how it affects the students. Based on the current research discoveries on Phantom Touch in VR (please refer to section *Phantom Touch in VR Research*), there is great potential in evoking the touch sensations among students, as well as using the method to decrease pulse, hence creating tranquility in individuals—which could effectively be used for student wellbeing and mental health care. Nonetheless, for this to be fully confirmed, future studies and experiments should be carried out in collaboration with institutions to understand how Phantom Touch in VR can aid students.

Furthermore, it is possible that some of those virtual experiences must be made bespoke for some universities based on their needs, although currently there is ongoing research on making virtual programs which could target more institutions with similar requirements. Therefore, some limitations of this study include the lack of experiment work done on Phantom Touch and VR in schools. Currently, the findings made on both VR in education, social touch, and Phantom Touch suggest the probable utilities in the field—nonetheless, firsthand trials of this technology must be done. Future work will include implementing virtual reality and Phantom Touch in academia, focusing on the most promising aspects, such as mental health, as touch-based experiences in VR could bring substantial wellbeing benefits to students. All other mentioned areas, such as immersive learning, require greater understanding of the institute's needs and will be explored on an individual basis.

Should you be interested in discussing the potential of Phantom Touch in VR, how to develop it, and how it can be implemented in education or other fields, please contact Sasha Alexdottir on sa@phantomtouchvr.com. Collaborations and partnerships from all over the world are very encouraged.

References

1. Skwiot, S.: Virtual Reality in Education: How Schools Are Using VR—42West. 42 West, the Adorama Learning Center, Dec. 07, 2022. https://www.adorama.com/alc/?p=90526. Accessed 26 Jan 2023
2. Kamińska, D., et al.: Virtual reality and its applications in education: survey. Information **10**(10), Art. no. 10, Oct. 2019. https://doi.org/10.3390/info10100318

3. Philippe, S., et al.: Multimodal teaching, learning and training in virtual reality: a review and case study. Virtual Real. Intell. Hardw. **2**(5), 421–442 (2020). https://doi.org/10.1016/j.vrih.2020.07.008
4. Herrera-Aliaga, E., Estrada, L.D.: Trends and innovations of simulation for twenty first century medical education. Front. Public Health **10**, 619769 (2022). https://doi.org/10.3389/fpubh.2022.619769
5. Sp, O., et al.: Past Present and Future of Simulation in Internal Medicine, Nov. 2019, Accessed: Jan. 05, 2023. [Online]. Available: https://europepmc.org/article/NBK/nbk549883
6. Graham, A.C., McAleer, S.: An overview of realist evaluation for simulation-based education. Adv. Simul. **3**(1), 13 (2018). https://doi.org/10.1186/s41077-018-0073-6
7. Hannaford, C.: Smart Moves: Why Learning Is Not All in Your Head. Great Ocean Publishers, Inc (1995)
8. Cascio, C.J., Moore, D., McGlone, F.: Social touch and human development. Dev. Cogn. Neurosci. **35**, 5–11 (2019). https://doi.org/10.1016/j.dcn.2018.04.009
9. Gray, P.H.: Henry Dwight Chapin: pioneer in the study of institutionalized infants. Bull. Psychon. Soc. **27**, 85–87 (1989). https://doi.org/10.3758/BF03329906
10. Cohen, S., Janicki-Deverts, D., Turner, R.B., Doyle, W.J.: Does hugging provide stress-buffering social support? A study of susceptibility to upper respiratory infection and illness. Psychol. Sci. **26**(2), 135–147 (2015). https://doi.org/10.1177/0956797614559284
11. Fishman, E., Turkheimer, E., DeGood, D.E.: Touch relieves stress and pain. J. Behav. Med. **18**(1), 69–79 (1995). https://doi.org/10.1007/BF01857706
12. Parker, A., Hosking, G., Pascoe, M., Dash, S.: Miss hugs? Touch forms bonds and boosts immune systems. Here's how to cope without it during coronavirus. The Conversation, May 13, 2020. http://theconversation.com/miss-hugs-touch-forms-bonds-and-boosts-immune-systems-heres-how-to-cope-without-it-during-coronavirus-137612. Accessed 26 Jan 2023
13. Light, K.C., Grewen, K.M., Amico, J.A.: More frequent partner hugs and higher oxytocin levels are linked to lower blood pressure and heart rate in premenopausal women. Biol. Psychol. **69**(1), 5–21 (2005). https://doi.org/10.1016/j.biopsycho.2004.11.002
14. Gallace, A., Spence, C.: The science of interpersonal touch: an overview. Neurosci. Biobehav. Rev. **34**(2), 246–259 (2010). https://doi.org/10.1016/j.neubiorev.2008.10.004
15. Guéguen, N.: Nonverbal encouragement of participation in a course: the effect of touching. Soc. Psychol. Educ. **7**(1), 89–98 (2004). https://doi.org/10.1023/B:SPOE.0000010691.30834.14
16. Owen, P.M., Gillentine, J.: Please touch the children: appropriate touch in the primary classroom. Early Child Dev. Care **181**(6), 857–868 (2011). https://doi.org/10.1080/03004430.2010.497207
17. Steward, A.L., Lupfer, M.: Touching as teaching: the effect of touch on students' perceptions and performance. J. Appl. Soc. Psychol. **17**(9), 800–809 (1987). https://doi.org/10.1111/j.1559-1816.1987.tb00340.x
18. Wheldall, K., Bevan, K., Shortall, K.: A Touch of Reinforcement: the effects of contingent teacher touch on the classroom behaviour of young children. Educ. Rev. **38**(3), 207–216 (1986). https://doi.org/10.1080/0013191860380301
19. von Mohr, M., Kirsch, L.P., Fotopoulou, A.: Social touch deprivation during COVID-19: effects on psychological wellbeing and craving inter-personal touch. R. Soc. Open Sci. **8**(9), 210287 (2021). https://doi.org/10.1098/rsos.210287
20. Mitchell, R.L.C., Phillips, L.H.: The psychological, neurochemical and functional neuroanatomical mediators of the effects of positive and negative mood on executive functions. Neuropsychologia **45**(4), 617–629 (2007). https://doi.org/10.1016/j.neuropsychologia.2006.06.030
21. Press Center—With an Assist from Oculus Quest 2, 2022 AR/VR Device Shipments Revised Up to 14.19 Million Units, Says TrendForce | Trend-Force—Market research, price trend of DRAM, NAND Flash, LEDs, TFT-LCD and green energy, PV. TrendForce. https://www.trendforce.com/presscenter/news/20220216-11122.html. Accessed 9 Dec 2022
22. 'VRChat'. VRChat Inc, Jan. 16, 2014

23. Islam, M.S., Lim, S.: Vibrotactile feedback in virtual motor learning: a systematic review. Appl. Ergon. **101**, 103694 (2022). https://doi.org/10.1016/j.apergo.2022.103694
24. Pamungkas, D., Ward, K.: Electro-tactile feedback system to enhance virtual reality experience. Fac. Eng. Inf. Sci. - Pap. Part A, 465–470 (2016). https://doi.org/10.7763/IJCTE.2016.V8.1090
25. Sun, C., Nai, W., Sun, X.: Tactile sensitivity in ultrasonic haptics: do different parts of hand and different rendering methods have an impact on perceptual threshold? Virtual Real. Intell. Hardw. **1**(3), 265–275 (2019). https://doi.org/10.3724/SP.J.2096-5796.2019.0009
26. Parida, K., Bark, H., Lee, P.S.: Emerging thermal technology enabled augmented reality. Adv. Funct. Mater. **31**(39), 2007952 (2021). https://doi.org/10.1002/adfm.202007952
27. Price, S., Jewitt, C., Yiannoutsou, N.: Conceptualising touch in VR. Virtual Real., vol. 25, Sep. 2021. https://doi.org/10.1007/s10055-020-00494-y
28. Sinha, A.K., Goh, G.L., Yeong, W.Y., Cai, Y.: Ultra-low-cost, cross-talk-free, fast-responding, wide-sensing-range tactile fingertip sensor for smart gloves. Adv. Mater. Interfaces **9**(21), 2200621 (2022). https://doi.org/10.1002/admi.202200621
29. van Wegen, M., et al.: An overview of wearable haptic technologies and their performance in virtual object exploration. Sensors **23**(3), Art. no. 3, Jan 2023. https://doi.org/10.3390/s23031563
30. Fitzek, F., et al.: Communications and control, pp. 249–275 (2021). https://doi.org/10.1016/B978-0-12-821343-8.00023-X
31. Perret, J., Vander Poorten, E.: Touching virtual reality: a review of haptic gloves. In: ACTUATOR 2018; 16th International Conference on New Actuators, Jun. 2018, pp. 1–5
32. I.-I. B. Ltd.: The phenomenology of virtual reality and phantom sensations. Stud. Univ. Babes-Bolyai—Philos. **55**(3), 81–96 (2010). Accessed: 10 Dec 2022. [Online]. Available: https://www.ceeol.com/search/article-detail?id=130749
33. Sykownik, P., Masuch, M.: The experience of social touch in multi-user virtual reality. In: 26th ACM Symposium on Virtual Reality Soft-ware and Technology, in VRST'20, pp. 1–11. Association for Computing Machinery, Nov. 2020, New York, NY, USA. https://doi.org/10.1145/3385956.3418944
34. Kato, Y., Sugimoto, M., Inami, M., Kitazaki, M.: Communications in virtual environment improve interpersonal impression. In: 2021 IEEE Conference on Virtual Reality and 3D User Interfaces Abstracts and Workshops (VRW), Mar. 2021, pp. 613–614. https://doi.org/10.1109/VRW52623.2021.00189
35. Alexdottir, S., Yang, X.: Phantom Touch phenomenon as a manifestation of the Visual-Auditory-Tactile Synesthesia and its impact on the users in virtual reality. In: 2022 IEEE Inter-national Symposium on Mixed and Augmented Reality Adjunct (ISMAR-Adjunct), pp. 727–732 (2022). https://doi.org/10.1109/ISMAR-Adjunct57072.2022.00218.
36. Drescher, V.M., Gantt, H.W., Whitehead, W.E.: Heart rate response to touch1. Psychosom. Med. **42**(6), 559–565, Nov. 1980, Accessed: Dec. 10, 2022. [Online]. Available: https://journals.lww.com/psychosomaticmedicine/Citation/1980/11000/Heart_Rate_Response_to_Touch1.4.aspx
37. Okolicsanyi, H.: Depression and anxiety on the rise among UK students. https://www.symplicity.com/blog/depression-and-anxiety-on-the-rise-among-uk-students. Accessed 27 Jan 2023
38. Soong, I.: Mental health of students in Singapore | Ingel Soong. Delivers Results, May 19, 2022. https://ingelsoong.com/mental-health-of-students-in-singapore/. Accessed 27 Jan 2023
39. 50 current student stress statistics: 2023 data, analysis & predictions. Research.com, Oct. 26, 2020. https://research.com/education/student-stress-statistics. Accessed 25 Jan 2023
40. 'TRIPP VR'. TRIPP, Inc., 2022. [Online]. Available: https://www.tripp.com/
41. 'International students'. Migration data portal. https://www.migrationdataportal.org/themes/international-students. Accessed 16 Jan 2023
42. 'ZOOM'. Zoom video communications, Qumu Corporation, Sep. 10, 2012
43. 'Microsoft Teams'. Microsoft, Mar. 14, 2017
44. 'Prezi'. Prezi Inc, Apr. 05, 2009
45. Lack of student engagement within online learning | education paper example. Chalky Papers. https://chalkypapers.com/lack-of-student-engagement-within-online-learning/. Accessed 28 Jan 2023

46. James, P.: "VR Chat" used to deliver one of the first university lectures in virtual reality. Road to VR, Nov. 28, 2014. https://www.roadtovr.com/vr-chat-helps-deliver-first-virtual-university-lecture/. Accessed 31 Jan 2023
47. 'How virtual reality technologies are revolutionising STEM learning—study International', Mar. 07, 2019. https://www.studyinternational.com/news/how-virtual-reality-technologies-are-revolutionising-stem-learning/. Accessed 28 Jan 2023
48. 'Bodyswaps'. Bodyswaps, 2022
49. Sorokowska, A., et al.: Affective interpersonal touch in close relationships: a cross-cultural perspective. Pers. Soc. Psychol. Bull. **47**(12), 1705–1721 (2021). https://doi.org/10.1177/0146167220988373
50. McDaniel, E., Andersen, P.A.: International patterns of interpersonal tactile communication: a field study. J. Nonverbal Behav. **22**(1), 59–75 (1998). https://doi.org/10.1023/A:1022952509743
51. Shams, L., Seitz, A.R.: Benefits of multisensory learning. Trends Cogn. Sci. **12**(11), 411–417 (2008). https://doi.org/10.1016/j.tics.2008.07.006
52. Dorros, S., Hanzal, A., Segrin, C.: The Big Five personality traits and perceptions of touch to intimate and nonintimate body regions. J. Res. Personal. **42**(4), 1067–1073 (2008). https://doi.org/10.1016/j.jrp.2007.11.004

La Scuola di Atene: Immersive Virtual Reality for Science Education

Mao-Ren Zeng and Mei-Hung Chiu

Abstract During the last few decades, virtual reality has developed rapidly as relevant technology and wearable devices become available. Many studies have investigated the educational benefit of immersive virtual reality (I-VR) in science education. However, the increasing use of virtual reality in education reveals little research conducted on I-VR's effectiveness for middle and high school students. This study aims to explore the possibility of learning with I-VR among ninth, eleventh and twelfth grades students in Taiwan. We have chosen abstract scientific concepts and embedded them in a virtual learning environment, then developed curricula based on the national standard. Moreover, we built a virtual art museum called the Scuola di Atene along with a laboratory and designed hands-on tasks to engage students in learning. To test the efficacy of Scuola di Atene, we recruited middle school students ($n = 27$) and high school students ($n = 72$) and asked for their feedback via self-report. The results indicated that most students were engaged in the activities through I-VR. However, a few students commented that they lacked opportunities to interact with their peers and felt uncomfortable with the headset. In addition, we also found that the hearing-impaired students enjoyed and responded positively to the tool they used and actively collaborated on solving the problems in the I-VR environment. In conclusion, this study has created six I-VR learning curricula and environments for middle and high school students. The I-VR curricula not only allowed students to visualize abstract concepts (e.g., the structure of molecular), it also allowed students to engage in learning activities to develop their experimental skills (e.g., operating the experiment). The implications of this study suggest that it is crucial to pay attention to students' individual needs and attitudes toward I-VR while designing an immersed-interactive I-VR curriculum; it is also crucial to empower teachers to make good use of the tool to engage students in learning tasks.

Keywords Curriculum-based · Immersive virtual reality · Science education

M.-R. Zeng · M.-H. Chiu (✉)
Graduate Institute of Science Education, National Taiwan Normal University, Taipei, Taiwan
e-mail: mhchiu@gapps.ntnu.edu.tw

Y. Cai et al. (eds.), *Virtual and Augmented Reality Technology-Enhanced Learning*, Gaming Media and Social Effects, https://doi.org/10.1007/978-981-96-2332-7_16

1 Introduction

Virtual reality (VR) has proliferated over the past few decades due to the development of relevant technology and the economic feasibility of the device, such as computer, projection, and VR headsets [1–3]. VR has been identified as very useful in representing abstract knowledge, increasing motivation, problem-solving, and deep engagement in the authentic environment [4, 5]. In education, Villena-Taranilla et al. [6] argued that VR fosters immersion and a sense of presence. Immersion refers to the degree to which a user feels absorbed in the virtual environment. In other words, immersion is the objective condition for the VR system such as graphics and sound [7]. In contrast, presence is defined as the subjective experience that indicates the difference between the user's feelings in the virtual environment and the real-world [8]. Therefore, we can classify the technological attributes of VR based on the degree of immersion. Offering the highest degree of immersion, immersive virtual reality (I-VR) allows users to navigate within a virtual environment by themselves and enhances the sense of reality for users through the head-mounted-display (HMD) [9, 10]. In other words, I-VR gets rid of the screen by using HMD and provides more opportunities for users to engage in the learning activities, enhancing the learning experience, and interacting with virtual environments [2, 9, 11]. Thus, many review [1, 6, 9, 11] and empirical studies [12–14] have investigated the educational benefit of I-VR on learning outcomes such as specific declarative knowledge, emotion and attitude, and skill.

Although science learning is the major area of educational I-VR research, many researchers have ascertained the effects of I-VR without emphasizing the needs of curriculum and instruction design for specific learning contents [11]. Moreover, there has been far less research on the instruction with I-VR for middle and high school students [1, 15, 16]. In other words, based on previous studies, there is little doubt on the effectiveness of I-VR for student learning. Instead, we should focus on the needs of different learning content and the diverse audience when we design I-VR-based curricula. In addition, research has shown that students are lacking motivation and interest in learning science. This is a universal phenomenon, especially in high academic performance countries (such as Japan, Korea, and Taiwan) [17]. As such, how we engage students in their learning and support their ownership as an agent while learning with I-VR have become important considerations when we designed the study.

This study aims to explore the possibility of learning with I-VR among middle and high school students in Taiwan. To do so, we have developed I-VR curricula based on the science standards in Taiwan, which aimed to arouse students' curiosity on scientific phenomenon and cultivate students' motivation for active learning [18]. Then, we collected feedback from students after implementing the I-VR curricula and aimed to answer these three research questions based on the data collected:

1. From the perspective of learning promotion, how did the students respond to the I-VR activities?

2. From the motivation enhancing perspective, how did the students respond to the I-VR activities?
3. From an interaction perspective, how did the students respond to the I-VR activities?

2 The Educational Benefit and the Rationale of Immersive Virtual Reality in Science Learning

Over the past few years, I-VR has become an increasingly popular technological tool in science learning. Many studies showed educational benefits in cognitive, skill, and affective domains [1]. Taking the cognitive domain as an example, students use I-VR to better understand specific declarative knowledge. In another aspect, Lamb et al. [19] asked students to manipulate multiple representations of DNA with different scales (e.g., microscale and macroscale) in the virtual environment. Moreover, research also indicated that students with lower levels of self-efficacy may have been more immersed in I-VR environments and held even stronger positive learning attitudes [20]. However, the effectiveness of I-VR varied and is dependent on the learning context and the learning material [1]. For instance, Allcoat and von Mühlenen [21] stated that the benefit of I-VR would not be significant if students are unfamiliar with the I-VR environments.

To reap the educational benefits of I-VR, Matovu et al. [11] argued, researchers should align with the rationale of I-VR while developing the I-VR curriculum. Based on the studies reviewed [1, 9, 11, 16], we divided the rationales for using I-VR into two parts. The first part related to students' understanding and practices of science concepts such as visualization and practical skills development. I-VR is a visualization technology that helps students visualize abstract science concepts in a virtual environment. For instance, in chemistry and biology, researchers presented structures of molecules and proteins in 3D or at the microscopic level. In addition, researchers could ensure safety and eliminate demands of physical space by building virtual laboratories for students to practice doing experiments.

The second part focused on students' engagement in science learning including the enhancement of learning experiences, virtual field trips, and first-personal experiences. On enhancing learning experiences, I-VR enhanced students' learning experiences by improving students' engagement and motivation via game-like strategies such as setting rules, granting rewards, and providing immediate feedback [11]. As such, researchers designed game-based I-VR to engage students in learning by allowing students to feel free to manipulate the objects and test or evaluate their ideas [1]. Moreover, students can explore magnificent natural landscapes or historic landmarks, such as volcanos [22] and Pearl Harbor [12], in the virtual world. Lastly, teaching empathy or a positive attitude toward a specific community is difficult for teachers because students lack the experience or the skill of perspective-taking. Therefore, I-VR provides first-personal experiences for students such as inhabiting the body of animals [23].

While the success of encouraging students to engage in learning with I-VR tools is based upon learning theories, such as embodied cognition, situational theory, and media richness theory, however, research based upon other cognitive theories like cognitive load, cognitive theory of multimedia theory indicated negative effect of students learning because of the limited capacity of memory working space that allow learners to be [24–26].

3 The Immersive Virtual Reality Curriculum Design in Science Learning

Based on the empirical studies in I-VR, Suh and Prophet [27] proposed two main research streams. The first stream emphasizes how I-VR improves students' performance, including learning effectiveness and outcomes. There is a general agreement that learning theories (e.g., flow theory, situated cognition theory, and technology acceptance model) play a critical role in I-VR activity designs [1, 2, 11]. The second stream focuses on what immersive features could enhance students' immersion in I-VR activities. Therefore, the studies highlight immersive technological features such as visual effects, audio, and interactivity with virtual objects.

3.1 Learning Theory Approach in Science Learning via Immersive Virtual Reality

Matovu et al. [11] advocated that learning theories can support researchers in the design, implementation, and evaluation of I-VR activities and help researchers align their activities with the learning goals. For example, the cognitive theory of multimedia learning (CTML) provides researchers with foundations for developing, evaluating, and revising the learning materials for effectiveness, such as by reducing students' cognitive load and providing reflection with peers after I-VR activities [11]. Moreover, Hamilton et al. [1] proposed two major learning theory approaches in I-VR: generative learning and situated learning which also provide students with constructive opportunities for active participation in linking new declarative and procedural knowledge with the existing framework.

In the generative learning approach, students would integrate their prior knowledge and experience to create new knowledge. Therefore, researchers tend to focus on student's cognitive processes (e.g., multiple representations and cognitive load) in I-VR activities. Based on the information process models, Johnstone [28] proposed the chemistry triplet to explain why chemistry is difficult and how students learn chemistry based on the information process models. Considering the nature of the concepts in chemistry, Johnstone considered that students pull out nothing from long-term memory because concepts are abstract. Furthermore, the excess information from the learning environment (e.g., instructions and techniques) could cause the overload of working memory to reduce students' high-level performance opportunities (e.g., interpretation and knowledge synthesis) during learning. In the chemistry triplet, it is important to integrate macroscopic, microscopic, and symbolic representations in learning complex and abstract concepts [28–31]. The macroscopic level stands for the observable phenomenon, while the microscopic level refers to the unobservable characteristics of an entity, such as ions or atoms. The third level is the symbolic representation of the entity. Therefore, while explaining the phenomenon, students should emphasize the transformation among the three representations rather than using just one representation [28, 30]. Thus, to provide the appropriate information for students, the chemistry triplet plays a crucial role in leveraging sufficient representations and decreasing cognitive load for students while designing the I-VR curricula.

3.2 Immersive Design Features

Various approaches to designing I-VR tools are proposed. For instance, a number of researchers identified the key features of I-VR 3D as visualization [32–34], interactive experience with the virtual objects [4, 9], and social interaction [2, 33]. Other systematic review studies extracted and categorized the design elements in I-VR, such as passive observation, immediate feedback, and virtual rewards [2, 35]. Among the proposed design approaches, Matovu et al. [11] used four immersive design features that were based on Dede [36] as a framework (e.g., sensory, actional, narrative, and social immersion) for designing sense-making I-VR activities.

Taking the immersive design features as an example, firstly, sensory immersion refers to the user's immersion in the virtual environment via sensory inputs. For instance, visuals (e.g., 3D as representations and duplicating the specific environment), audio effects (e.g., background audio instructions or sound from peers), and haptic feedback (e.g., vibrations or pressure) can create a more immersive experience with HMD. Secondly, actional immersion depends on the I-VR system tracking participants' movement and updating the scene to allow students to experience I-VR smoothly with wearable devices, such as walking around in the virtual environment and interacting with virtual objects. Thirdly, narrative immersion is related to how the designer organizes the learning content rather than technology. In other words, the I-VR designer is a storyteller who engages students in activities by creating a distinctive character, a complex storyline, and various challenges. Lastly, social immersion involves interactions with peers, teachers, or pedagogical agents and presents collaborative learning features. However, most studies were designed for individual users to experience the virtual environment by themselves. Thus, we need more empirical studies to better understand the effects of social immersion.

Researchers could design the I-VR learning environment using various immersive design features. For instance, we created the virtual scenarios with the features of sensory immersion, such as cleaning up the painting with the alcohol in a spray or manipulating the chemical equipment to conduct the experiment and examining the properties of gases coming out of the tube in our study. Experiencing the discovery processes allowed the students to appreciate how science theories were developed and validated.

4 Methods

4.1 Procedure

To answer our research questions, we need to develop and implement curriculum-based I-VR in science classroom so we can explore the possibility of learning with I-VR among middle and high school students in Taiwan. Based on the high cost and complexity of I-VR, we followed the iterative design and revision process to maintain the quality of I-VR. We organized the procedure into three stages, namely, development, implementation, and reflection. First, in the development stage, we had to generate a prototype of I-VR curricula, so we recruited engineers and teachers to design the script of the I-VR, including the learning content (e.g., the function of the living body and the structure of the molecules), the scenes (e.g., La Scuola di Atene virtual art museum and Lavoisier's laboratory), the representations (e.g., text, simulation, atom), and the manual for using I-VR (e.g., guideline, prompts, and instruction).

Secondly, we validated the prototype of our I-VR curricula via subject experts, teachers, and students to verify the accuracy and operability of the prototype in the implementation stage. In the beginning, we invited three professors as subject experts to use the I-VR curriculum prototype and comment on the accuracy of learning content, the pertinency of instruction design (e.g., the guideline and assessment in each task), and the rationale of I-VR (e.g., the level of immersion and presence). Then, we collected the expert teachers' suggestions on the learning content and instruction design after they have experienced using I-VR. Lastly, 12 university students who majored in science provided feedback on how smooth the I-VR program is running, such as the continual update of scenes while the body moved. After revising the prototype based on the comments received, we conducted a field test in middle and high schools. In this study, we investigated students' satisfaction with the questionnaire, that was adapted and revised from the Ministry of Education website (https://moevrar.tku.edu.tw/download.cshtml). Moreover, we interviewed students about their experiences with the I-VR curriculum.

Finally, in the reflection stage, we adjusted the I-VR curriculum based on the results of the field test. For example, we added a detailed manual and warning information for students who are using I-VR for the first time. Overall, the I-VR curriculum is a complex and multidimensional system that also needs to consider the opinions of its many audiences (e.g., subject experts, teachers, and students) and follow the iterative design process for evaluation and revision.

4.2 Participants and Wearable Devices

This study was conducted in middle and high schools in Taipei, Taiwan. We recruited ninth grade students ($n = 27$), eleventh grade students with hearing impairment ($n = 7$), and twelfth grade students ($n = 65$). All students had learned the science concepts and were taking the I-VR-based course for the first time. During the I-VR course, we provided two wearable devices (HTC VIVE) to students to complete the task during the class. Moreover, the students were informed about the research purpose and could withdraw from the course at any time.

4.3 Questionnaire of Student' Satisfaction Toward Learning with Immersive Virtual Reality

To measure students' satisfaction with I-VR-based learning, we adapted the self-report questionnaire from the Ministry of Education website (Table 1). This questionnaire aimed to show students' satisfaction on how the I-VR promote their learning, enhance their motivation, and support the interaction among the user, learning environment, and peers. The questionnaire consisted of 11 items using a five-point Likert scale that ranged from one (strongly disagree) to five (strongly agree), and one open question. We divided the items into three aspects based on the potential benefits of I-VR, namely promoting learning, enhancing motivation, and interaction with the scientific objects and phenomenon. First, there were four items for promoting student learning via I-VR (e.g., I can learn scientific concepts efficiently via I-VR activities), four items for enhancing student's motivation during the I-VR activities (e.g., I can maintain my motivation in learning sciences via I-VR activities), and three items for the degree of interaction among students, virtual environments, and peers in I-VR (e.g., I can manipulate the experiment easily in the I-VR activities). All students were invited to fill out this questionnaire after they have taken the I-VR-based course.

4.4 Curriculum-Based Immersive Virtual Reality: La Scuola di Atene

We developed six I-VR curricula for middle and high schools based on the curriculum guidelines of natural science of the Ministry of Education, Taiwan [18]. This curriculum guidelines emphasize three goals in science learning, including (1) providing opportunities for students to develop their ability of scientific inquiry; (2) supporting students to understand how scientific knowledge is developed and obtain the attitude toward the nature of science; (3) guiding students to learn the core concepts of scientific knowledge. Thus, we developed the curriculum-based I-VR, which intergraded the learning performance (e.g., scientific inquiry skills and attitude) and learning content (e.g., scientific knowledge), to support students in science learning.

Meanwhile, we chose biology and chemistry as the subject and focused on the representation of molecules or atoms. Therefore, we divided the six I-VR curricula into two categories based on the learning domain [1]. The first category refers to the cognitive domain, which is related to teaching science concepts (e.g., elements and compounds have specific chemical symbol representations) and consists of the three curricula: element identification, the periodic table, and molecular structures. The second category emphasizes the cognitive and procedural domain for learning content, such as students conducting laboratory experiments and interpreting the information to solve problems. Therefore, we developed the virtual laboratory and the art museum in the second group: conservation of artistic works, Lavoisier's laboratory, and Davy's laboratory.

Table 1 Items of the questionnaire assessing student satisfaction

Items
Promoting learning
I learn scientific concepts efficiently via I-VR activities
I remember the topics learned after I-VR activities
I achieve the learning goals quickly via I-VR activities
I familiarize myself with the learning content through appropriate tasks via I-VR activities.
Enhancing motivation
I enjoy using I-VR to learn chemistry and other related scientific concepts
I am willing to use I-VR activities to learn about chemistry and related scientific concepts
Learning with I-VR activities is interesting
I can maintain my motivation to learn science via I-VR activities.
Interaction
I enjoy interacting with the learning environment via I-VR activities
I can manipulate the experiment easily in I-VR activities
I have more opportunities to discuss the concepts with my classmates and understand the relevant chemical knowledge via I-VR activities.

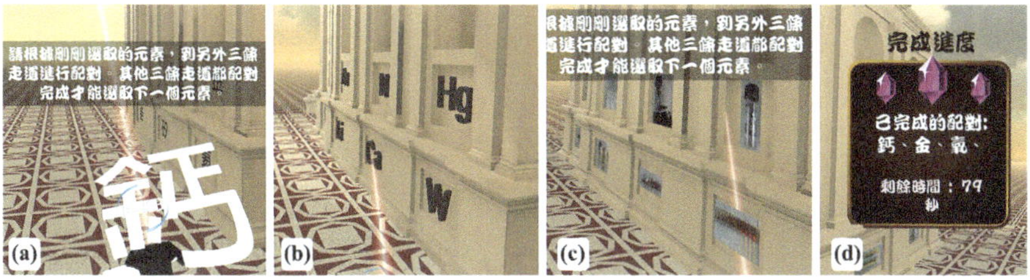

Fig. 1 Interface of the periodic table

4.4.1 Learning Theories in Curriculum-Based Immersive Virtual Reality

Based on the learning content, we applied the learning theories (such as learning flow [37]; Johnstone's triangle for chemistry education [29]; multiple representations [26, 38]) to design the curriculum and the instructional approach to support students' learning in I-VR. From the learning content perspective, we used the chemistry triplet [28, 29] as the framework to emphasize the different representations of the same substance and promote students' learning of chemical knowledge [30, 31]. For instance, we prepared the chemical symbol (symbolic) (Fig. 1a, b) and the color (macroscopic) (Fig. 1c) of the elements in the periodic table. Then, we asked students to match them correctly within a limited amount of time.

From the learning engagement perspective, we believe that students would engage in activities and meaningful learning when they are immersed in the virtual environment. To enhance students' engagement, we designed the learning material based on students' interests and skills and used task-based activities as challenges to assess students' performance. For example, we created an appealing virtual environment (e.g., laboratory and art museum) and drafted guideline to serve as the scaffolding to help students complete the task. Then, we set time limits for each task and ranked their performance (Fig. 1d).

4.4.2 The Features of Immersion in Le Scuola Di Atene

To immerse students in I-VR, we adopted several approaches to immersion and evaluated the I-VR curricula accordingly. Specifically, we classified these approaches into three categories based on their immersive design features [11], namely, sensory, actional, and narrative. For example, we built a virtual art museum (sensory) called the Scuola di Atene along with a laboratory in the storylines (narrative). Students needed to manipulate the objects to collect and interpret the information (actional). Moreover, students answered a series of questions to show their understanding of the scientific concepts in the context (actional). We introduce the two activities in La Scuola di Atene as follows.

Fig. 2 Interface of conservation of artistic works

Conservation of artistic works. Students played the role of a conservator repairing art pieces (sensory and narrative). To clean up mold on paintings, students could freely choose from a wide selection of cleaning agents (e.g., 75% alcohol, bleach, and laser) and experiment (actional) (Fig. 2a). During the cleaning process, students can use the super-magnifier to observe the amount of the mold (Fig. 2b) and record the time elapsed to compare the effectiveness of the cleaning agents (actional). Moreover, students can read the principle of sterilization (Fig. 2c) and complete the assessment.

Lavoisier's and Davy's laboratories. We designed the virtual laboratory to show students how Lavoisier identified oxygen in the eighteenth century (narrative). Furthermore, we replicated all the laboratory apparatus (e.g., glass retort and furnace) (Fig. 3a, b) and embedded the simulation of the experiment with microscopic representations to explain the process (sensory) (Fig. 3c). As such, students can use their prior knowledge to conduct experiments (actional), such as replacing the heating resource, recording the height of the mercury, and watching the film to understand the mechanism of the oxidation reaction from the microscopic perspective. Due to students' diverse backgrounds in science, we also prepared guidelines and prompts to help students complete the tasks. Additionally, we also replicated Davy's laboratory, which engaged students in generating oxygen and hydrogen and examining their properties through electrolysis (Fig. 3d, e). Videos about hydrogen combustion were produced for students to experience the flames and burst sound (Fig. 3f).

5 Results

In the following paragraphs, we have provided the quantitative and qualitative data to address each of our research questions. The students' receptiveness toward the I-VR curricula was measured through the questionnaire. Moreover, we conducted semi-structured interviews to better understand user experiences, specifically with regard to user immersion and smoothness of I-VR.

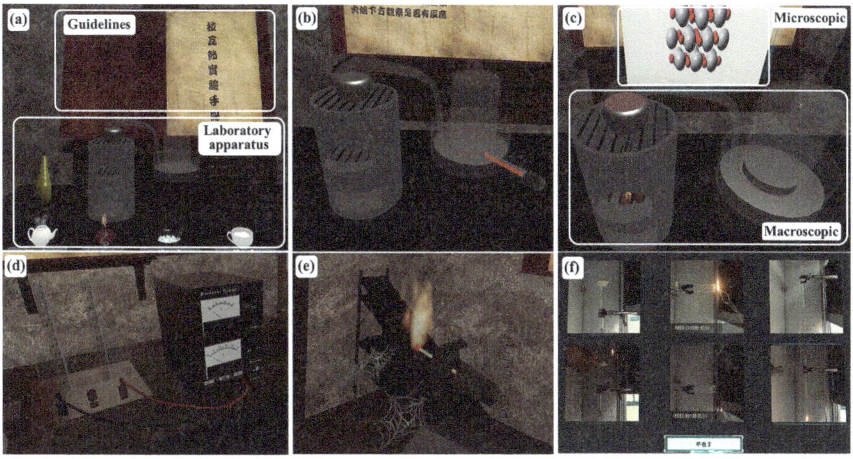

Fig. 3 Interface of Lavoisier's and Davy's laboratory

The overall satisfaction is above three in all grades and indicates that students are satisfied with La Scuola di Atene, as shown in Table 2. In addition, the eleventh grade ($M = 4.27$) and twelfth grade ($M = 4.38$) students were highly satisfied with La Scuola di Atene. Students also provided positive feedback on immersive features and their interest in La Scuola di Atene during their interviews. For example, middle school students thought the virtual environment was fascinating and were willing to use I-VR for learning in the future. Moreover, the high school students were fully engaged in I-VR activities and were able to understand the science concepts.

> It's really refreshing. I believe there will be more and more interesting ones for teaching in the future. (9th grade student)
>
> I learned a lot and have been interested in experiments via IVR activities. (12th grade student)
>
> You may not be good at chemistry, but you will learn more about chemistry principles via IVR activities. (11th grade student)

On the aspect of learning promotion, while all students acknowledge the benefit of I-VR in promoting learning, some middle school students posted negative comments on I-VR activities, including learning efficiency and user experience. For instance, students mentioned that learning with I-VR activities took more time compared to didactic instruction. In addition, the 3D virtual environment and HMD make students feel dizzy during the I-VR activities.

> Although I have never used VR, I do not think it is suitable for learning because it is time-consuming. (9th grade)
>
> I felt dizzy. It is fun, but it is very inefficient. (9th grade)
>
> The HTC Vive VR helmet is too heavy, and I feel dizzy. (11th grade)

Moreover, all students exhibited high agreement with regard to the aspect of enhancing motivation (Table 2). However, it is more difficult for ninth grade students

Table 2 Items of questionnaire for user satisfaction

Items	Ninth grade ($n = 27$)	Eleventh grade ($n = 7$)	Twelfth grade ($n = 65$)
Overall	3.81	4.27	4.38
Promoting learning	3.59	4.21	4.22
1[a]	3.50	4.29	4.00
2	3.66	4.29	4.23
3	3.44	4.15	4.26
4	3.81	4.15	4.39
Enhancing motivation	4.01	4.39	4.48
5	4.11	4.29	4.43
6	4.07	4.43	4.37
7	4.03	4.43	4.62
8	3.81	4.43	4.49
Interaction	3.82	4.19	4.46
9	4.04	4.15	4.45
10	3.77	4.14	4.48
11	3.66	4.29	4.46

Note Eleventh grade students are hearing-impaired
[a]Item IDs

($M = 3.81$) to remain motivated in I-VR activities compared to eleventh ($M = 4.43$) and twelfth ($M = 4.49$) grade students. In other words, I-VR activities can enhance students' motivation but are not a good long-term learning approach, especially for middle school students. Finally, as shown in Table 2, the interaction design with the virtual environment, such as the virtual laboratory and the operation process, is appropriate for high school students but interaction with peers (item 11) was lacking for middle school students ($M = 3.66$). Therefore, while La Scuola di Atene could facilitate interactions between the user and the virtual environment well, it still needs to be improved in terms of facilitating peer interaction.

Overall, all the students acknowledged the benefits of I-VR curricula and held positive attitudes toward using I-VR as a learning tool. However, students also raised their own concerns on learning efficiency and social interaction with peers via I-VR.

6 Discussion and Conclusions

In conclusion, the results indicated that most students were engaged in I-VR and learned the science concepts through immersion. These results aligned with previous studies on the educational benefit of I-VR [1, 9, 11, 16]. In other words, we have developed meaningful I-VR curricula for middle and high school students based on the science standard in Taiwan. However, a few students provided negative comments based on their experiences, specifically, they complained about the lack of opportunities to interact with their peers and feeling dizzy as a result of wearing the VR devices. The data also revealed many different findings on user experiences, out of which the following three are most noteworthy:

6.1 Developing I-VR Based on Learning Theories and the National Science Standard

To develop I-VR curricula, we applied learning theories and the national science standard to design, evaluate, and revise the learning activities. Meanwhile, to accentuate the advantages of I-VR, we chose the chemistry triplet [28, 29] and focused on the transformation among the macroscopic, microscopic, and symbolic representations in science learning. Also, we used the elements of optimal learning moments [39] to design the learning activities. As such, students can visualize abstract concepts in the virtual environment (e.g., the structure of molecular), engage in learning activities within interesting scenarios, get hints to support their learning and feel challenged with the limited time allotted. The present study not only shows the educational benefit of I-VR but also implements Matovu et al.'s [11] emphasis on aligning I-VR activities with school curricula.

6.2 Designing I-VR Activities with the Immersive Features

In this study, most of the students were highly satisfied with I-VR. This result may be explained by considering the immersive features [11, 36] while we designed the I-VR activities. We co-design the I-VR activities with engineers and teachers. The engineers provided the technological support to transform teachers' scripts into I-VR activities and created the immersive features. For sensory immersion, I-VR would generate audio effects when students answered correctly; haptic feedback through vibration was also given. Moreover, we allowed students to walk around the virtual environment and interact with virtual objects for actional immersion. Since middle school students asked for peer interactions in their feedback, more efforts should be made to develop social immersion in the future. Finally, for narrative immersion, students played different roles to complete their tasks and experience the different storylines in each activity.

6.3 Emphasizing the Needs of Students at Different Learning Stages of I-VR Activities

In the self-report questionnaire of user satisfaction, all the students believed La Scuola di Atene could help them understand science concepts, engage in learning activities, and interact with the virtual environment. However, students have reservations about the effectiveness of learning and the user experience via I-VR activities at different learning stages. According to the results of the interview, the more likely explanation rests in the needs of individual users.

First of all, students believed I-VR plays a different role in promoting science learning. For example, high school students showed high agreement on the efficiency of learning (item 1) and achieving the learning goal (item 3). In other words, high school students believed that I-VR as a learning tool could help them achieve their learning goals. However, some middle school students believed that I-VR activities were inefficient for learning but were suitable for content review. In other words, middle school students believe that I-VR can support them in integrating their existing knowledge but not obtaining new knowledge. Therefore, the significant gap between the task goal and students' expectations would negatively impact user experience and learning outcomes [11]. In addition, several researchers [40, 41] used the CTML [24, 25] to explain the effects and effectiveness of leaning with visualization technology [1]. The CTML said that high-quality images and animations would increase the cognitive load and become a distraction in learning activities. Thus, to develop meaningful I-VR curricula, researchers and designers should consider the rationale of representation and their learning goals to reduce learners' cognitive load.

In addition, this study presents that students experienced physical discomfort (or cybersickness), including dizziness and headache, as a result of engaging in I-VR activities. Some studies confirmed cybersickness through behavior (e.g., heart rate and skin conductance at the forehead) and self-report questionnaire measurements (e.g., Simulator Sickness Questionnaire) of VR users [42]. Weech et al. [43] also showed a negative correlation between presence and cybersickness. Reduced sensory mismatch, sensory conflict conditions, and stereoscopy were named as the key factors in reducing cybersickness. Moreover, middle school students in this study stated that it is difficult to remain motivated in I-VR activities. This finding aligns with Villena-Taranilla et al. [6], who argued that the shorter the usage time with I-VR activities is, the more effective the learning outcome would be for young students.

7 Limitation and Implication

Even though this study has the merit of offering valuable insights into the I-VR design for middle school students in science learning, it has some limitations. It is important to emphasize that the methodological problems in the research design which limited our interpretations. The first limitation concerns the specific subject used in the current study. That is, the results cannot be generalized to other science topics as well as other subjects. Also, it should be noted that this study recruited participants in a small size sampling with similar cultural backgrounds. Thus, the generalization of the results to other populations with different educational backgrounds might not be suitable. Further research is needed to conduct the experiment with participants with various educational backgrounds and on different science subjects to ensure the generalizability of the effectiveness of I-VR. Finally, the systematic data collections, before, during, and after the use of the intervention of I-VR, should be carried out to deeply understand the strength, weakness, challenges and opportunities on implementation of the I-VR in school science practice.

In spite of the aforementioned limitations, there are still several promising findings in this study that could potentially make a difference in science education. One might assume I-VR is only for ordinary students and not for hearing-impaired students who might have limited capacity to follow the instruction given by teachers. However, this study found that even hearing-impaired students can benefit from the use of I-VR in science learning. Moreover, active engagement in the activities and collaboration among students were observed among the students while using I-VR.

Also, adopting historical discoveries of chemistry in I-VR learning environment allows students to interact with scientists' instruments and experience how science could be discovered. Also, providing games (matching structural configurations and properties of chemical elements) to enhance student motivation and create linkages between different representations of chemical elements was another benefit of the use of I-VR. Moreover, we agree with Makransky et al. [13] about cognition overloading caused by the various features of I-VR that might foster users' sense of presence but also distract them from their learning. However, this kind of data (such as electroencephalogram, EEG, used in Makransky et al.'s study [13]) was not collected in this study. Researchers might take a cognitive load into account to ensure the net effect of I-VR is to be positive in supporting learning. Finally, we found several features in this I-VR learning environment to be promising. For instance, students were able to test their hypothesis of discovering oxygen, and if they failed, they could watch the instructional videos repeatedly.

In sum, choosing appropriate concepts for I-VR is essential to making I-VR a success. However, assessing the impact on students' motivation and cognition was challenging. An appropriate assessment tool to measure students' motivation in learning chemistry and the longitudinal impact of such learning requires more attention from researchers and educational practitioners when the I-VR is implemented.

Acknowledgements The authors of this article would like to express their appreciation to the Ministry of Education, Taiwan, for supporting this study. Also, thanks go to Hsiao-Lan Chung, Wei-Chun Tang, Jin-Wei Hsu, Jiaying Lin, and Ta-Min Hung for their assistance on the project.

References

1. Hamilton, D., McKechnie, J., Edgerton, E., Wilson, C.: Immersive virtual reality as a pedagogical tool in education: a systematic literature review of quantitative learning outcomes and experimental design. J. Comp. Educ. **8**(1), 1–32 (2021). https://doi.org/10.1007/s40692-020-00169-2
2. Radianti, J., Majchrzak, T.A., Fromm, J., Wohlgenannt, I.: A systematic review of immersive virtual reality applications for higher education: design elements, lessons learned, and research agenda. Comput. Educ. **147** (2020). https://doi.org/10.1016/j.compedu.2019.103778
3. Wu, B., Yu, X., Gu, X.: Effectiveness of immersive virtual reality using head-mounted displays on learning performance: a meta-analysis. Br. J. Edu. Technol. **51**(6), 1991–2005 (2020). https://doi.org/10.1111/bjet.13023
4. Araiza-Alba, P., Keane, T., Chen, W.S., Kaufman, J.: Immersive virtual reality as a tool to learn problem-solving skills. Comput. Educ. **164** (2021). https://doi.org/10.1016/j.compedu.2020.104121
5. Checa, D., Bustillo, A.: A review of immersive virtual reality serious games to enhance learning and training. Multimedia Tools Appl. **79**(9–10), 5501–5527 (2019). https://doi.org/10.1007/s11042-019-08348-9
6. Villena-Taranilla, R., Tirado-Olivares, S., Cózar-Gutiérrez, R., González-Calero, J.A.: Effects of virtual reality on learning outcomes in K-6 education: a meta-analysis. Educ. Res. Rev. **35** (2022). https://doi.org/10.1016/j.edurev.2022.100434
7. Slater, M.: A note on presence terminology (2003)
8. Gaggioli, A., Bassi, M., Fave, A.D.: Quality of experience in virtual environments. In: Riva, G., Davide, F., IJselsteijn, W.A. (eds.), Being there: concepts, effects and measurements of user presence in synthetic environments. IOS Press (2003)
9. Di Natale, A.F., Repetto, C., Riva, G., Villani, D.: Immersive virtual reality in K-12 and higher education: a 10-year systematic review of empirical research. Br. J. Edu. Technol. **51**(6), 2006–2033 (2020). https://doi.org/10.1111/bjet.13030
10. Slater, M., Sanchez-Vives, M.V.: Enhancing our lives with immersive virtual reality. Front. Robot. AI **3** (2016). https://doi.org/10.3389/frobt.2016.00074
11. Matovu, H., Ungu, D.A.K., Won, M., Tsai, C.-C., Treagust, D.F., Mocerino, M., Tasker, R.: Immersive virtual reality for science learning: design, implementation, and evaluation. Studies Sci. Educ., 1–40 (2022). https://doi.org/10.1080/03057267.2022.2082680
12. Cheng, K.-H., Tsai, C.-C.: A case study of immersive virtual field trips in an elementary classroom: students' learning experience and teacher-student interaction behaviors. Comput. Educ. **140** (2019). https://doi.org/10.1016/j.compedu.2019.103600
13. Makransky, G., Petersen, G.B., Klingenberg, S.: Can an immersive virtual reality simulation increase students' interest and career aspirations in science? Br. J. Edu. Technol. **51**(6), 2079–2097 (2020). https://doi.org/10.1111/bjet.12954
14. Sankaranarayanan, G., Wooley, L., Hogg, D., Dorozhkin, D., Olasky, J., Chauhan, S., Fleshman, J.W., De, S., Scott, D., Jones, D.B.: Immersive virtual reality-based training improves response in a simulated operating room fire scenario. Surg. Endosc. **32**(8), 3439–3449 (2018). https://doi.org/10.1007/s00464-018-6063-x
15. Çeken, B., Taşkın, N.: Multimedia learning principles in different learning environments: a systematic review. Smart Learn. Environ. **9**(1) (2022). https://doi.org/10.1186/s40561-022-00200-2
16. Freina, L., Ott, M.: A literature review on immersive virtual reality in education: State of the art and perspectives. eLearning and Software for Education (eLSE), Bucharest (Romania) (2015). https://www.itd.cnr.it/download/eLSE%202015%20Freina%20Ott%20Paper.pdf
17. OECD.: PISA 2015 Results (Volume I) (2016). https://doi.org/10.1787/9789264266490-en
18. Ministry of Education.: Curriculum guidelines of 12-year basic education for elementary, junior high schools and general senior high schools—Natural Sciences (2018). Retrieved from https://www.naer.edu.tw/eng/PageSyllabus?fid=148

19. Lamb, R., Antonenko, P., Etopio, E., Seccia, A.: Comparison of virtual reality and hands on activities in science education via functional near infrared spectroscopy. Comput. Educ. **124**, 14–26 (2018). https://doi.org/10.1016/j.compedu.2018.05.014
20. Cheng, K.-H., Tsai, C.-C.: Students' motivational beliefs and strategies, perceived immersion and attitudes towards science learning with immersive virtual reality: a partial least squares analysis. Br. J. Educ. Technol. **51**(6), 2140–2159 (2020). https://doi.org/10.1111/bjet.12956
21. Allcoat, D., von Mühlenen, A.: Learning in virtual reality: Effects on performance, emotion and engagement. Res. Learn. Technol. **26**(0) (2018). https://doi.org/10.25304/rlt.v26.2140
22. Pinto, T., Pinto, A., Maia-Lima, C., Barbot, A.: Learning about volcanoes using virtual field trips: a study in a Portuguese elementary school. In: ICERI2016 Proceedings (2016)
23. Ahn, S.J.G., Bostick, J., Ogle, E., Nowak, K.L., McGillicuddy, K.T., Bailenson, J.N.: Experiencing nature: embodying animals in immersive virtual environments increases inclusion of nature in self and involvement with nature. J. Comput.-Mediat. Commun. **21**(6), 399–419 (2016). https://doi.org/10.1111/jcc4.12173
24. Mayer, R.E.: Multimedia learning. In: Psychology of Learning and Motivation, vol. 41, pp. 85–139. Academic Press (2002). https://doi.org/10.1016/S0079-7421(02)80005-6
25. Mayer, R.E.: Cognitive theory of multimedia learning. In: The Cambridge Handbook of Multimedia Learning, pp. 43–71 (2014). https://doi.org/10.1017/cbo9781139547369.005
26. Kozma, R.B.: Technology and classroom practices. J. Res. Technol. Educ. **36**(1), 1–14 (2003). https://doi.org/10.1080/15391523.2003.10782399
27. Suh, A., Prophet, J.: The state of immersive technology research: a literature analysis. Comput. Hum. Behav. **86**, 77–90 (2018). https://doi.org/10.1016/j.chb.2018.04.019
28. Johnstone, A.H.: Why is science difficult to learn? Things are seldom what they seem. J. Comput. Assist. Learn. **7**(2), 75–83 (1991). https://doi.org/10.1111/j.1365-2729.1991.tb00230.x
29. Johnstone, A.H.: Teaching of chemistry—logical or psychological? Chem. Educ. Res. Practice **1**(1), 9–15 (2000). https://doi.org/10.1039/a9rp90001b
30. Taber, K.S.: Revisiting the chemistry triplet: drawing upon the nature of chemical knowledge and the psychology of learning to inform chemistry education. Chem. Educ. Res. Pract. **14**(2), 156–168 (2013). https://doi.org/10.1039/c3rp00012e
31. Talanquer, V.: Macro, submicro, and symbolic: the many faces of the chemistry "triplet." Int. J. Sci. Educ. **33**(2), 179–195 (2010). https://doi.org/10.1080/09500690903386435
32. Alharbi, H.F., Reddy, N., Patel, A., Tsaramirsis, K., Randhawa, P.: Virtual and augmented reality for mechatronics based applications. In: 2022 4th International Conference on Applied Automation and Industrial Diagnostics (ICAAID) (2022)
33. Merchant, Z., Goetz, E.T., Cifuentes, L., Keeney-Kennicutt, W., Davis, T.J.: Effectiveness of virtual reality-based instruction on students' learning outcomes in K-12 and higher education: a meta-analysis. Comput. Educ. **70**, 29–40 (2014). https://doi.org/10.1016/j.compedu.2013.07.033
34. Limniou, M., Roberts, D., Papadopoulos, N.: Full immersive virtual environment CAVETM in chemistry education. Comput. Educ. **51**(2), 584–593 (2008). https://doi.org/10.1016/j.compedu.2007.06.014
35. Wohlgenannt, I., Fromm, J., Stieglitz, S., Radianti, J., Majchrzak, T.A.: Virtual reality in higher education: preliminary results from a design-science-research project (2019)
36. Dede, C.: Immersive interfaces for engagement and learning. Science **323**(5910), 66–69 (2009). https://doi.org/10.1126/science.1167311
37. Csikszentmihalyi, M.: Motivation and creativity: toward a synthesis of structural and energistic approaches to cognition. New Ideas Psychol. **6**(2), 159–176 (1988). https://doi.org/10.1016/0732-118x(88)90001-3
38. Gilbert, J.K., David, T.: Multiple Representations in Chemical Education (2009). https://doi.org/10.1007/978-1-4020-8872-8
39. Schneider, B., Krajcik, J., Lavonen, J., Salmela-Aro, K., Broda, M., Spicer, J., Bruner, J., Moeller, J., Linnansaari, J., Juuti, K., Viljaranta, J.: Investigating optimal learning moments in U.S. and finnish science classes. J. Res. Sci. Teach. **53**(3), 400–421 (2015). https://doi.org/10.1002/tea.21306

40. Parong, J., Mayer, R.E.: Learning science in immersive virtual reality. J. Educ. Psychol. **110**(6), 785–797 (2018). https://doi.org/10.1037/edu0000241
41. Parong, J., Mayer, R.E.: Cognitive and affective processes for learning science in immersive virtual reality. J. Comp. Assist. Learn. **37**(1), 226–241 (2021). https://doi.org/10.1111/jcal.12482
42. Mazloumi Gavgani, A., Hodgson, D.M., Nalivaiko, E.: Effects of visual flow direction on signs and symptoms of cybersickness. PLoS ONE **12**(8), e0182790 (2017). https://doi.org/10.1371/journal.pone.0182790
43. Weech, S., Kenny, S., Barnett-Cowan, M.: Presence and cybersickness in virtual reality are negatively related: a review. Front. Psychol. **10**, 158 (2019). https://doi.org/10.3389/fpsyg.2019.00158

A Cyber Physical Architecture for Symbiotic Multi-robot Fleet Management

Daniel Mitchell, Yue Gu, Jamie Blanche, Samuel Harper, Blair Archibald, Michele Sevegnani, and David Flynn

Abstract The Metaverse is transforming the way we interact with robots, sensors, environments, and humans, via the creation of a digital domain that is inter-operable, unified, and co-created by many collaborators. Several sectors have identified that robots are valuable for optimizing current processes. However, robotic platforms face challenges in the environments they are deployed in, requiring a diverse range of capabilities and roles. A Multi-robot (MR) fleet presents opportunities via the leverage of attributes of diverse robot types. Symbiotic interactions are established via collaborative robotics for the completion of objectives and ensuring mission resilience. The Metaverse provides distinct opportunities for a human-in-the-loop to engage with an MR-fleet via cyber physical platforms including simulation and mixed reality. This chapter presents a Symbiotic Multi-Robot Fleet (SMuRF) and Operational Decision Support Interface (ODSI) which co-ordinates the orchestration and deployment of an MR team within an autonomous inspection, maintenance and repair mission, and where the ODSI provides a human-in-the-loop with a holistic overview of the robotic platforms at run-time. In addition, we demonstrate how formal methods, in the form of probabilistic models to address mission success probabilities, can be verified during run-time alongside the SMuRF. This improves

D. Mitchell (✉) · Y. Gu · J. Blanche · S. Harper · B. Archibald · M. Sevegnani · D. Flynn
University of Glasgow, Glasgow, Scotland
e-mail: D.Mitchell.5@research.gla.ac.uk

Y. Gu
e-mail: yue.gue@Glasgow.ac.uk

J. Blanche
e-mail: Jamie.Blanche@Glasgow.ac.uk

S. Harper
e-mail: Samuel.Harper@Glasgow.ac.uk

B. Archibald
e-mail: Blair.Archibald@Glasgow.ac.uk

M. Sevegnani
e-mail: Michele.Sevegnani@Glasgow.ac.uk

D. Flynn
e-mail: David.Flynn@Glasgow.ac.uk

Y. Cai et al. (eds.), *Virtual and Augmented Reality Technology-Enhanced Learning*, Gaming Media and Social Effects, https://doi.org/10.1007/978-981-96-2332-7_17

overall situational awareness and validates decisions made by the human operator in real-time and in advance of an autonomous mission to provide foresight.

Keywords Cyber-physical system · Multi-robot fleet · simulation · mixed reality

1 Introduction

Humanity is responding to several challenges which can be tackled if we improve the productivity, safety and resilience for autonomous systems. These challenges include: (A) the global climate crisis where robots are being designed to inspect, maintain and repair the complex renewable assets via autonomous deployments. (B) Operation and maintenance challenges in nuclear facilities and decommissioning of legacy facilities to ensure the safety of human workforce and examine/explore areas which may be subject to radioactive chemicals. (C) How the metaverse can be used to improve procedures and evaluation in healthcare. (D) How customers can improve their interactions with products within the Metaverse and real world. Several opportunities are created within cyber physical systems (CPS) to improve current procedures that exist within the software (cyber component) and hardware (physical component) of the systems.

Of the previously discussed overhead challenges, next we discuss more focused challenges inhibiting regular deployments. Current robot deployments include individual platforms within specific short-term inspection and maintenance activities all within controlled environments. Whilst the deployment of robotics in the field is a valid next step, this brings challenges due to the requirement for a dedicated team of at least two deployed human engineers per robot. As presented in Table 1, a human may be required for teleoperation, to intervene when challenges are faced and also in the analyzation of engineering results [1–3]. These different tasks all represent different skillsets which can be trained but also require years of experience (e.g. validation of inspection). Intervention may seem like an easy solution; however, when the robots are positioned in dangerous environments, this creates substantial risks [2]. For example, if a robot faces a challenge in a confined space and a human is required to rectify sensed and unsensed problems which inhibit mission completion. A risk assessment would be required for the confined space and robot, a human would be required to intervene and an additional human would be required to be on standby to help if anything went wrong. However, these dangerous scenarios can also include working at height and radioactive environments too. To summarize, the deployment of robots should not present an additional operational hazard, which detracts from human productivity, therefore, autonomy within inspection missions require operational governance. This is important for even the most basic mission in the real world alongside the ability for a robot to complete a range of functional tasks according to its capability.

The Metaverse unlocks opportunities that can benefit from dynamic, real-time and remote interaction across a host of different digital models and technologies

Table 1 List of different human roles required during a single teleoperated robotic inspection mission

Human role	Description	Problem
Teleoperation	To teleoperate the robot into position to complete inspection	This leads to three distinct roles which require different training for the deployment of a robot (1) Learning how to teleoperate a robot safely (2) Years of experience learning about inspection faults (3) Safe human intervention and accessing potentially dangerous areas (confined space training)
Engineering expertise	To validate the inspection or re-pair process via their engineering knowledge	
Intervention	To assist the robot in overcoming challenges when these are faced. These can include trip hazards, reliability failure, mapping failure, etc.	

[4]. This creates an ecosystem which is agnostic to digital technology, models and methods of interaction with the aim to enhance individual and collective human-team experiences [5, 6]. Digital twins is a method currently in development within the robotics community which can be further enhanced by the metaverse [7]. There are different types including digital model, shadows and twins [8]. Within this, we also have different levels of functionality and intelligence which is all driven by the capability challenge in addressing operational decision-making, planning and knowledge capture [9]. In summary, the digital twin acts as an information integrator to enable and enhance operational decision-making for a digital representation of an in design or real-world asset. This can improve validation, verification, reduce costs and risks whilst maximizing the exploration of different factors.

In relation to robotics, the metaverse promotes opportunities to further develop the relationships which occur across humans, digital twins and robotic platforms. In the example where there is a facility where robots are used for operation and maintenance. The metaverse enhances the digital version of the real asset where a human can interact and view real data of an asset and compare this with the digital version as shown in Fig. 1. This allows the human observer to 'teleport' to different areas of the digital plant seamlessly to view different areas with adaptability between a holistic and close-up view of the facility. The digital twin also allows for increases in productivity. Current procedures can be easily adapted to validate and verify different approaches to view for productivity, efficiency and safety improvements. This allows for these approaches to be verified ahead of running them onboard the real asset.

The connection of robots to the metaverse promotes agile systems and approaches, enabling smart nuclear, smart warehouses, novel medical applications and agile energy infrastructure. How each sector utilizes the metaverse will differ in terms of lists of priorities. However, as robots and autonomous systems are iteratively introduced into current procedures, the key metrics will remain the same. These include the following:

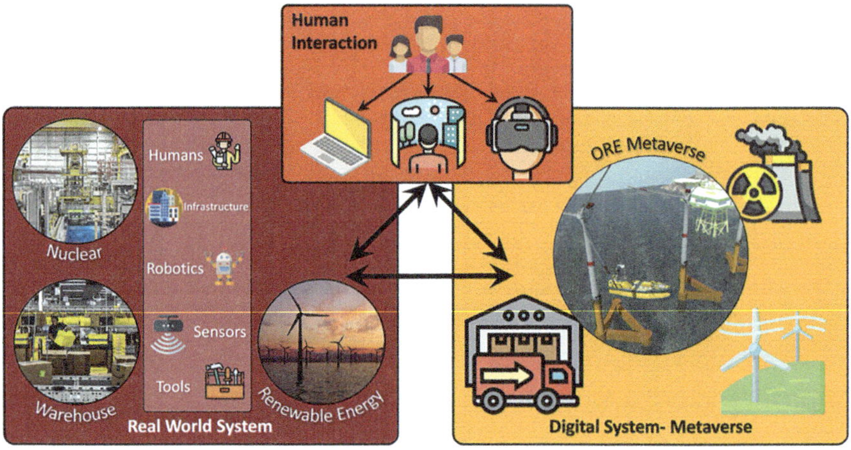

Fig. 1 Representation of the key components linked to the digital system in the Metaverse consisting of real-world system, digital system and human interaction

Level of current capability—The ability of a system to complete a designed task, e.g. inspection, maintenance or repair.
Type of deployment—Levels of autonomy from fully autonomous to teleoperation [10].
Metrics to gauge value—Validation of the value created for operators when handing over systems for operations can be measured in cost reductions, safety, resilience and productivity improvements.

This paper aims to enable for an improved understanding of the state-of-the-art to identify how the cyber and physical roadmaps converge to become a cyber physical system. Enable for improved understanding of the concepts required to improve multi-robot fleets and the types of technology being applied within several sectors including nuclear, logistics, military and renewables. The chapter subheading structure can be viewed within Fig. 2.

1.1 Trends and Imperatives

The robotics market is a market which will positively disrupt sectors due to economic benefits and reduction in the intensiveness of labour [11]. Globally, the robotics market is expected to grow at a Compound Annual Growth Rate (CAGR) of 22.8% between 2021 and 2030 to reach $214.8 billion. It is expected that this growth will be established due to the adoption of artificial intelligence and robotics across defence, security, manufacturing, electronics, automotive, and healthcare domains. The industrial share of the robotics market is expected to grow at a CAGR of 14.3% from 2020 to 2027 to reach a value of $30.8 billion. In addition, within the healthcare sector,

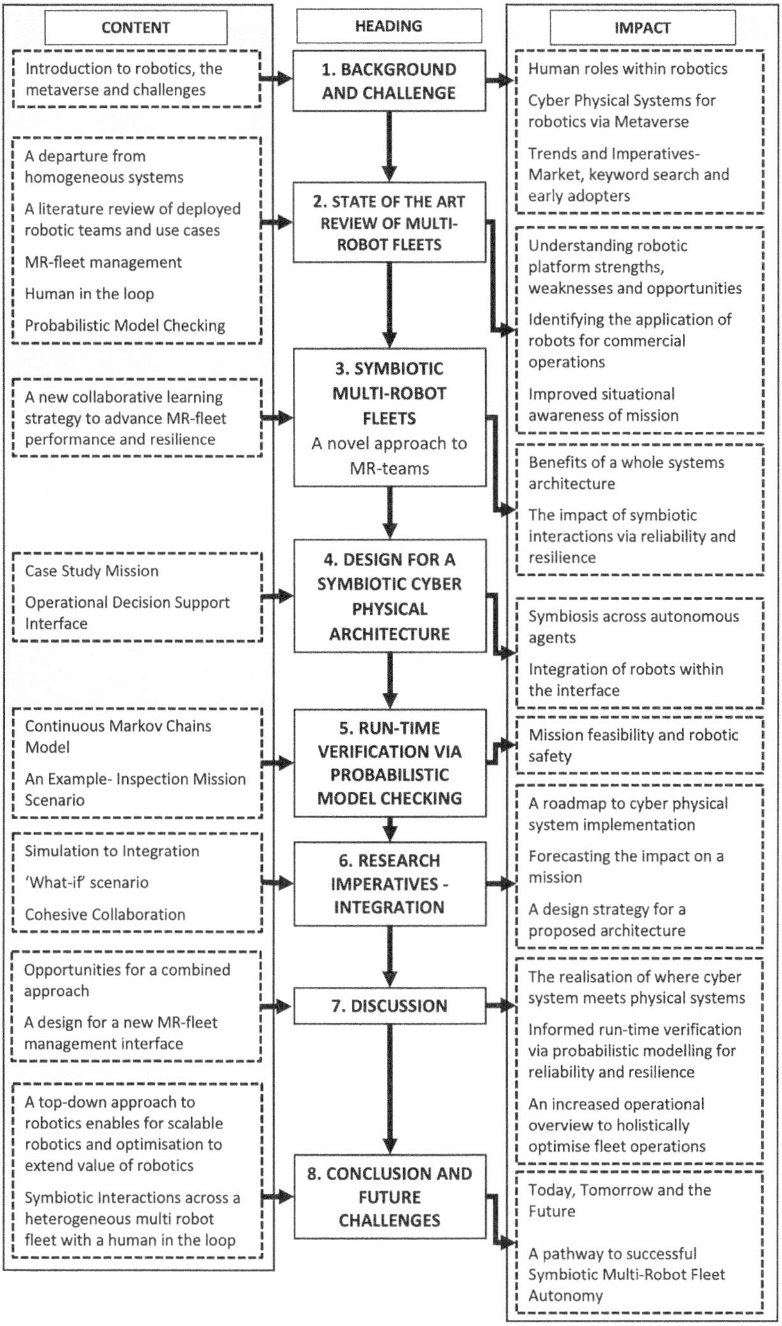

Fig. 2 Paper structure highlighting content and impact from each subheading in the chapter

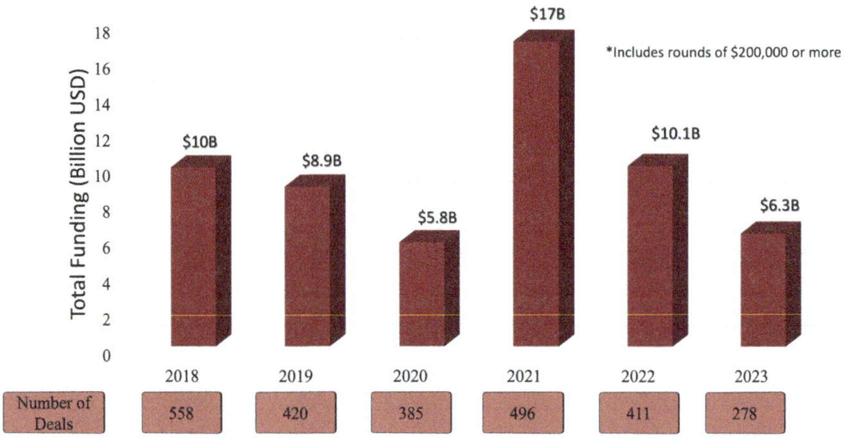

Fig. 3 Funding in robotics including pre-seed, seed and all venture rounds (last updated November 2023) [14]

surgical robots are expected to grow by 19.3% within 2022–2030 to reach $18.2 billion [12], with key developments in orthopaedics, neurology, urology and gynaecology [13]. In 2022, funding in robotics reached record highs with over $17 billion being invested in venture capital robotic start-ups. This tripled the investment when compared with 2020 as displayed within Fig. 3 [14].

A keyword search was conducted within robotics to provide valuable insights into the target applications and trends which exist within the last decade. When conducting the keyword search, Scopus was utilized to search for words within the title, abstract and keywords sections of publications. The terms included "*Robots*" AND "*keyword*", where the keywords can be identified within Fig. 4 ('*Team*', '*Swarm*', '*Multi-Robot*', '*Heterogeneous*' and '*Homogeneous*'). 'Swarm' is typically used within nature to describe a large or dense group of flying insects, however, '*swarm robotics*' is used within research as a collective term for multiples of the same ground, aerial or subsea robots, and with an approach to the coordination of the robots as a system [15]. The results of the search indicate a rise in publications in the last 5 years for '*Multi-Robot (MR)*', '*team*', '*swarm*' and '*heterogeneous robotics*'. This represents research focusing on gaps which exist in creating scalable robot teams that have value for operators to coordinate, collaborate and optimize partnerships with robots, humans and facilities. Whilst heterogeneous and homogeneous keywords are used less, this is mainly due to a collective consensus in the preference of terms which are more easily defined, especially by the wider public. Current research is oriented towards homogeneous robotics however, we identify a gap in research that exists across a diverse MR-fleet, which can include commercial off the shelf and bespoke platforms connected with a human operator to oversee activities. This will operate as human–machine teams, where the operator can designate tasks to robots and the robots can work together to complete tasks whilst updating the human-in-the-loop and allowing them to focus on more complex tasks.

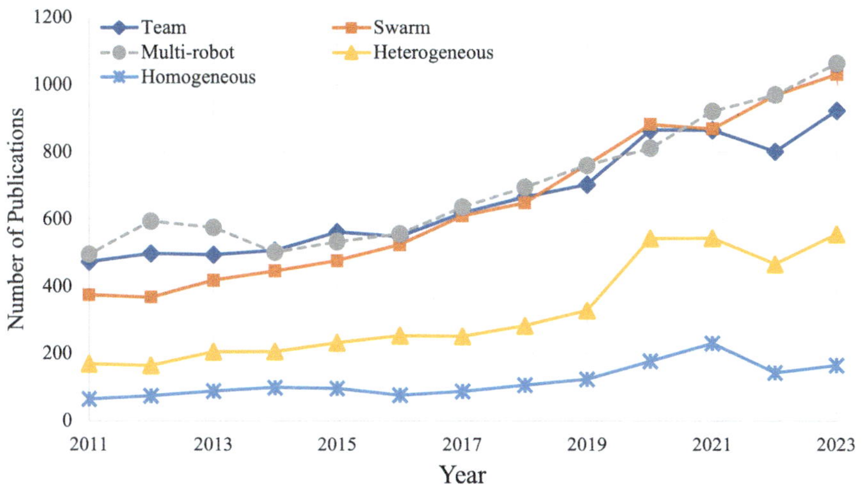

Fig. 4 Number of publications each year identified within Scopus using keyword search terms 'robot and keyword term' (last updated April 2024)

Early adopters within industry are taking advantage of ground-up capabilities from robots, where an MR-fleet approach is allowing them to excel within their operations. As identified in the introduction, there is a shift from single robot deployments to robots that can operate within a facility more regularly. For persistent autonomy to occur, a facility operator must remove the current requirement for a robot to be operated by a team of human operators as this removes the key value from having a robot on site in the first place. This creates opportunities in fleet management as, once a single robot is implemented within a digital twin, it can be easier to scale to multiple robots of the same type. This allows for a single dashboard where a diverse MR-fleet can be operated by a single operator to coordinate the team. Hence, the reason for an increase in publications using these keyword terms. The remainder of this subsection discusses three use cases: inspection of energy facilities, warehouse operations and the Ministry of Defence UK.

Energy Robotics

Energy Robotics represent a company which helps operators of oil and gas, chemical and power and utilities facilities to capture key information from a robotic fleet. Their aim is to improve inspection via the creation of a digital dashboard to display data (Fig. 6a) from a range of ground-based robots, including Boston Dynamics SPOT, ExR-2 and Rover robots (Fig. 6b), to complete a range of inspection tasks. These robots have a range of standard, well-proven sensors onboard, such as LiDAR, visual cameras, thermal cameras, gas sensors and microphones. Whilst each robot supports different combinations of payloads, the key objective of their service is to collect high-quality data and present this within a user interface to digitalize inspection. Case studies include the inspection of a Shell plc operated Oil Refinery in the Netherlands, where a SPOT and ExR-2 are used to complete regular routine inspection around well-defined routes [16]. This allows humans to focus on more complex tasks. Advantages of their digitalization tool are as follows:

1. Robot control: A user can manually control robots live during a mission (teleoperate).
2. Robot health status: Key status elements such as robot latency, battery %, mode of operation (driving, stopped etc.), types of sensors implemented and emergency stop.
3. Teachable inspection missions: Robots can be manually teleoperated around a route with sensors aimed at different locations throughout a route. The robot can then repeat this mission and feedback data to the digital tool.
4. Inspection reporting: An automatic report is generated for a human operator with a visual representation of robot, map and key inspection points visualized in graphs or images.

This CPS highlights key elements which are required for a human operator at a facility to extract the key value from robotics and automated systems. This identified approach highlights the key requirement for a multi-robot fleet and the management required as a fleet of robots increases around a facility. This highlights the need of the research approach within this chapter where key improvements in future could include robots which can autonomously work together to complete objectives of the mission, e.g. by sharing sensing payloads. In addition, the inclusion of more robotic platforms such as unmanned aerial vehicles (UAVs) can improve the holistic inspection of a site.

Amazon Fulfilment Centre

In 2012, Amazon announced that they would purchase Kiva Systems-which interconnect hardware and software to streamline the process of picking, packaging and shipping e-commerce products for $775 million [17]. In June 2022, Amazon announced their first autonomous warehouse robot named Proteus (Fig. 6c). This robot uses a powerful hydraulic lift to convey tall, wheeled cages containing packages around a warehouse. This allows for efficient and effective delivery of items to personnel to be sealed into cardboard boxes before postage. The robotic platforms, which are more effectively defined as swarm ground robots (several of the same robots working together to a common objective) are built and tested alongside their own diagnostics with the aim to support growth which enables more jobs due to the humans which build, support, interact and deploy them [18].

In terms of fleet management, robots and cages are spread out at random locations on the warehouse floor. There is no assigned cage specifically for each item. This reduces the amount of time it takes to get an item to the picking station on the warehouse floor. The algorithms also monitor popular items which are being sold and moves the cages with these items closer to the picking station location to allow for a reduction in time. To keep track of the robots there is a grid system on the ground using a series of QR codes which allows the units to work together to clear paths for each other to retrieve items as efficiently as possible. Cameras onboard the robots identify if any items fall out of the cages and notify a human to retrieve the objects. When things go wrong, a robot technician is permitted to enter the warehouse floor provided they are wearing a vest which sends a signal to nearby robots to stop. This allows a human to collect items which may have fell onto the shop floor or complete maintenance on the robotic platforms [19]. Whilst this may be implemented, there is no mention of how issues in robotic fleet maintenance are identified. Algorithms in future could be used to self-certify a robotic platform and ensure it can extend its remaining useful life ahead of key maintenance periods.

Ministry of Defence: British Army

The British Army released their approach to robotics and autonomous systems in 2022, which included a vision to 2035 where the central idea consists of creating an operational advantage for land forces by integrating adaptable robotic and autonomous systems within human machine teams, with the aims to generate mass and tempo (as defined in Fig. 5) whilst reducing risk [20]. This can be used to create an operational advantage in real time on the battlefield and within mission support to aid decision-making. Human machine teaming is a key element of the deployment of robots alongside soldiers (Fig. 6d, e) [21]. This will include a large amount of real-time fleet management and coordination via increased mass whilst reducing human density in close combat roles.

In September 2022, a successful deployment of swarm UAVs was conducted alongside military exercises where the soldiers themselves deployed multiple live drones. The first key result included an improvement in the regulation of UAVs as this was the first time the Military Aviation Authority issued a categorization of this

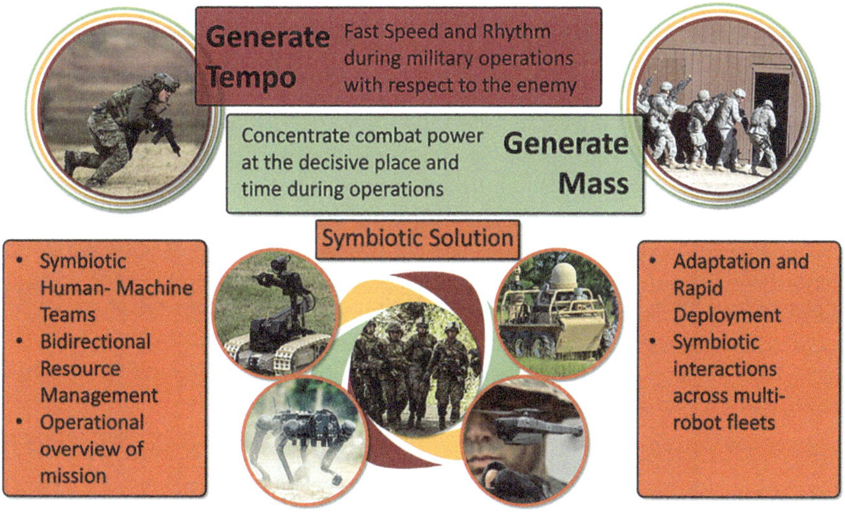

Fig. 5 Overview of key military definitions and how a symbiotic robotic approach can improve military operations

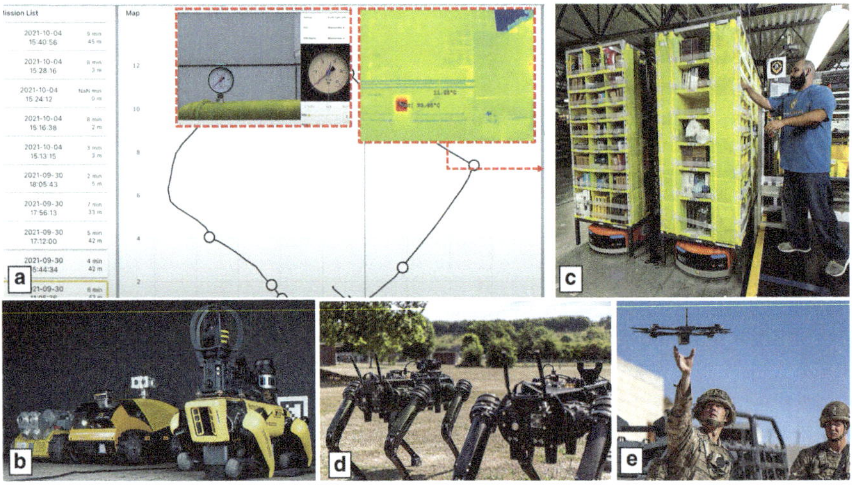

Fig. 6 Early adopters of swarms and multi-robot fleets highlighting Energy Robotics with their software interface for multi-robot fleet management (**a**) and multi-robot fleet (**b**), Amazon warehouse robots (**c**) and early tests of robotics within the British Army for reconnaissance (**d**, **e**) [16, 22–24]

type of swarm to be operated by a single operator. Of the UAVs and equipment demonstrated, the company Atlas, presented the capability for a single operator to control four drones using a tablet with the creation of manual mission taskings. Elbit Systems UK Ltd. performed a second demonstration, which included one operator having the ability to task 6 drones via autonomous missions. This allowed for an operator to task multiple drones with the same mission or various missions for the swarm [25].

Of key importance to the military is trust and explainable AI, whilst the robot is what we see visually, there are multiple algorithms, AI and data sharing occurring in the background, which allows for these robots to become effective trustworthy partnerships within a Human–Machine Team. This is where the value is created from robotics and autonomous systems. In future, we shall see more iterations of demonstrations which include autonomous ground vehicles for transport of ammunitions, UAVs for reconnaissance missions, quadruped robots completing reconnaissance ahead of soldiers and rescue evacuations via large drones [22, 23, 26–28].

2 State-of-the-Art Review of Multi-robot Fleets

2.1 Departing Homogeneity

The characteristics that define a swarm are: decentralized control, limited inter-robot communication, use of local information streams, and the emergence of global behaviour patterns. These have typically consisted of several simple robotic platforms, with swarming techniques beginning to be applied to more complex robots and fleets [29]. Applications of swarm robotics to date, however, are limited to proof-of-concept studies under highly controlled conditions. These swarms are also characterized by homogeneity of the robotic platform [30]. An MR-fleet shares many of the attributes of a swarm, but with the added advantage of the inclusion of more than one type of robotic platform. This results in a larger degree of flexibility in the MR-fleet due to the differing capabilities of the platforms used, allowing for the leveraging of capabilities for maximum benefit. With swarms and MR-fleets exhibiting relative merits, the need for resilient and trustworthy autonomy demands flexibility in operations, a criterion that is better suited to the varied capabilities of an MR-fleet.

2.2 Applications of Multi-robot Fleets

Warehouses

The use of robotics in the industrial and commercial sectors is well-established, where facilities are advancing further in warehouses designed to accommodate the needs of a robotic fleet [31]. Warehouses are excellent examples of use cases for wheeled robotic agents due to smooth concrete flooring and due to the predictable and managed operating conditions, which are optimized further to facilitate the robot operations. However, the practical application of robotic systems in unmanaged dynamic environments require run-time path planning capable of identifying a safe route for navigation through areas less suitable for robotic operations [32]. Accounting for environmental dynamism in path planning for Robotic Autonomous Systems (RAS), while an established field, places a focus on object detection and collision avoidance as part mobile Simultaneous Location And Mapping (SLAM) operations [33]. An identified area of improvement includes an emphasis on research to identify the safe passage of a robot through an area with uncertain ground conditions that would otherwise impede RAS operations.

Termed "*Robotic Mobile Fulfilment Systems*" (RMFS), robotically operated smart warehouses play a key role in maintaining the efficiency and smooth running of the logistics sector worldwide. Typical robot activities undertaken include lifting cargo pods from a pick-up area and delivery to a manual sorting station, with these missions including a MR-fleet working towards the same goals. Ensuring this efficiency drive is MR task allocation, which assigns tasks to each platform and is aimed at reducing the total cost of tasks, while also minimizing the number of robots used. Task allocations can assign multiple tasks to a single robot for consecutive execution or, for more complex objectives, instruct the platforms to form coalitions where required. At run-time, the addition of new tasks due to dynamic operational requirements can extend mission parameters and result in reallocations/re-tasking as a situation develops [34].

Offshore

The offshore energy sector is operationally complex, with high capital and operational expenditure costs associated with both the extraction/generation and transmission of energy to markets. With the energy markets currently supporting the offshore renewables sector, the offshore environment will remain a key operational area for many decades to come, with research being undertaken to optimize robotic platforms for safe and resilient offshore operations. Robotics in the offshore sector remains an emergent field, with the vast majority of research conducted focused on submarine systems that are designed to inspect and repair submarine structures, which is beyond the scope of this chapter. This section will focus on non-aquatic systems operating above the waterline, commonly termed as "*topside*".

Taurob: Designed with ATEX (equipment for potentially explosive environments) compliant electronics, Taurob has been developed, in partnership with Total, since 2019. A tracked unmanned ground vehicle (UGV) design, Taurob (Fig. 7a) was designed from inception as a mobile robotic platform for offshore platform environments for the O&G sector and is capable of climbing stairways up to 45°, while carrying a range of environmental sensors [35]. The Taurob brand has since diversified into the chemical, power generation, emergency response, security and surveillance sectors.

MIMREE/Bladebug: The Multiplatform inspection, Maintenance and Repair in Extreme Environments (MIMREE) project utilized a multi-robot system to automate the delivery and recovery of robotic inspection devices for the maintenance of offshore wind farms. The delivery system comprised of an autonomous surface vessel (ASV) and UAV, which is deployed once the ASV arrives at the offshore site.

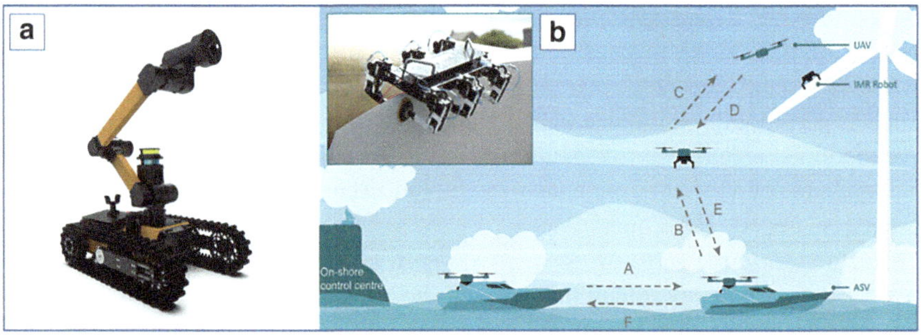

Fig. 7 **a** Taurob ATEX compliant robotic platform [36]. **b** MIMREE/Bladebug; A vacuum-based inspection robot that is delivered to a WT blade via a combination of UAV/ASV carrier robots. The Bladebug (inset) moves around the blade on an inspection mission. Adapted from [37]

The UAV, carrying the inspection robot as a payload, then delivers the payload to the site of interest. The inspection robot, Bladebug (Fig. 7b inset), is an insect-like design that utilizes vacuum pads for adhesion and locomotion. Figure 7b shows the operational process as A) the ASV and UAV deploy with the payload robot, as commanded by a remote operations centre. B) The UAV deploys with the payload robot, where it is delivered to the asset at point C). Following completion of the inspection mission, the payload robot is collected from the wind turbine blade surface at point D) and returned to the ASV at point E). The ASV is then reassigned or returns to base at point F) [37]. In the duration of this project several challenges were addressed, however, since the close of the project, Bladebug is the only outcome which has continued development and deployment [38, 39].

The ANYbotics ANYmal robot (Fig. 8) is a quadrupedal autonomous system used for proof-of-concept robotics for offshore substation/platform operations. Intended for residency on an offshore site, this system (and the similar SPOT system from Boston Dynamics) has been tested in environmental operating analogues and has also been deployed in an offshore environment [40, 41]. Though clear advancements have been made, with demonstrations given, in late 2019 to industry representatives at Port of Blyth, of quadruped, UAV, AUV, ASV and wheeled robotic designs, it is clear that these systems have not yet reached the level of non-intervention required for extended periods of autonomous offshore service [42]. Anybotics recently improved their robotic platform releasing an ANYmal X which includes ATEX-certified equipment allowing it to be operated within hazardous and potentially explosive environments. The robot has the ability to inspect using visual and thermal cameras with the ability to navigate over unstructured, complex terrains.

Consequently, in spite of the hazardous environment and high intensity labour requirements imposed by the offshore energy sector, the use of robotics systems offshore is in its infancy and limited to the inspection of offshore, subsea structures via remotely operated vehicles (with some automation but mainly teleoperation by a human operator) [36]. Ground-up improvements include improving manipulation

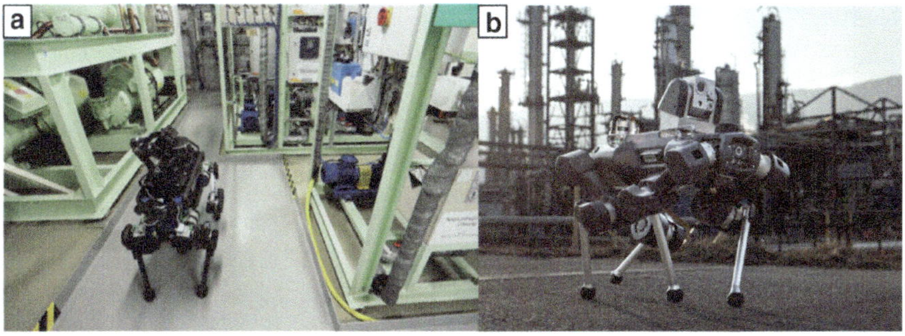

Fig. 8 ANYmal: **a** Quadrupedal system used for proof-of-concept robotics for offshore substation/platform operations [36]. **b** Anymal X Ex-certified robot [43]

activities which robots can complete alongside robots being used further for maintenance and repair activities. While research continues towards MR-fleets, resilient operations and trusted autonomy in beyond visual line of sight environments, with promising results, the nature of the high consequence environment offshore means that progress is slow. This is due to regulation in ensuring safe deployment of robotic platforms within facilities and with the capabilities from both ground-up and top-down challenges.

Nuclear

Robotics in a nuclear environment is a complex topic due to the severity of reactions when procedures are not followed. The nuclear sector has clear and well-documented origins in the wake of the worst nuclear accidents/incidents in the history of the sector [44]. It is therefore necessary to limit this discussion to mobile robotic platforms that utilize autonomy, rather than teleoperation, and to systems that return data other than visual/camera feeds.

The state-of-the-art in edge dynamic path planning has resulted in autonomous mobile robots that can account for uncertainty in the conditions of its operational environment, with navigational decisions made based on information from a suite of sensors [45, 46]. Costmaps, grid map representations of risk used in 2D path planning and navigation, can provide semi-autonomous systems with awareness of the position of static and mobile obstacles in an environment using sensors, such as Lidar or depth cameras [47]. Path planning algorithms then use this information to avoid areas where a collision may occur [48]. This costmap approach provides robots with a spatially resolved representation of risk, which can be expanded to a range of sensing modalities and threats. An example is the integration of ionizing radiation sensing with UGVs to map radiation levels in deployment areas where human presence is denied due to the radioactive environment [49]. This places greater emphasis on edge sensing to both detect and map high-risk areas, in addition to safeguarding the robotic platform by preventing a condition of irretrievable failure. The consequent evasion of highly localized radiation levels improves mission resilience and advances the RAS capacity for non-intervention in a hazardous environment.

2.3 Fleet Management

A multi-robot fleet is a relatively new concept due to the approaches from roboticists and engineers when addressing challenges. Many approaches look to design methods for ground-up capabilities which include aspects of the capabilities of robots, such as traversing difficult terrain, manipulation of valves, identification of data within gauges and teleoperation of robotic platforms. However, a new approach is required which creates additional value when these ground-up capabilities are regularly deployed within a facility, which includes fleet management and task coordination of a diverse multi-robot fleet to capture the true value and optimization.

This can be especially difficult as designers of robotic platforms often work in independent silos, making it difficult to orchestrate robots in shared environments and monitor multiple robots within a single systems view [50–52].

The operation of multiple robots within a mission space requires coordination and monitoring. These tasks are typically performed by a Fleet Management System (FMS), which is a centralized control system that issues commands and processes communications to and from robotic elements within a fleet or swarm [53, 54]. The role of the FMS (and similar systems) is to enable optimized task allocation to the fleet/swarm to reduce cost, energy expenditure and increase productivity for a given set of mission objectives [34, 55–57]. As discussed earlier, Energy Robotics utilize an FMS where data can be visualized within this system alongside the deployment of the robotic platforms also. Formant also presents an FMS for robotic platforms which can access robotic software development kits to allow for autonomous missions and teleoperation of a robotic fleet. Their interface can be summarized under operation (remote control of a robot), analytics (data management) and observe [58, 59]. However, in general for all RAS, it can be said that autonomy in its current form is unable to solve all problems and often requires human intervention when circumstances leave the autonomy envelope. This leads to human-in-the-loop optimization.

2.4 Human in the Loop

The human-in-the-loop (HITL) represents a pathway to the real world, where a robot (or fleet) is operating in a particular environment to perform a task or set of tasks, with a virtual world model (a digital twin) of the operating area detailing information gathered by the robot on mission parameters and the operating space. Under a fully autonomous paradigm, the robotic fleet will be given instructions via the Digital Twin (DT), and any amendments to mission planning or task allocation are issued by the DT [60].

The realities of commercial and industrial deployments of robotic fleets in warehouses, such as those optimized for robotic systems at Amazon, Nimble, Plus One, Waymo and Zoox, are such that impediments to task progress, and where robots are deemed to be at risk, require human intervention by a teleoperator/HITL to rectify. Using machine learning and AI, it is possible to train a robot fleet to improve its operational efficiency by self-correcting for common problems without the need for repetitive human interventions. This optimizes the use of HITL time and focus for less common errors [61].

As such, optimization of HITL time has become a priority in autonomy research and incorporates advances in perception capability to improve the accuracy and fidelity of human–robot interaction. The future of industrial autonomy will require significant levels of fleet automation to remain flexible and adaptable enough to maintain task efficiency, even when conditions depart those optimized for operations [62]. This is especially important to enable market growth by increasing the speed of

MR-fleet operation, while lowering operating costs. Thus, collaborative autonomy in dynamic and less predictable operating conditions, including the presence of humans, will continue to be a major research imperative for robotics [63, 64].

Whilst a human would be able to validate the success of a mission, as MR-fleets become larger, it is more effective for probabilistic modelling to validate and verify mission success in real time. This allows a human operator to focus on more complex tasks and as artificial intelligence is better suited to cognitively validate the rate of success of a mission. Additional benefits also include human factors where a human may disregard the state of a robotic platform to perform a mission (even though it is damaged etc.) to ensure that they meet targets, etc. Therefore, probabilistic modelling ensures that a fleet is used properly and responsibly.

2.5 *Probabilistic Model Checking*

Despite the advantages of multi-robot fleets, the increased complexity in unknown environments makes it difficult to reason about and guarantee the properties of the overall system when deploying robots in critical missions (e.g. inspections). Human operators are often paired with the robot fleets to improve the decision-making but only if the operator maintains situational awareness—understanding of the current state—as well as anticipating future states of the system and the environment [65]. Previous studies have shown that a strong awareness significantly improves human performance [66], while humans with poor awareness may have problems detecting and intervening in abnormal behaviours of robots [67, 68].

Formal methods that use mathematical models for the analysis and verification of systems have been widely applied to support the development of software and hardware systems, such as data refinements, code verification, and industrial applications [69]. Model checking is a formal technique to assess whether a given system satisfies certain properties. It has been used to provide insights into multi-robot systems (e.g. swarms) and analyzes the robustness [70] (e.g. fault-tolerance in robotic swarms) and security [71] (e.g. verification of security protocols in multi-agent systems). Considering the probabilistic nature of real systems, such as unpredictable failures, probabilistic model checking is designed to formally verify stochastic systems against quantitative properties. A typical approach to solving the probabilistic model checking problem involves the construction of mathematical models of the system (i.e. possible system states, transitions between states and the likelihood of these transitions) and an iterative examination of all possible executions of the system model to compute the exact numerical measure [72].

Many tools have been developed to perform probabilistic model checking. For example, PRISM [73] supports different types of mathematical models, including discrete-time Markov chains (DTMCs), Markov decision processes (MDPs) and continuous-time Markov chains (CTMCs). It is applicable and expressive enough to perform an exhaustive analysis of the system under study [73]. A core component of PRISM is a probabilistic model checker that quantifies all possible behaviours of a system, e.g. asking the probability that a system succeeds. It has been successfully used in a wide range of domains, from real-time probabilistic communication protocols (e.g. IEEE 802.3 CSMA/CD [74]) to biological systems (e.g. complex biological pathways [72].

Instead of verifying a system before deployment, run-time monitoring [75], also known as run-time verification, uses formal methods to dynamically analyze the execution traces that a system generates at run-time. It has grown actively in the last decades and numerous practical tools have emerged. Falcone et al. compare and classify 60 tools in different dimensions, including their specifications and deployment [76]. Current tools rely on two main activities to perform the verification: creating monitors from specifications and receiving observations (system state updates) from the monitored system [77]. Monitors are static, once created and only the observations are updated at run time. These tools are broadly applied to verify agent-level properties, e.g. safety, but it is challenging to analyze mission-level properties, such as the success of human-in-the-loop system, where the system evolves over human interactions.

In this work, we investigate using run-time probabilistic model checking to model and analyze mission-level properties when human operators are in the loop. We allow formal models to evolve at run time to account for system dynamics and human interactions. The models are then used to perform model checking against properties of interest. By modelling, we don't only monitor the process, but also predict the possible effects of interactions to improve situational awareness. Our previous studies have applied similar techniques to predict failures in a multi-sensor lift system [78] and to provide valuable insight into the delivery mission of the UAV swarm [79]. We specifically focus on the human factors: how human interventions might affect the running system (and hence model checking results) and how the formal information aids human operators in making decisions. In our latest work, a user study was conducted on the UAV swarm scenario and showed that the formal information increased human-swarm performance without affecting the workload and system usability [80].

3 Symbiotic Multi-robot Fleets

As identified in the trends and imperatives section, there exist trends to reach regular autonomous deployment. These include MR teams and increased human–robot interaction via agile cyber physical systems. Robots are currently being deployed as single robots for single use cases such as within an inspection role. As facilities continue to realize the potential of robotics, there will be an increasing number of robots and types of robots being deployed within operational facilities sector-wide. To further enable the value from robotics this will require several methods for humans to interact with robotic platforms, autonomous systems and other types of sensors. Cyber physical systems will allow for digital tools to: synchronously allow for query-based logic, adapt missions and work alongside robots to increase productivity by allowing humans to work on more complex tasks.

To achieve this, key steps include MR-fleets which can be considered as a wide range of robots (quadruped, wheeled, UAV, amphibious vehicles etc.), deployed to complete O&M activities as individual robots or partnered together. To realize this ambition, there will be necessary steps for resilient robots which can autonomously overcome day-to-day challenges due to environmental conditions such as terrain, weather conditions and mission profiles. Reliability will also be a key component of MR-fleet management, where robots are capable of self-certification of onboard systems to validate safety procedures. Additionally, MR-fleet management allows for humans to allocate tasks and coordinate the autonomous MR-fleet to ensure productivity [81].

A Symbiotic Multi-Robot Fleet (SMuRF) provides a new collaborative learning strategy to advance robot team performance and resilience for autonomous missions. This also features improvements in cyber physical coordination, management and deployment of MR-fleets via a holistic system of systems approach. A whole systems architecture is presented within Fig. 9, which includes three key layers: RAS, data and digitalization and personnel. People are core to the success of technology, this is mainly due to how they interact, communicate and ensure safe operation of both robotics and humans. Digitalization of a cyber physical system allows for personnel to visualize and co-create, resulting in a partnerships across human–machine teams. This also establishes increased visualization, where remote operators can access key information to visualize the robot within the environment and to interact effectively with autonomous missions. Data is a powerful tool within a CPS, and where inaccurate data can result in inaccuracies within a system of systems approach, especially as this feeds into autonomy. Within the RAS layer, this can be summarized alongside infrastructure, sensors, autonomy, AI and robotics. These systems have been paired together as they all require similar architectures and all feed data into the lower layers within Fig. 9. In most cases, it is often the autonomy and AI (software) which lets down the physical robotics platforms performance. Therefore, resilience and reliability within autonomy are especially important in future to ensure problems which robots face can be overcome when in an autonomous mode of deployment. Currently, many problems which autonomous systems face simply require teleoperation to overcome, therefore this would minimize this. Symbiotic interactions feature across all three segments allowing for the whole system architecture to co-create and mutually optimize missions. Examples of symbiotic interactions are discussed in the following segment.

A requirement for an effective SMuRF includes a diverse MR-fleet, where an example is displayed in Fig. 10, representing a nuclear facility. The environment displays an area which could not be fully inspected via a single robotic platform. This represents the first key motivation supporting a SMuRF, where a ground robot can be used to inspect low areas, aerial vehicles can inspect high areas and robots can work together to complete inspection objectives. To expand on teamwork, symbiosis represents the second motivation, where robots mutually benefit from cooperating with each other to complete mission objectives. This can include requesting different

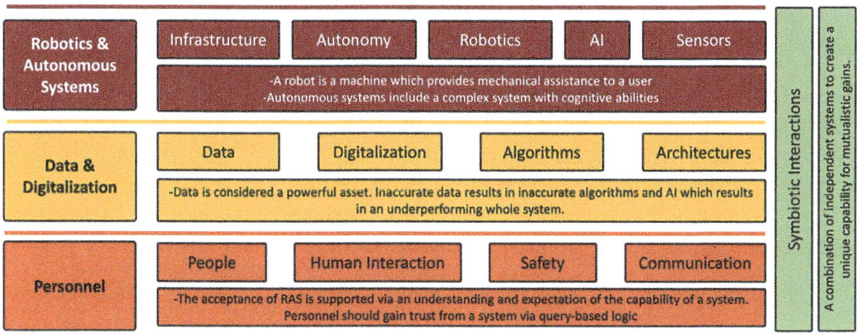

Fig. 9 Whole systems architecture featuring symbiotic interactions

robots due to different payloads to ensure the most effective inspection of an area. The third motivation includes fleet management and digitalization where robots can be coordinated via a HITL and information transfers to the human about the SMuRF missions.

Fig. 10 Composite image of a SMuRF within a nuclear setting, consisting of ground-based robots (Clearpath Husky, Boston Dynamics SPOT), aerial platforms (Intel Falcon, DJI Tello) and infrastructural sensors (Limpet sensor) [82, 83]

4 Design for a Symbiotic Cyber Physical Architecture

The transition from semi to persistent autonomy requires robots working alongside humans. These resident robots must demonstrate safety compliance and decision-making to preserve safety and maintain the trust of humans in shared and dynamic workspaces. Designing for a symbiotic CPS allows a MR-fleet to perform missions while increasing the chance of mission success through considering factors which affect the resilience and reliability of the fleet. An example of such a mission is given in the case study which follows.

4.1 Case Study Mission

This case study reflects a real-world scenario involving autonomous interactions in a remote, dynamic and enclosed environment. A SMuRF is tasked with a routine inspection mission on an offshore asset. This is described graphically in Fig. 11. The grey baseline mission shows this routine inspection, involving an MR-fleet inspecting different areas depending on their capabilities (e.g., at height for the UAV). Symbiosis is demonstrated through an unplanned fault deviation from the baseline, this is shown by green line, and the level of symbiosis demonstrated by the increase on the Y-Axis.

The deviation begins when a wheeled UGV experiences a low battery warning. The operator is informed of this, then the rest of the fleet moves to assist autonomously. The UAV completes a short mission to corroborate a clear path via the onboard camera, then a legged robot UGV, equipped with a manipulator arm, moves to pick up and deposit a battery before all robots return to their starting points. This interaction is designed to prove the increase in resilience provided by a SMuRF, which increases the chance of mission success.

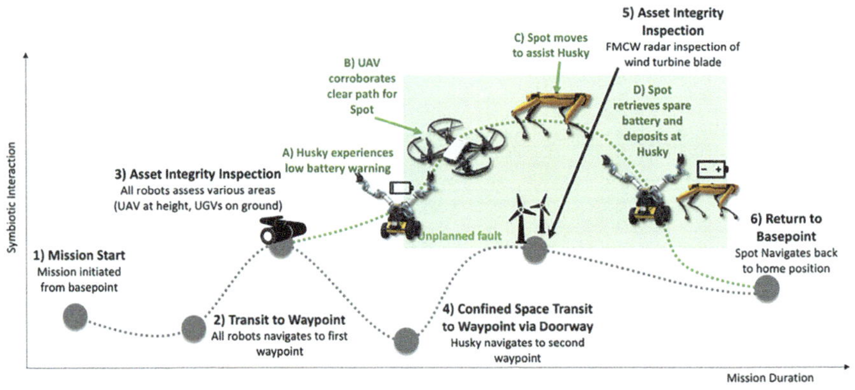

Fig. 11 Description of the MR mission plan. The grey line indicates the baseline mission, and the green line indicates the deviation upon fault detection

4.2 Operational Decision Support Interface

The Operational Decision Support Interface (ODSI) gives an operator a hyper-enabled overview of a SMuRF and allows for orchestration of the fleet. It is intended to be used as a tool to enable a human-in-the-loop to complete a successful autonomous mission with a SMuRF, via the provision of information and diagnostics on the robots in the fleet. To ensure cross-platform support across multiple devices, operating systems and input methods, the ODSI was created in Unity 2020.3.11f1. Unity allows the creation of an interface which can be adapted per-platform via layouts, text and buttons.

The ODSI allows for multimodal communications between devices and platforms, using USB, Wi-Fi and Ethernet connections. For this case study, the assets connected were as follows: DJI Tello (UAV), Boston Dynamics Spot (Quadruped UGV), Clearpath Husky (Wheeled UGV), 3 × webcams and Limpet (Multimodal sensor). A graphic depicting how these different assets were connected is shown in Fig. 12. More detail on each of these assets can be found in their relevant subsections.

The ODSI is shown in Figs. 13 and 14. All features are annotated A-J and all controls 1–15. The definitions of these are as follows:

ODSI Features:

A. Status of Husky (uptime, motor currents, battery voltages, driver temperatures, battery capacity and charge estimate)

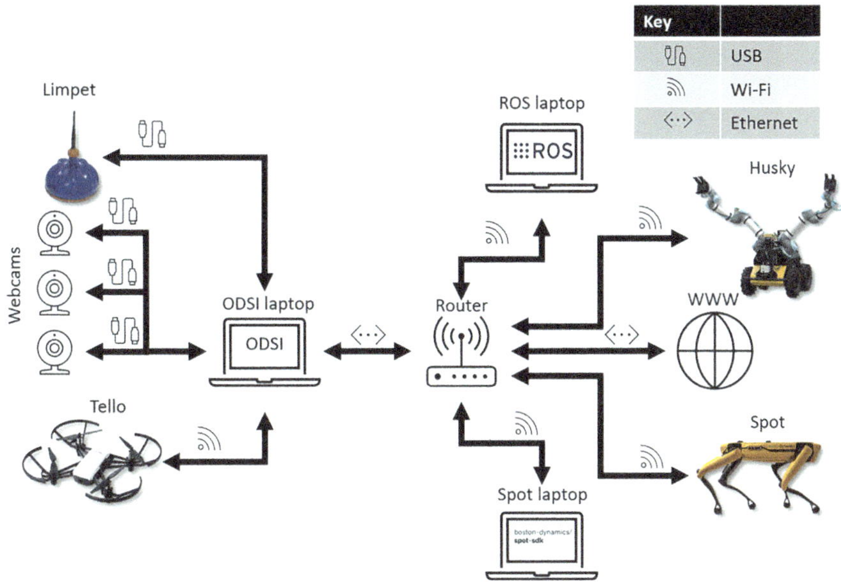

Fig. 12 Network layout of demonstration

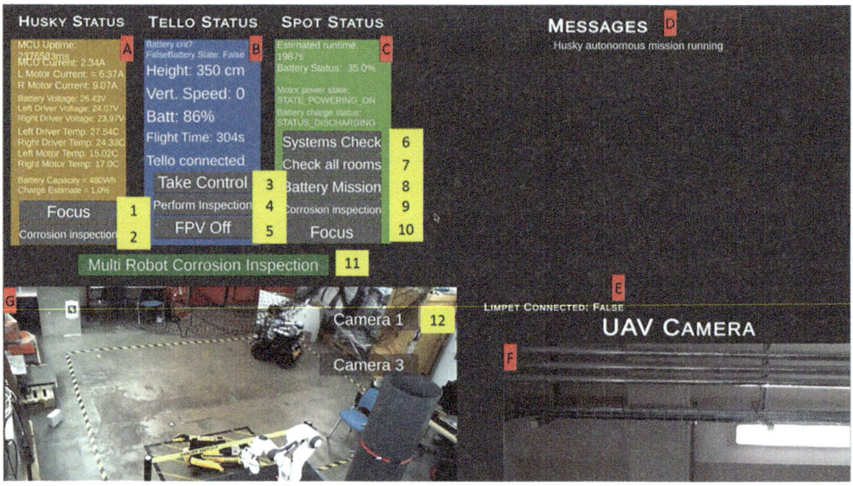

Fig. 13 Main control interface of the ODSI

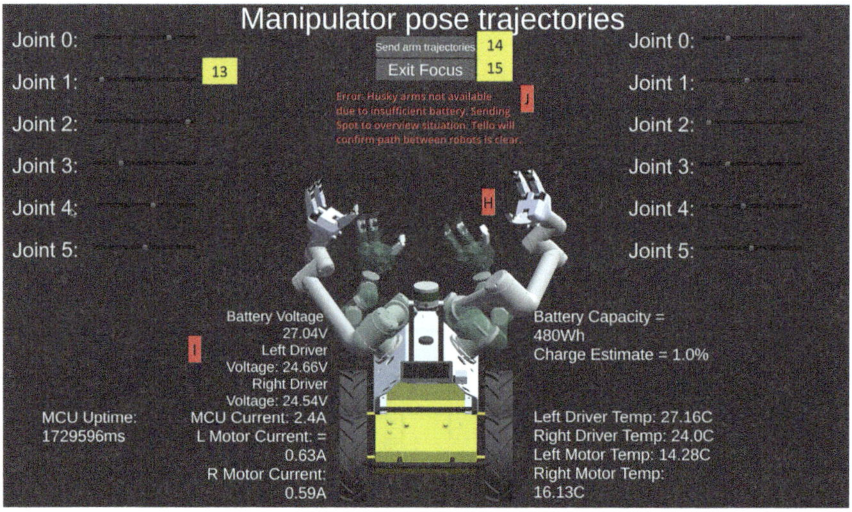

Fig. 14 Focus page of ODSI for the Husky robot

B. Status of Tello (battery criticality, battery state, height, vertical speed, battery charge, flight time and connection status)
C. Status of Spot (remaining run-time, battery charge, motor power state and battery status)
D. Messages from connected devices
E. Limpet connection status
F. Tello camera live feed

G. Webcam live feed
H. 3D model of Husky with ghost (current positions) and solid (projected positions) models
I. Status display similar to A
J. Error message sent from Husky.

ODSI Controls:

1. Focus screen for Husky
2. Perform Husky solo autonomous mission
3. Show teleoperation commands for Tello
4. Perform Tello solo autonomous mission
5. Toggle UAV Camera display (F.)
6. Perform Spot start-up sequence
7. Perform Spot room exploration mission
8. Perform Spot-Husky assistance mission
9. Perform Spot solo autonomous mission
10. Focus screen for Spot
11. Perform autonomous mission on all robots
12. Toggle camera feed
13. Manipulator joints desired positions, sliders control individual joint angles
14. Send arm trajectories to Husky
15. Exit focus screen.

4.3 Aerial Vehicle Within the SMuRF

Using a UAV within a SMuRF enables access to areas which cannot be reached by ground robots, including areas at height and over rough terrain. Additionally, the camera on a UAV allows the human operator to teleoperate and gain an overview of the work area. For use within this demonstration, a commercially available DJI Tello drone was chosen [84]. This platform has advantages in its low cost and small size allowing it to be flown indoors.

However, issues are presented with the limitations presented by the closed operating system onboard Tello and the size of the device. It is too small for any sensor payload beyond what is present on the device, consequently, the only practical application is for providing overviews of an environment through the onboard camera. Additionally, the lack of inbuilt sensors on the platform (only a height sensor and forward-facing camera) limits platform ability to perform autonomous missions that are more complex than hovering at a set height. There is also the requirement for a direct Wi-Fi communication between the operator and the UAV. Therefore, to implement the Tello within a SMuRF, accommodations must be made in the design of the network layout (see Fig. 12).

As the ODSI is built upon Unity, the '*TelloForUnity*' framework was cloned from GitHub [85]. This framework enables critical functions within ODSI for Tello, such as run-time telemetry, teleoperation and a live camera feed. It also handles communications between ODSI and the Tello through Universal Data Protocol messages.

4.4 Autonomous Ground Vehicles Within the SMuRF

A legged platform within a SMuRF allows access at ground level on uneven terrain or up/down staircases. Spot from Boston Dynamics was chosen as a commercially available platform for this operation. It is a quadrupedal robot capable of carrying a payload on its back, in this case a robotic arm which is used for manipulation of sensors and other objects [86, 87].

Limitations exist in its software accessibility—without a separate ROS PC, access to the systems of Spot is limited. A Python language Software Development Kit is provided for programming autonomous missions and functions, this was used to enable run-time communications of telemetry and mission status.

Limitations also exist in the creation of autonomous missions; Spot is capable of autonomous navigation but relies on fiducial markers placed in the environment and must be trained on a specific route through an environment before it is capable of traversing this route autonomously. Additionally, Spot must be able to locate at least one fiducial at any given time, which introduces challenges when assigning an autonomous route that passes through multiple rooms, e.g. all doors must be open for the mission. Spot is also capable of autonomous pick and place style operations with the robotic arm, but it must first be trained on the object which is to be manipulated through image recognition techniques. Each model is very large (> 5 GB) and only ensures detection in particular circumstances—the object must be at a similar angle with similar lighting in the same environment and position as the original training model.

Through these identified limitations, the scope to run a dynamic mission with Spot is narrow. Despite these limitations, the platform can still be used for routine inspection missions and assistance of other robots, e.g. through battery recovery, hence its application in this SMuRF.

To handle communications between Spot and the ODSI, a laptop running the Spot SDK in Python was connected to Spot through Wi-Fi. This laptop was, in turn, running a Python TCP client to send information to the TCP server in the ODSI. This allowed the ODSI to send commands to Spot and receive run-time telemetry information.

Wheeled platforms within SMuRFs allow access to confined spaces at ground level and have enhanced payload capability for sensors and manipulators. For this application, a Clearpath Husky was chosen as a commercially available platform. The Husky is equipped with a variety of positional and visual sensors, as well as two Universal Robotics UR-5 manipulators for various tasks [88].

Husky has advantages in terms of having an open software base, using the standard ROS system on its onboard computer. This gives the platform more flexibility in how it is implemented, and as a result means complex autonomous missions can be planned and executed. An additional advantage of this is it is possible to simulate failure modes to trigger recovery missions. The limitations of the platform are physical—it is unable to cross uneven terrain or inspect at height, additionally it is the heaviest of the three robots.

To enable autonomous navigation for this study, a laptop was connected to Husky via Wi-Fi to act as the ROSCore, to avoid making possible breaking changes to the platform's onboard computer. This laptop was in turn connected to the same Wi-Fi network as the ODSI. Communications between ROS and the ODSI were handled by a custom Python script. Unity has the capability of accepting a ROS connection, however the messages are large and frequent and contain information not necessary for the operations performed in this demonstration. Therefore, the custom client script will only send relevant robotic telemetry and mission started/stopped indicators.

4.5 Providing Overview for a Human in the Loop

To enable operator overview of the mission environment, USB webcams were deployed. The video feed from these can be switched in the ODSI.

A critical aspect of any SMuRF is the inclusion of infrastructural sensing. This can inform robotic decision-making through the early detection of adverse conditions. Limpet is an ROS-enabled integrated sensor platform for infrastructural condition monitoring, and was designed for mass production at low cost, allowing for deployment in an operating area in large collectives to monitor the environmental conditions on offshore installations [89]. Typically affixed to infrastructure via magnets, the Limpet has a robust protective housing to enable operation in harsh environments. It contains sensors for; time-of-flight, sound, pressure, temperature, humidity, hall-effect, acceleration, angular velocity and optics. The sensing module is inside a weather and impact proof shell for environmental protection and has an integrated magnet and sticking strip for attaching to surfaces. Consequently, the Limpet platform, albeit not a mobile robot, is useful for signal relay, navigation fidelity and situational/environmental awareness.

For this study, the Limpet is connected to the ODSI via a USB cable and communicates over a serial connection. This enables environmental effects to impact the SMuRF decision-making.

5 Run-Time Verification via Probabilistic Model Checking

Given the case study above, we focus on an example scenario to show how the probabilistic model checking can reason about the system and provide valuable insights into the mission at run time. We model the scenarios of each individual robot in the fleet and the operator's actions. As illustrated in Fig. 15, Husky operates to explore

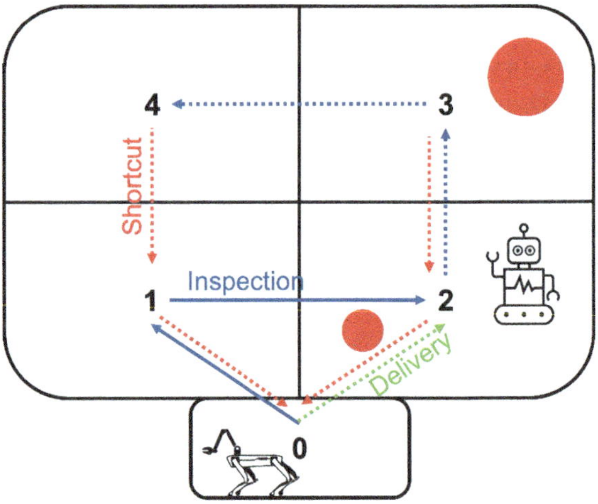

Fig. 15 Example scenario of the abstracted inspection mission: The continuous environment is discretized into a **2 × 2 grid** with numbered cells (0 being the **base** station, where robots are on standby.) Each cell might contain obstacles (**red** objects) with different sizes. Robots follow the pre-designed paths to perform **inspection**, **delivery** new batteries and return to base via **shortcuts**

the environment and identify obstacles whilst Spot is on standby to deliver batteries for Husky when there is a battery failure, or the operator asks to do so. We count the inspection as completed when Husky inspects all regions within the environment and all robots in the fleet safely return to the base station.

5.1 Continuous Markov Chains Model

As the multi-robot fleet operates in continuous environments, we take the form of (labelled) Continuous Markov chains (CTMCs) as our model. A CTMC is a triple $C = (S, R, L)$, where S is a finite set of states with a designated initial state, $R : S \times S \rightarrow \mathbb{R}_{\geq 0}$ is a rate matrix that represents a set of non-negative real numbers, and $L : S \rightarrow 2^{AP}$ is a label function that assigns atomic predicates with states, e.g. $s \mapsto \{\text{success}\}$. A transition occurs between two states s and s' only when rate $\lambda\big(s, s'\big) > 0$, in which case the probability of this transition being triggered within t time units equals to $1 - e^{-\lambda(s,s')\cdot t}$. Thus, transitions with higher rates occur more often. If $\lambda\big(s, s'\big) > 0$ holds for more than one state s', there exists a race between outgoing transitions from state s.

Rather than directly build a CTMC, we use the PRISM modelling language [73], a state-based language based on reactive modules [90], that supports high-level specification of processes. Processes are represented by modules consisting of non-deterministic choice over action-labelled guarded commands (which denote transitions); modules are composed of all common actions.

5.2 Properties

Properties of interest in PRISM are represented as an extension of the Continuous Stochastic Logic (CSL) [91], which is a temporal logic with probabilistic operators. In this work, we focus on the *eventually* temporal operator, that examines, for all paths, if a state labelled where is true can be eventually reached. PRISM allows us to quantify the properties via an operator that determines, not if a path exists where is true, but the *likelihood* of this being the case. A time boundary can also be added to require to be true within time units (domain-dependent, e.g. hours, minutes or seconds) through.

An essential property of the mission is *feasibility*, that is, how likely the inspection is to succeed based on the current state of the MR-fleet and the environment. Another important property to be concerned about is the *safety* of the robots, such as the likelihood of the robot failure over the next time units. Given the complexity and stochasticity involved during the mission, it is always difficult for operators to reach a conclusion about the feasibility and safety information by merely observing.

5.3 Implementation

As is typical in modelling, it is often necessary to abstract specific details of the scenarios, e.g. by discretizing continuous values, to maintain a meaningful trade-off between performance and accuracy. For our model, we aim to determine the likelihood that all robots make it from the base station to spatial regions and back to the base safely, where the failure can come from dead batteries or background failure (collisions etc.). Each robot has a battery level that we discretize to four values: *high*, *mid*, *low* or *critical*, similar to the method used in [92]. Batteries are charged at a fixed rate at the base. When robots are not at the base, their batteries drain at fixed rates corresponding to the current level, e.g. batteries near critical levels might discharge faster than those fully charged. A robot with a critical battery cannot continue with its current operation and needs to return to the base, and then resumes after its battery is fully charged. Meanwhile, the battery fails at a fixed rate based on its current level when robots operate outside the base, e.g. batteries near critical levels might fail more often than those fully charged. When its battery fails, a robot's movement also terminates.

As CTMCs are discrete state models (but with continuous transitions), they are not well-suited to modelling the continuous space. Therefore, to capture the movement of robots, we discretize the environment into a grid of cells, e.g. a 2×2 grid as shown in Fig. 15, allowing each cell to be identified using integers (with 0 being the base). When Husky operates inspection, it follows a pre-designed path (e.g. $0 \rightarrow 1 \rightarrow 2 \rightarrow 3 \rightarrow 4$) through all cells, and takes *shortcuts* on its way back (e.g. $4 \rightarrow 1 \rightarrow 0$, or $3 \rightarrow 2 \rightarrow 0$). Spot always moves following the shortcuts to ensure batteries can be delivered on time. Note that this can be determined ahead of time/calculated by planning tools before running the model.

To each cell, except the base, we associate a background failure rate that captures unknown failures within that cell, e.g. hazards, and collisions, which is updated during the inspection of Husky, e.g. higher rates of failure if large obstacles are detected. Furthermore, a rate of movement determines how quickly robots move between adjacent cells, assuming higher rates for Spot than Husky in all cells.

During the inspection of Husky, the operator determines when and whether to send out Spot for delivering new batteries. Their action is modelled as the synchronization of the Husky module and the Spot module, where the process of Husky is paused to wait for the delivery process of Spot to be finished and resumed after that. We assume this synchronization always happens when the battery on Husky fails.

5.4 Example Scenario

Cells might contain obstacles with different sizes that are unknown to Husky and Spot before the inspection. The background failure rate associated with each cell is updated with knowledge of the obstacles acquired by Husky during the inspection. To show how our model can aid the operator in making decisions on whether to send out Spot to deliver new batteries for Husky, we present the following scenarios. Note that decisions can be made at any time step and the results below compare the effects with and without operator interference.

Scenario 1: Husky currently performs the inspection in cell 2, as shown in Fig. 15, and has a low battery level. Figure 16 shows how the success probability of the mission and the safety of robots change within the next time units with/without the operator request to send out Spot.
Scenario 2: Similar to Scenario 1, if Husky is currently in cell 3, the probabilities are presented in Fig. 16.

In both scenarios, sending out Spot increases the probability that Spot fails during the delivery. Especially in Scenario 2, Spot might face a high risk to fail as Husky detects a large obstacle in cell 3 and the background failure rate is increased accordingly. Therefore, sending out Spot might not be a smart choice as the success probability is already high in Scenario 2. However, it might be worth taking a small risk of losing Spot to improve the success probability in Scenario 1.

6 Research Imperatives—Integration

The previous example scenario in Sect. 5.2 serves as a template for verification of a SMuRF at run time, which means multiple robots can be involved in different tasks. As for the integration, the physical SMuRF can, at any point, provide all known

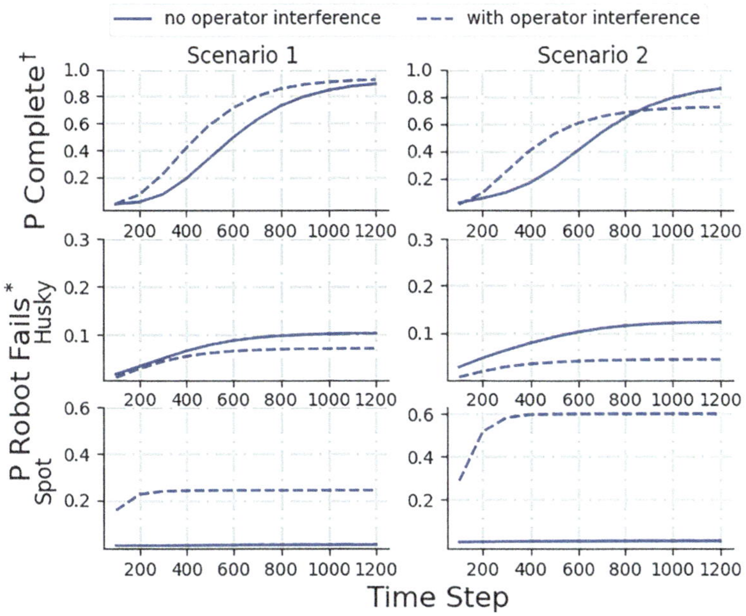

Fig. 16 Mission feasibility and robot safety: the probability changes within the next time units. Operator may interfere and ask Spot to deliver a new battery for Husky. † The probability of Husky completing the inspection and all robots being safe. * The probability of the robot failing due to the battery failure and background failures

information, including the knowledge of obstacles, and the status of robots (e.g. locations, battery levels), allowing the PRISM model to be constructed and updated accordingly and run for analysis in parallel. A key benefit of having a formal model is that we can run 'what-if' scenarios to give the operator more information before making decisions. A 'what-if' scenario can be described as a forecasted speculation about how a given situation (list of events during an autonomous mission), might occur by having a positive or negative impact on the key performance indicators of the mission. This can be displayed to an operator graphically in the interface, for example, an operator might query the trade-off between the feasibility of the mission and the safety of robots based on Fig. 16. Similarly, different what-if scenarios can be analyzed and large sets of properties can be abstracted by the upper and lower probability bounds, i.e. their envelopes of behaviour, to improve the situational awareness [93].

Future research imperatives include a symbiotic system of systems approach to enable for these technologies to work cohesively and collaboratively together [94, 95]. An overview of the systems required to achieve this is displayed in Fig. 17 where a human operator has access to a SMuRF deployed in the field. Moving from right to left, the human operator must have access to all elements of a robotic team whether deployed in the field or within simulation. To enable this at the hardware

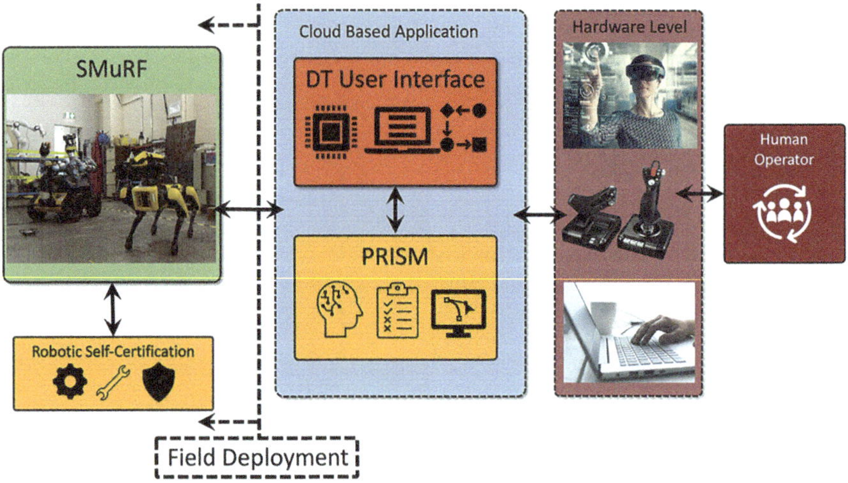

Fig. 17 Proposed design strategy of the positions of systems required where the SMuRF is deployed out in the field

level the human operator must have access to virtual reality headsets, joysticks and laptop to access the correct information. A virtual reality/augmented reality headset can enable for humans to position themselves in the digital world which can act as a representative of the real world. This can lead to improved decision-making during operations where a SMuRF is positioned remote in the field, the headsets can also act as effective tools in training and for presenting information to the human-in-the-loop in a more interactive approach [96–98]. A laptop can be used for any data analytics which is required to be done including accessing overviews from missions, completed objectives, errors, accessing cameras and other information from sensors. Finally, joysticks/controllers are effective tools for in the case where a robot is required to be teleoperated. This would be aided from a display or virtual reality headset for remote control of an autonomous vehicle [99–101].

The cloud-based application layer is where we would strategically position our digital twin in a future research imperative direction. This would be to enable several engineers to access the digital twin and to allow for add ons to be easily implemented. However, there are several definitions of digital twins available which can lead to confusion in the case of autonomous operations. There will certainly be different key performance indicators depending on the application and owner of the digital twin. As a reminder we define a digital twin as 'a software which acts as an information integrator to enable and enhance operational decision making for a digital representation of an in design or real-world asset'. The digital twin would include bidirectional information where the sensor data which the robots deployed in the field collect are transferred in real time to the digital twin to enhance the operational overview of the human operator via the relative hardware methods. This would allow for access to data related to inspection, environment and robot performance. Several add ons such as PRISM can enhance a digital twin to enable for forecasting of different events to further improve the verification and validation of a SMuRF deployed in the field.

Finally, we discuss the robots deployed in the field. Since, wireless communications in robotic missions cannot always be guaranteed, robots should have their own self-certification protocols onboard to safeguard against autonomous operation above its own physical limits, alongside cybersecurity in the case of a local threats [81, 102–105]. The robot team deployed in the field must strategically share information through each layer to the digital twin to minimize bandwidth issues, alongside inter and intra communication amongst the fleet to solve tasks related to inspection.

7 Discussion

To allow humans to focus on complex tasks, their role must adapt as robotic systems are scaled in the workplace. To date, up to three humans can be required to conduct an inspection using a mobile service robot due to the requirements to (1) teleoperate, (2) provide engineering expertise, and (3) intervene when things go wrong. This is disadvantageous due to both the cost of personnel, and increased risks from having more people working in a hazardous environment. Even as robotic platforms and autonomy improves, human roles will still be required in some scenarios, but with different approaches via different levels of autonomy, and improvements in persistent and trusted autonomy [10]. Full autonomy would be the first choice for inspection where, if the human-in-the-loop was not satisfied with results, they could teleoperate the robot to gain an improved understanding of the inspection area. This would be further enriched by metaverse applications to improve the perspective for the human via data insights [106, 107].

This chapter has investigated early adopters of multi-robot teams in industrial applications and described research areas and robotic platforms via our state-of-the-art literature review. We identified a shift away from homogeneity for inspection missions, due to the wide range of requirements that are faced in industrial environments. In many cases it is non-optimal to have a single robot to complete all inspection tasks, and the inspection capability from a diverse multi-robot fleet has substantial benefits to industrial facility operators. Human-in-the-loop remains an important feature within robotics due to the requirement to feedback information from inspections and for the human to understand the inspection mission. Probabilistic modelling was identified as an approach to allow measurements and predictions of key mission parameters which negatively impact or stop the mission. These can be simulated in advance of the real-world mission and compared against the real robot to ensure the success and recovery of a robot. A mixed SMuRF becomes even more important when considering ground, air, and subsea inspection modes.

An opportunity exists where a combination of both SMuRF and probabilistic modelling addresses resilience and reliability challenges for robots. This can be evaluated via simulation and physical robotic tests to comparatively evaluate the success rates of the robotic fleet to a number of variables which challenge the success of the mission. With a human-in-the-loop, the operator should be aware of the situation of robots and their operational environment which present critical hazards, such as working at height, weather conditions or radioactive dangers. Probabilistic modelling enables increased levels of awareness using validation and verification to present the human operator insight into the risk of future system failure. This should not overbear the operator requiring high levels of cognitive loads allowing them to easily identify whether the mission is achievable or not.

A combination of both approaches (cyber and physical) will be enhanced via the metaverse. Fleet management will enable information to be fed to advanced technologies such as virtual reality, mixed reality and augmented reality. This will reduce the requirement for laptops and mobile phones but still enable complete access. Digital twins already exist but will see significant advancement by giving humans the ability to teleport into a space for visualization purposes to oversee a mission. This data can be fed into a simulation that is representative of the environment and enhanced via real-time footage of robots corroborating the actions of the fleet.

Increased processing capabilities both onboard the robots, and via the cloud, enables a digital twin which uses informed run-time verification to inform on decision-making of autonomous systems. For example, systems should be designed so that if the mission success chance is above a set safety or reliability threshold the mission can continue automatically. Only if it falls below a set success chance, then the human-in-the-loop is notified (the mission continues, but the operator has the option to stop the mission) and if the threshold drops further then the human-in-the-loop is notified that the mission is being aborted. This enables mission critical updates to a human operator and can allow for the redesignation of tasks across a multi-robot fleet to ensure the mission is completed by another robot.

A multi-robot fleet creates opportunities for several types of missions across a myriad of sectors. However, the human-in-the-loop requires in-depth knowledge and information about each diverse robot to perform the necessary computations. This also includes the human operator having the knowledge to know which robots to deploy for different missions and know what their robotic capabilities are at the start and during the mission.

Coordinating a multi-robot fleet with a human operator requires an operational overview of facility, robotic assets and areas to be inspected. These can be accessed via a laptop or metaverse application overseen remotely by a human. Key performance indicators of the dashboard would include the following:

Task assignment: The ability for a human-in-the-loop to designate tasks to different robotic platforms via waypoints. These waypoints would include the positions of areas needing inspection. Although a human operator designates tasks, the robotic and autonomous systems would create their own decisions on how best to inspect the areas and if they are the most suitable robots to inspect the area (as in some cases the human may have deployed the wrong robot for the job).

Predicting and mitigation of failures: Robots face challenges daily which they may be able to or unable to overcome. Therefore, the autonomy in the system should be able to predict when a failure is going to occur and try to prevent the error or overcome the error once it has occurred.

Productivity: Comparisons of how other autonomous missions could be completed in parallel to the real-time autonomous mission. This would ensure that robots are operating optimally and productivity levels are maximized.

We propose a novel MR-fleet management system which could be created in the metaverse and used to oversee future robotic operations. The dashboard would incorporate sensing, robotics, infrastructural and environmental data, where the human operator can access desirable data to ensure the efficient operation of a facility. Additionally, it can display logbooks of historical results, warnings, and reliability issues within the fleet with different tabs. Figure 18 displays the design of a dashboard where the home page is displayed presenting a Clearpath Husky, Boston Dynamics SPOT and DJI Tello drone. The home screen allows for a rapid overview of the robots completing their different tasks alongside different tabs which can be opened via the top left. This displays an image of each robot, battery status, thumbnail of the navigation map/camera feed and overall mission updates.

The navigation tab is presented in Fig. 19. On the left-side a summarized view is displayed for both UAV and SPOT robots (these views can be expanded). The right-side displays a more detailed view for the Husky robotic mission. The human-in-the-loop has access to a mission summary, live feed from cameras and the option

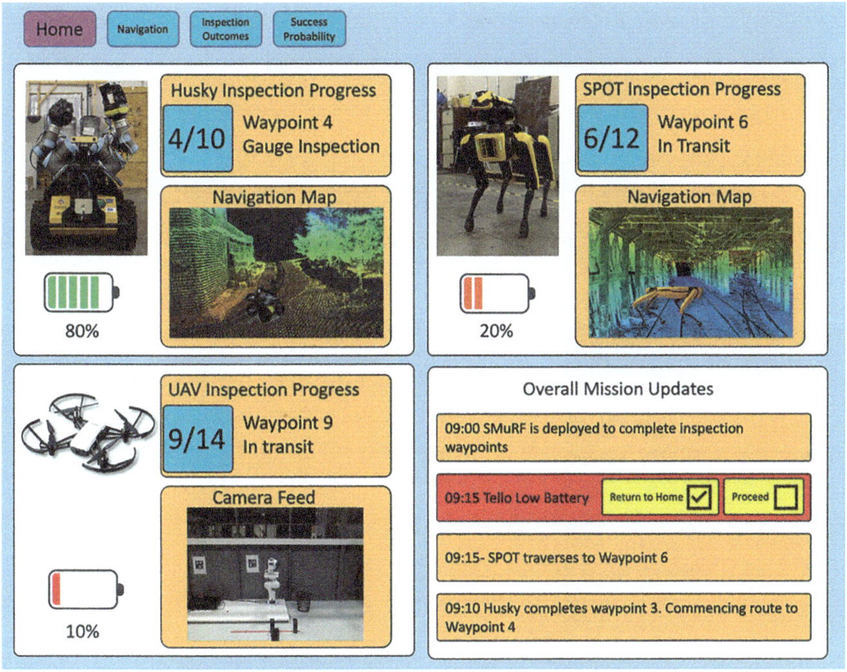

Fig. 18 Home screen tab displaying three robotic platforms and key information [108, 109]

to teleoperate if necessary. For safety purposes, each robot has an emergency stop (E-STOP) button.

The inspection tab is presented in Fig. 20 where a detailed overview of condition of inspections can be identified. The tab offers the opportunity to view warnings in a 3D environment where photographs of the inspection can be accessed. The metaverse allows for engineers to use virtual reality to inspect areas in the 3D model to ensure safe procedures can be in place ahead of rectifying issues in the real world.

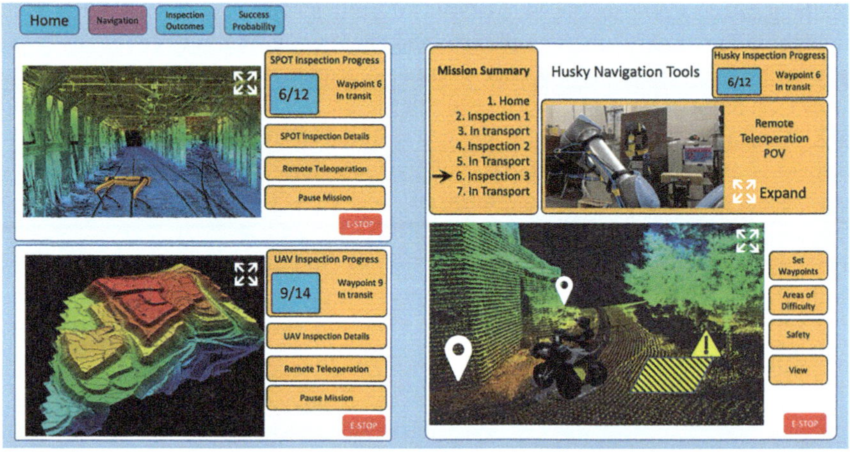

Fig. 19 Navigation tab displaying the navigation results for all of the conceptualized robotic missions [110]

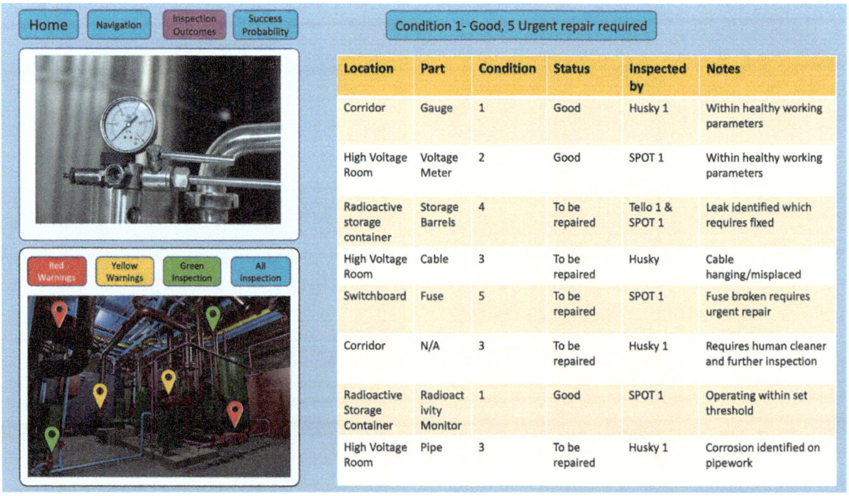

Location	Part	Condition	Status	Inspected by	Notes
Corridor	Gauge	1	Good	Husky 1	Within healthy working parameters
High Voltage Room	Voltage Meter	2	Good	SPOT 1	Within healthy working parameters
Radioactive storage container	Storage Barrels	4	To be repaired	Tello 1 & SPOT 1	Leak identified which requires fixed
High Voltage Room	Cable	3	To be repaired	Husky	Cable hanging/misplaced
Switchboard	Fuse	5	To be repaired	SPOT 1	Fuse broken requires urgent repair
Corridor	N/A	3	To be repaired	Husky 1	Requires human cleaner and further inspection
Radioactive Storage Container	Radioact ivity Monitor	1	Good	SPOT 1	Operating within set threshold
High Voltage Room	Pipe	3	To be repaired	Husky 1	Corrosion identified on pipework

Fig. 20 Inspection outcomes tab highlighting the results from sensing methods onboard robotic platforms [111]

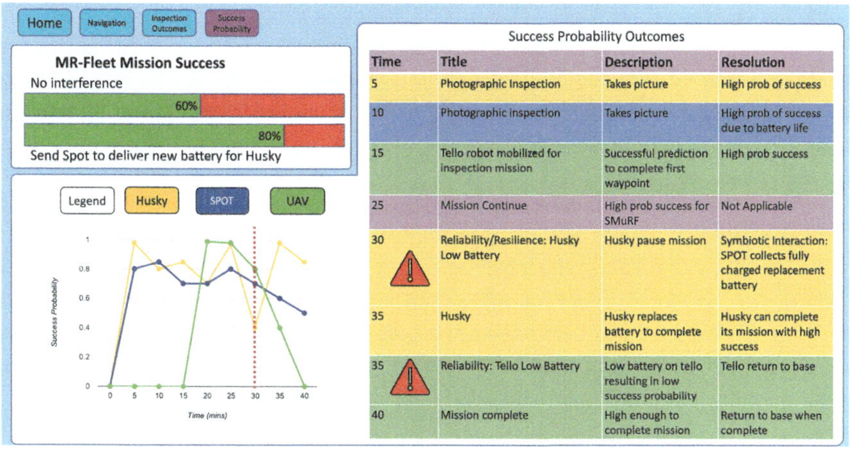

Fig. 21 Success probability tab presenting the success probabilities where the red dashed line indicates a symbiotic interaction which takes place for the Husky robot to increase the success probability of the robotic mission

The success probability tab presented in Fig. 21 gives an overview probability of the SMURF completing the mission with and without operator interferences (e.g. sending another robot for assistance). This is predicted ahead of the current process and updated at run time. Each tab for the robot presents the success probability over time using line graphs and highlights the symbiotic interaction points and other important events for the operator in a tabulated overview. Overall, the success probability dashboard gives the human-in-the-loop reassurance that the robot can complete a mission or that the robot needs task reassignment. Secondly, pressure to meet deadlines in the workplace can lead to human error via strict deadlines. This can lead to robotic failure and loss of an asset due to irresponsible management of inspection tasks resulting in extended timelines and costs. In this case, the success probability predicts the effects of possible interferences to assist the operator in making decisions.

8 Conclusions and Future Challenges

To conclude, robotics and autonomous systems are advancing rapidly in addressing improvements in capabilities required to conduct inspection (manipulation, sensing and autonomous missions). However, this improvement is limited for facility operators if they are unable to scale a couple of robotic platforms to a coordinated multi-robot fleet. This will allow for a wide range of inspections to be conducted leading to productivity improvements for both robots and human engineers who can focus on more complex tasks. Digitalization and metaverse applications are a necessary requirement in the coordination of multi-robot fleet deployments as a human must have access to high levels of interaction to gain the information they require and must not be limited by wireless communications or robot platform agility/mobility in accessing an area, e.g. attaining the required photograph or limited teleoperation due to communication bandwidths. Whilst ground-up applications of single robotic platforms are developed and deployed for different applications, the scaling includes a step to multi-robot fleets where a human-in-the-loop can coordinate a wide range of individual robotic platforms. A route towards optimized symbiotic multi-robot autonomy for today, tomorrow and the future is presented within Fig. 22. Currently, robotic platforms are teleoperated in the field with large teams of engineers to ensure they operate as intended. If a range of robots were deployed in a state-of-the-art facility, this would require a team for each robot which would not be feasible. The next advancement includes a meeting of the top-down and ground-up capability challenges where we can trust robots to operate as intended and coordinate them from a centralized interface where data can be transceived to each robot individually. The future advancement includes the symbiotic multi-robot fleet where the cognitive load on the human-in-the-loop is reduced via symbiotic interactions. This occurs when robots can request assistance to overcome challenges with resilience, reliability or safety during run-time in a mission. The creation of symbiotic multi-robot fleets where individual robotic platforms can adapt with respect to each other's mission profiles to provide assistance and improve overall mission quality. Many research and commercial parties are realizing the potential of a scaled robotic team for different applications within the nuclear and energy sectors, therefore a review on the emergent technologies and multi-robot fleets was required to improve the understanding, inform knowledge in robotics, and realize the potential in coordinating remote robots. Finally, we contribute to the state-of-the-art via the discussion of a symbiotic multi-robot fleet deployed in an offshore substation analogue scenario alongside the opportunities for probabilistic modelling to improve the reliability and resilience of autonomous missions.

Artificial intelligence is more widely being known to rapidly overcome challenges via its advantages of becoming highly adaptable in different scenarios. Some research is being conducted on robots which can learn from human engineers such as within small-scale agriculture [106] and applying artificial intelligence to robots that provide a service via communication and interaction [112]. This also has benefits for probabilistic modelling and a symbiotic multi-robot fleets. In future, this could enable to autonomous learning of how to do tasks, how to improve task accuracy and overcome resilience challenges by using the intelligence from the robotic platform. In addition to understanding and detecting cognitive overload by the human operator leading to shared control of information and key information fed to the human-in-the-loop.

This chapter provides a guide of how the metaverse can be applied to cyber physical systems for symbiotic multi-robot fleet management. An in-depth review of the state-of-the-art and how the cyber and physical roadmaps converge is discussed. Key concepts and technologies are discussed and how these can be applied to real-world applications are discussed too. Probabilistic modelling for the self-certification of robotic platforms is discussed alongside validation of mission success.

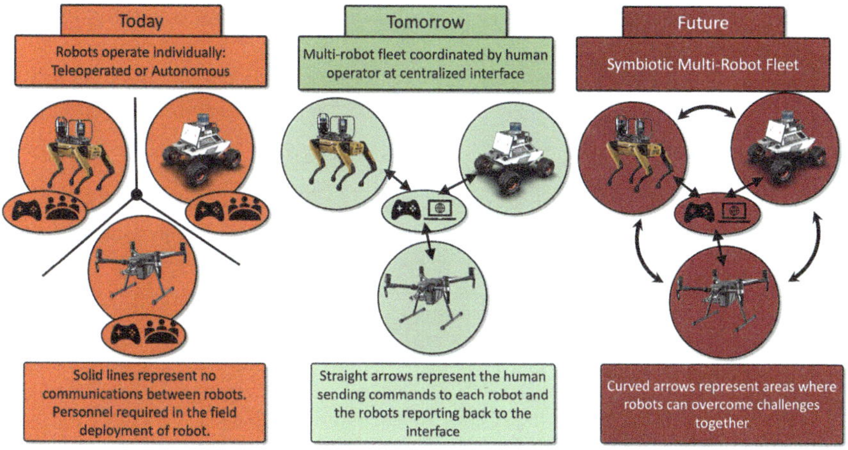

Fig. 22 Today, tomorrow and future pathway to optimized inspection and autonomy via robotics and autonomous systems

The contribution of this work provided a timely, academic-led review of cyber physical architectures for multi-robot fleets and a pathway for further development for robots conducting inspection activities in commercial, operational environments. Firstly, the background and challenge are presented where human roles within robotics and cyber physical systems via the metaverse alongside trends and imperatives for robotics are discussed. Secondly, a state-of-the-art review of robotic platforms for commercial activities and multi-robot fleets are discussed. A new approach named a symbiotic multi-robot fleet for inspection, maintenance and repair is introduced which features the whole systems architecture and discusses the key benefits of symbiotic interactions to improve reliability and resilience challenges in autonomous missions. In Sect. 4, the design for a symbiotic cyber physical architecture is presented alongside the robots which are part of the symbiotic multi-robot fleet and mission environment. Next probabilistic modelling with run-time verification is presented discussing mission feasibility and robot safety. The chapter then highlights key analysis from the chapter via a research imperatives section which focusses on the integration and a route to a cyber physical system which is deployed as a cohesive package. The discussion section is presented in Sect. 7 where the design for a novel multi-robot fleet management tool is presented to enable for multi-robot fleet coordination and overviews of autonomous missions. Finally, we present the conclusions of the book chapter.

Our work shows that most projects focus on the ground-up applications of robotic platforms (autonomous mission, manipulation, showcasing data to a human-in-the-loop) however, there is a need to realize that robots will require scaling and coordinated with a top-down view of the multi-robot fleet. To minimize cognitive overload, many decisions to overcome safety, resilience and reliability issues can be made by the autonomous systems themselves which leads to symbiotic interactions across a multi-robot fleet where information can still be fed to the human-in-the-loop via mission priorities.

References

1. Potenza, A., Kiselev, A., Loutfi, A., Saffiotti, A.: Towards sliding autonomy in mobile robotic telep-resence-a position paper. ACM Ref. Format 2, 17 (2017). https://doi.org/10.1145/nnnnnnn.nnnnnnn
2. Crestani, D., Godary-Dejean, K., Lapierre, L.: Enhancing fault tolerance of autonomous mobile robots. Rob. Auton. Syst. **68**, 140–155 (2015). https://doi.org/10.1016/J.ROBOT.2014.12.015
3. Paluch, S., Wirtz, J., Kunz, W.H.: Service robots and the future of services. Marketing Weiterdenken, pp. 423–435 (2020). https://doi.org/10.1007/978-3-658-31563-4_21
4. Meng, Z., She, C., Zhao, G., De Martini, D.: Sampling, communication, and prediction co-design for synchronizing the real-world device and digital model in metaverse. IEEE J. Sel. Areas Commun. **41**(1), 288–300 (2023). https://doi.org/10.1109/JSAC.2022.3221993
5. Azar, T., Barretta, R., Mystakidis, S.: Metaverse. Encyclopedia **2**(1), 486–497 (2022). https://doi.org/10.3390/ENCYCLOPEDIA2010031
6. Pozniak, H.: Could engineers work in the Metaverse? Eng. Technol. **17**(4) (2022). https://doi.org/10.1049/ET.2022.0408
7. Kong, L.C.W., Harper, S., Mitchell, D., Blanche, J., Lim, T., Flynn, D.: Interactive digital twins framework for asset management through internet. In: 2020 IEEE Global Conference on Artificial Intelligence and Internet of Things, GCAIoT 2020, Dec 2020. https://doi.org/10.1109/GCAIOT51063.2020.9345890
8. Huang, Z., Shen, Y., Li, J., Fey, M., Brecher, C.: A survey on AI-driven digital twins in Industry 4.0: smart manufacturing and advanced robotics. Sensors **21**(19), 6340 (2021). https://doi.org/10.3390/S21196340
9. Dani, A.A.H., Supangkat, S.H.: Combination of digital twin and augmented reality: a literature review. In: 9th International Conference on ICT for Smart Society: Recover Together, Recover Stronger and Smarter Smartization, Governance and Collaboration, ICISS 2022—Proceeding (2022). https://doi.org/10.1109/ICISS55894.2022.9915160
10. Mitchell, D., et al.: A review: challenges and opportunities for artificial intelligence and robotics in the offshore wind sector. Energy and AI **8**, 100146 (2022). https://doi.org/10.1016/J.EGYAI.2022.100146
11. Acemoglu, D., Restrepo, P.: Robots and jobs: evidence from US labor markets> SSRN Electron. J. Jun 2017. https://doi.org/10.2139/SSRN.2940245
12. Pistilli, M.: How to Invest in Robotics (Updated 2022). Accessed: 19 Dec 2022. [Online]. Available: https://investingnews.com/daily/tech-investing/emerging-tech-investing/robotics-investing/how-to-invest-in-robotics/
13. Surgical Robots Market Size, Share & Growth Report, 2030. Accessed: 30 Mar 2023. [Online]. Available: https://www.grandviewresearch.com/industry-analysis/surgical-robot-market
14. The Robots Are Still Rising, Albeit With Less Funding. Accessed: 8 Apr 2024. [Online]. Available: https://news.crunchbase.com/robotics/ai-global-us-venture-funding-figure/
15. Schranz, M., Umlauft, M., Sende, M., Elmenreich, W.: Swarm robotic behaviors and current applications. Front. Robot. AI **7** (2020). https://doi.org/10.3389/frobt.2020.00036
16. Software For Autonomous Inspection | Energy Robotics. Accessed: 19 Dec 2022. [Online]. Available: https://www.energy-robotics.com/robotic-inspection-software#fleet-management
17. Amazon Acquires Robot-Coordinated Order Fulfillment Company Kiva Systems For $775 Million In Cash | TechCrunch. Accessed: 15 Dec 2022. [Online]. Available: https://techcrunch.com/2012/03/19/amazon-acquires-online-fulfillment-company-kiva-systems-for-775-million-in-cash/?guccounter=1

18. Meet the robots at Amazon—YouTube. Accessed: 16 Dec 2022. [Online]. Available: https://www.youtube.com/watch?v=HSA5Bq-1fU4
19. Amazon's first fully autonomous warehouse robot is called Proteus—The Verge. Accessed: 16 Dec 2022. [Online]. Available: https://www.theverge.com/2022/6/21/23177756/amazon-warehouse-robots-proteus-autonomous-cart-delivery
20. British Army Approach to Robotics and Autonomous Systems: Generating Human-Machine Teams. Accessed: 16 Dec 2022. [Online]. Available: https://www.army.mod.uk/media/15790/20220126_army-approach-to-ras_final.pdf
21. What next for Army Robotics and Autonomous Systems? | The British Army. Accessed: 16 Dec 2022. [Online]. Available: https://www.army.mod.uk/news-and-events/news/2022/03/what-next-for-army-robotics-and-autonomous-systems/
22. Sit. Stay. Roll over? | The British Army. Accessed: 16 Dec 2022. [Online]. Available: https://www.army.mod.uk/news-and-events/news/2022/07/first-robotic-autonomous-systems-enhanced-brigade-combat-team/
23. Project Convergence 22: Rangers and Yorkshire Regiment lead the way | The British Army. Accessed: 16 Dec 2022. [Online]. Available: https://www.army.mod.uk/news-and-events/news/2022/11/project-convergence-22-rangers-and-yorkshire-regiment-lead-the-way/
24. Enabling Predictive Maintenance Through Robotic Inspection—Metrology and Quality News—Online Magazine. Accessed: 19 Dec 2022. [Online]. Available: https://metrology.news/enabling-predictive-maintenance-through-robotic-inspection/
25. British Army carries out successful Swarming Drone capability | The British Army. Accessed: 16 Dec 2022. [Online]. Available: https://www.army.mod.uk/news-and-events/news/2022/09/british-army-carries-out-successful-swarming-drone-capability/
26. Dstl acquires first fleet of autonomous ground vehicle systems—GOV.UK. Accessed: 19 Mar 2020. [Online]. Available: https://www.gov.uk/government/news/dstl-acquires-first-fleet-of-autonomous-ground-vehicle-systems
27. Soldiers test drones to prepare for complex future warfare | The British Army. Accessed: 16 Dec 2022. [Online]. Available: https://www.army.mod.uk/news-and-events/news/2022/11/soldiers-test-drones-to-prepare-for-complex-future-warfare/
28. DE&S Technology Office procures cutting-edge Nano UAVs—Defence Equipment & Support. Accessed: 16 Dec 2022. [Online]. Available: https://des.mod.uk/des-pocures-nano-uavs/
29. Arnold, R., Carey, K., Abruzzo, B., Korpela, C.: What is a robot swarm: a definition for swarming robotics. In: 2019 IEEE 10th Annual Ubiquitous Computing, Electronics & Mobile Communication Conference (UEMCON), pp. 74–81 (2019). https://doi.org/10.1109/UEMCON47517.2019.8993024
30. Dorigo, M., et al.: Swarmanoid: a novel concept for the study of heterogeneous robotic swarms. IEEE Robot. Autom. Mag. **20**(4), 60–71 (2013). https://doi.org/10.1109/MRA.2013.2252996
31. Azadeh, K., de Koster, R., Roy, D.: Robotized warehouse systems: developments and research opportunities. In: ERIM Report Series Research in Management Erasmus Research Institute of Management, May 2017
32. Ganeshmurthy, M.S., Suresh, G.R.: Path planning algorithm for autonomous mobile robot in dynamic environment. In: 2015 3rd International Conference on Signal Processing, Communication and Networking (ICSCN), pp. 1–6 (2015). https://doi.org/10.1109/ICSCN.2015.7219901
33. Zaffar, M., Ehsan, S., Stolkin, R., Maier, K.M.: Sensors, SLAM and long-term autonomy: a review. In: 2018 NASA/ESA Conference on Adaptive Hardware and Systems (AHS), pp. 285–290 (2018). https://doi.org/10.1109/AHS.2018.8541483
34. Oliveira, G.S., Röoning, J., Carvalho, J.T., Plentz, P.D.M.: Efficient task allocation in smart warehouses with multi-delivery stations and heterogeneous robots. In: 2022 25th International Conference on Information Fusion (FUSION), pp. 1–8 (2022). https://doi.org/10.23919/FUSION49751.2022.9841337
35. Taurob Inspector—The world's most versatile data collecting robot. Accessed: 22 Dec 2022. [Online]. Available: https://taurob.com/taurob-inspector/

36. Bogue, R.: Robots in the offshore oil and gas industries: a review of recent developments. Ind. Robot. **47**(1), 1–6 (2020). https://doi.org/10.1108/IR-10-2019-0207/FULL/PDF
37. Jiang, Z., et al.: A multirobot system for autonomous deployment and recovery of a blade crawler for operations and maintenance of offshore wind turbine blades. J. Field Robot. (2022). https://doi.org/10.1002/ROB.22117
38. BladeBUG: Robotic LPS inspection. Accessed: 22 Dec 2022. [Online]. Available: https://www.bladebug.co.uk/media/lps/
39. The Engineer—BladeBUG unveils new robot. Accessed: 22 Dec 2022. [Online]. Available: https://www.theengineer.co.uk/content/news/bladebug-unveils-new-robot/
40. Procyk, A.: Ground robots head offshore | Oil & Gas Journal. Accessed: 22 Dec 2022. [Online]. Available: https://www.ogj.com/drilling-production/article/14235414/ground-robots-head-offshore
41. Ramezani, M., Brandao, M., Casseau, B., Havoutis, I., Fallon, M.: Legged robots for autonomous inspection and monitoring of offshore assets. In: Proceedings of the Annual Offshore Technology Conference, vol. 2020-May (2020). https://doi.org/10.4043/30694-MS
42. ORCA Robotics—ORE catapult and ORCA hub join forces. Accessed: 22 Dec 2022. [Online]. Available: https://www.orcahub.org/articles/ore-catapult-and-orca-hub-join-forces
43. ANYmal X: Ex-Proof Inspection Robot—ANYbotics. Accessed: 2 Mar 2023. [Online]. Available: https://www.anybotics.com/anymal-ex-proof-inspection-robot/
44. Zhang, Q., et al.: Research progress of nuclear emergency response robot. IOP Conf. Ser. Mater. Sci. Eng. **452**(4), 042102 (2018). https://doi.org/10.1088/1757-899X/452/4/042102
45. West, A., et al.: Use of Gaussian process regression for radiation mapping of a nuclear reactor with a mobile robot. Scientific Reports **11**(1), 1–11 (2021). https://doi.org/10.1038/s41598-021-93474-4
46. Tsitsimpelis, I., Taylor, C.J., Lennox, B., Joyce, M.J.: A review of ground-based robotic systems for the characterization of nuclear environments. Prog. Nucl. Energy **111**, 109–124 (2019). https://doi.org/10.1016/j.pnucene.2018.10.023
47. Lu, D.V., Hershberger, D., Smart, W.D.: Layered costmaps for context-sensitive navigation. In: 2014 IEEE/RSJ International Conference on Intelligent Robots and Systems, pp. 709–715 (2014). https://doi.org/10.1109/IROS.2014.6942636
48. West, A., Wright, T., Tsitsimpelis, I., Groves, K., Joyce, M.J., Lennox, B.: Real-time avoidance of ionising radiation using layered cost maps for mobile robots. Front Robot AI **9**, 61 (2022). https://doi.org/10.3389/FROBT.2022.862067/BIBTEX
49. Groves, K., Hernandez, E., West, A., Wright, T., Lennox, B.: Robotic exploration of an unknown nuclear environment using radiation informed autonomous navigation. Robotics **10**(2) (2021). https://doi.org/10.3390/robotics10020078
50. Optimized Robotics Automation—AWS IoT RoboRunner Features—Amazon Web Services. Accessed: 15 Dec 2022. [Online]. Available: https://aws.amazon.com/roborunner/features/
51. Piacentini, C., Bernardini, S., Beck, J.C.: Autonomous target search with multiple coordinated UAVs. J. Artif. Intell. Res. **65**, 519–568 (2019). https://doi.org/10.1613/JAIR.1.11635
52. Welburn, E., Wright, T., Marsh, C., Lim, S., Gupta, A., Crowther, W., Watson, S.: A mixed reality approach to robotic inspection of remote environments. UK-RAS Conference on Embedded Intelligence. Loughborough, United Kingdom, (2019). https://pure.manchester.ac.uk/ws/portalfiles/portal/84610133/UKRAS_19_Mixed_Reality_Approach_to_Robotic_Remote_Inspection.pdf
53. Berndt, M., Krummacker, D., Fischer, C., Schotten, H.D.: Centralized robotic fleet coordination and control. In: Mobile Communication—Technologies and Applications; 25th ITG-Symposium, pp. 1–8 (2021)
54. What is fleet management in robotics? | Robotnik®. Accessed: 19 Dec 2022. [Online]. Available: https://robotnik.eu/what-is-fleet-management-in-robotics/
55. Borzone, T., Morărescu, I.-C., Jungers, M., Boc, M., Janneteau, C.: Hybrid framework for consensus in fleets of non-holonomic robots. In: 2018 Annual American Control Conference (ACC), pp. 4299–4304 (2018). https://doi.org/10.23919/ACC.2018.8430918

56. Carreno, Y., Pairet, È., Petillot, Y., Petrick, R.P.A.: Task allocation strategy for heterogeneous robot teams in offshore missions. In: IFAAMAS, vol. 9 (2020). Accessed: 19 Dec 2022. [Online]. Available: www.ifaamas.org
57. Garattoni, L., Birattari, M.: Autonomous task sequencing in a robot swarm. Sci Robot **3**(20) (2018). https://doi.org/10.1126/SCIROBOTICS.AAT0430/SUPPL_FILE/AAT0430_SM.PDF
58. Formant | Solutions. Accessed: Dec. 22, 2022. [Online]. Available: https://formant.io/solutions/
59. Kamps, H.J.: Formant is solving the robotic Tower of Babel with a unified platform | TechCrunch. Accessed: Dec. 22, 2022. [Online]. Available: https://techcrunch.com/2022/01/05/formant-series-a/
60. Roldán, J.J., et al.: Multi-robot systems, virtual reality and ROS: developing a new generation of operator interfaces. Stud. Comput. Intell. **778**, 29–64 (2019). https://doi.org/10.1007/978-3-319-91590-6_2/TABLES/7
61. Hoque, R., et al.: Fleet-DAgger: Interactive Robot Fleet Learning with Scalable Human Supervision (2022). https://doi.org/10.48550/arxiv.2206.14349
62. Swamy, G., Reddy, S., Levine, S., Dragan, A.D.: Scaled autonomy: enabling human operators to control robot fleets. In: 2020 IEEE International Conference on Robotics and Automation (ICRA), pp. 5942–5948 (2020). https://doi.org/10.1109/ICRA40945.2020.9196792
63. Jost, J., Kirks, T., Gupta, P., Lünsch, D., Stenzel, J.: Safe human-robot-interaction in highly flexible warehouses using augmented reality and heterogenous fleet management system. In: 2018 IEEE International Conference on Intelligence and Safety for Robotics (ISR), pp. 256–260 (2018). https://doi.org/10.1109/IISR.2018.8535808
64. Bonci, A., Cheng, P.D.C., Indri, M., Nabissi, G., Sibona, F.: Human-robot perception in industrial environments: a survey. Sensors **21**(5), 1571 (2021). https://doi.org/10.3390/S21051571
65. Endsley, M.R.: Toward a theory of situation awareness in dynamic systems. Human Factors: J. Human Factors Ergon. Soc. **37**(1), 32–64 (1995). Accessed: 19 Dec 2022. [Online]. Available: https://www.researchgate.net/publication/21019B492
66. Endsley, M.R.: Situation awareness global assessment technique (SAGAT). In: Proceedings of the IEEE 1988 National Aerospace and Electronics Conference, vol. 3, pp. 789–795 (1988). https://doi.org/10.1109/NAECON.1988.195097.
67. Riley, J.M., Strater, L.D., Chappell, S.L., Connors, E.S., Endsley, M.R.: Situation awareness in human-robot interaction: challenges and user interface requirements, pp. 191–212 (2016). https://doi.org/10.4324/9781315587622-18
68. Mitchell, D., Blanche, J., Harper, S., Lim, T., Flynn, D.: Microwave foresight sensing for safety compliance in autonomous operations. In: IEEE International Conference on Omni Layer Intelligent Systems (2023)
69. Woodcock, J., Gorm Larsen, P., et al.: Formal methods: practice and experience. https://doi.org/10.1145/1592434.1592436.
70. Lomuscio, A., Pirovano, E.: Verifying Fault-Tolerance in Probabilistic Swarm Systems (2020)
71. Boureanu, I., Kouvaros, P., Lomuscio, A.: Verifying security properties in unbounded multiagent systems. Accessed: Dec. 22, 2022. [Online]. Available: www.ifaamas.org
72. Kwiatkowska, M., Norman, G., Parker, D.: Probabilistic Model Checking: Advances and Applications
73. Kwiatkowska, M., Parker, D., Norman, G.: PRISM: probabilistic symbolic model checker. In: Field, T., Harrison, P.G., Bradley, J., Harder, U. (eds.), Computer Performance Evaluation: Modelling Techniques and Tools, pp. 200–204. Springer Berlin Heidelberg, Berlin, Heidelberg (2002)
74. Duflot, M., et al.: Probabilistic model checking of the CSMA/CD protocol using PRISM and APMC. Electron. Notes Theor. Comput. Sci. **128**(6), 195–214 (2005). https://doi.org/10.1016/j.entcs.2005.04.012
75. Leucker, M., Schallhart, C.: The Journal of Logic and Algebraic Programming A brief account of runtime verification. J Log Algebr Program **78**, 293–303 (2009). https://doi.org/10.1016/j.jlap.2008.08.004

76. Falcone, Y., Krstić, S., Reger, G., Traytel, D.: A taxonomy for classifying runtime verification tools. Int. J. Softw. Tools Technol. Transf. **23**, 255–284 (2021). https://doi.org/10.1007/s10009-021-00609-z
77. Sánchez, C., et al.: A survey of challenges for runtime verification from advanced application domains (beyond software). Form Methods Syst. Des. 54 (2019). https://doi.org/10.1007/s10703-019-00337-w
78. Xin, X., Keoh, S.L., Sevegnani, M., Saerbeck, M.: Run-time probabilistic model checking for failure prediction: a smart lift case study. In: IEEE 8th World Forum on Internet of Things (WF-IoT2022) (2022)
79. Gu, Y., Hunt, W., Archibald, B., Xu, M., Sevegnani, M., Soorati, M.D.: Successful swarms: operator situational awareness with modelling and verification at runtime. In: International Conference on Robot and Human Interactive Communication (RO-MAN) (2023)
80. Abioye, A.O., et al.: The effect of predictive formal modelling at runtime on performance in human-swarm interaction. In: Companion of the 2024 ACM/IEEE International Conference on Human-Robot Interaction, in HRI '24. Association for Computing Machinery, New York, NY, USA, pp. 172–176 (2024). https://doi.org/10.1145/3610978.3640725.
81. Robu, V., Flynn, D., Lane, D.: Train robots to self-certify their safe operation. Nature **553**(7688), 281. Nature Publishing Group (2018). https://doi.org/10.1038/d41586-018-00646-w
82. Mitchell, D., et al.: Lessons learned: symbiotic autonomous robot ecosystem for nuclear environments. In: IET Cyber Systems and Robotics (2023). https://doi.org/10.1049/csy2.12103
83. Mitchell, D., et al.: Cover image: lessons learned: symbiotic autonomous robot ecosystem for nuclear environments. IET Cyber-Syst. Robot. **5**(4), i–i (2023). https://doi.org/10.1049/csy2.12106
84. Tello | Ryze Tech—Feel the Fun. Accessed: 19 Dec 2022. [Online]. Available: https://store.dji.com/uk/product/tello?vid=38421
85. GitHub—comoc/TelloForUnity: Ryze Tech/DJI Tello application development resources for Unity. Accessed: 19 Dec 2022. [Online]. Available: https://github.com/comoc/TelloForUnity
86. Spot Specifications. Accessed: 26 Jul 2021. [Online]. Available: https://support.bostondynamics.com/s/article/Robot-specifications
87. Spot® | Boston Dynamics. Accessed: 28 Nov 2020. [Online]. Available: https://www.bostondynamics.com/spot#id_third
88. Husky UGV—Outdoor Field Research Robot by Clearpath. Accessed: 21 Jun 2022. [Online]. Available: https://clearpathrobotics.com/husky-unmanned-ground-vehicle-robot/
89. Sayed, M., Nemitz, M., Aracri, S., McConnell, A., McKenzie, R., Stokes, A.: The Limpet: a ROS-enabled multi-sensing platform for the ORCA hub. Sensors **18**(10), 3487 (2018). https://doi.org/10.3390/s18103487
90. Alur, R., Henzinger, T.A.: Reactive modules. Form Methods Syst. Des. **15**(1), 7–48 (1999). https://doi.org/10.1023/A:1008739929481
91. Baier, C., Haverkort, B., Hermanns, H., Katoen, J.-P.: Model-checking algorithms for continuous-time Markov chains. IEEE Trans. Software Eng. **29**(6), 524–541 (2003). https://doi.org/10.1109/TSE.2003.1205180
92. Zhao, X., et al.: Towards integrating formal verification of autonomous robots with battery prognostics and health management. Lecture Notes in Computer Science (including subseries Lecture Notes in Artificial Intelligence and Lecture Notes in Bioinformatics), vol. 11724 LNCS, pp. 105–124 (2019). https://doi.org/10.1007/978-3-030-30446-1_6/FIGURES/10
93. Calder, M., Sevegnani, M.: Stochastic model checking for predicting component failures and service availability. Accessed: Dec. 22, 2022. [Online]. Available: https://www.dcs.gla.ac.uk/~michele/papers/tdsc.pdf
94. Mitchell, D., et al.: Symbiotic system of systems design for safe and resilient autonomous robotics in offshore wind farms, Oct. 2021. https://doi.org/10.1109/ACCESS.2021.3117727
95. McConnell, A., et al.: The future workplace: a symbiotic system of systems environment. In: Cyber-Physical Systems, pp. 259–329 (2021). https://doi.org/10.1201/9781003186380-18

96. Makhataeva, Z., Varol, H.A.: Augmented reality for robotics: a review. Robotics **9**(2), 21 (2020). https://doi.org/10.3390/ROBOTICS9020021
97. Pérez, L., Diez, E., Usamentiaga, R., García, D.F.: Industrial robot control and operator training using virtual reality interfaces. Comput. Ind. **109**, 114–120 (2019). https://doi.org/10.1016/J.COMPIND.2019.05.001
98. Burghardt, A., Szybicki, D., Gierlak, P., Kurc, K., Pietruś, P., Cygan, R.: Programming of industrial robots using virtual reality and digital twins. Appl. Sci. **10**(2):486 (2020). https://doi.org/10.3390/APP10020486
99. Hoebert, T., Lepuschitz, W., List, E., Merdan, M.: Cloud-based digital twin for industrial robotics. Lecture Notes in Computer Science (including subseries Lecture Notes in Artificial Intelligence and Lecture Notes in Bioinformatics), vol. 11710 LNAI, pp. 105–116 (2019). https://doi.org/10.1007/978-3-030-27878-6_9/FIGURES/4
100. Kshetri, N.: The economics of the industrial metaverse. IT Prof **25**(1), 84–88 (2023). https://doi.org/10.1109/MITP.2023.3236494
101. Stączek, P., Pizoń, J., Danilczuk, W., Gola, A.: A digital twin approach for the improvement of an autonomous mobile robots (AMR's) operating environment: a case study. Sensors **21**(23):7830 (2021). https://doi.org/10.3390/S21237830
102. Zaki, O.F., et al.: Self-certification and safety compliance for robotics platforms. In: Society of Petroleum Engineers (SPE), 2020, p. OTC-30840-MS. https://doi.org/10.4043/30840-ms
103. Zaki, O., Dunnigan, M., Robu, V., Flynn, D.: Reliability and safety of autonomous systems based on semantic modelling for self-certification. Robotics **10**(1) (2021). https://doi.org/10.3390/robotics10010010
104. Fisher, M., et al.: Verifiable self-certifying autonomous systems. In: Proceedings—29th IEEE International Symposium on Software Reliability Engineering Workshops, ISSREW 2018, pp. 341–348, Nov. 2018. https://doi.org/10.1109/ISSREW.2018.00028
105. Lacava, G., et al.: Cybsersecurity issues in robotics. https://doi.org/10.22667/JOWUA.2021.09.30.001
106. Kaarlela, T., et al.: Towards metaverse: utilizing extended reality and digital twins to control robotic systems. Actuators **12**(6), 219 (2023). https://doi.org/10.3390/ACT12060219
107. Guan, J., Morris, A.: Extended-XRI body interfaces for hyper-connected metaverse environments. In: 2022 IEEE Games, Entertainment, Media Conference, GEM 2022 (2022). https://doi.org/10.1109/GEM56474.2022.10017701
108. Phillips, S., Charron, N., McLaughlin, E., Narasimhan, S.: Infrastructure mapping and inspection using mobile ground robotics. Struct. Health Monit., 3120–3127 (2019). https://doi.org/10.12783/SHM2019/32469
109. Complete Mobile Surveying of Amtrak's New York Penn Station | The RIEGL Newsroom. Accessed: May 18, 2023. [Online]. Available: https://newsroom.riegl.international/2019/02/18/complete-mobile-surveying-of-amtraks-new-york-penn-station/
110. Drone LiDAR in Mining: Go with the Workflow. Accessed: May 18, 2023. [Online]. Available: https://www.microdrones.com/en/content/drone-lidar-in-mining-go-with-the-workflow/
111. Construction Companies & CAD Design Services | The CAD Room. Accessed: May 18, 2023. [Online]. Available: https://www.thecadroom.com/why-should-construction-companies-seek-out-cad-design-services-companies/
112. Huang, M.-H., Rust, R.T.: Engaged to a robot? The role of AI in service. J. Serv. Res. **24**(1), 30–41 (2021). https://doi.org/10.1177/1094670520902266

GPSR Compliance

The European Union's (EU) General Product Safety Regulation (GPSR) is a set of rules that requires consumer products to be safe and our obligations to ensure this.

If you have any concerns about our products, you can contact us on ProductSafety@springernature.com

In case Publisher is established outside the EU, the EU authorized representative is:

Springer Nature Customer Service Center GmbH
Europaplatz 3
69115 Heidelberg, Germany

Batch number: 08589224

Printed by Printforce, the Netherlands